An Introduction to IoT Analytics

Chapman & Hall/CRC Data Science Series

Reflecting the interdisciplinary nature of the field, this book series brings together researchers, practitioners, and instructors from statistics, computer science, Machine Learning, and analytics. The series will publish cutting-edge research, industry applications, and textbooks in data science.

The inclusion of concrete examples, applications, and methods is highly encouraged. The scope of the series includes titles in the areas of Machine Learning, pattern recognition, predictive analytics, business analytics, Big Data, visualization, programming, software, learning analytics, data wrangling, interactive graphics, and reproducible research.

Published Titles

Feature Engineering and Selection
A Practical Approach for Predictive Models
Max Kuhn and Kjell Johnson

Probability and Statistics for Data Science
Math + R + Data
Norman Matloff

Introduction to Data Science
Data Analysis and Prediction Algorithms with R
Rafael A. Irizarry

Cybersecurity Analytics
Rakesh M. Verma and David J. Marchette

Basketball Data Science
With Applications in R
Paola Zuccolotto and Marcia Manisera

JavaScript for Data Science
Maya Gans, Toby Hodges, and Greg Wilson

Statistical Foundations of Data Science
Jianqing Fan, Runze Li, Cun-Hui Zhang, and Hui Zou

Explanatory Model Analysis
Explore, Explain, and, Examine Predictive Models
Przemyslaw Biecek, Tomasz Burzykowski

For more information about this series, please visit: https://www.routledge.com/ Chapman--HallCRC-Data-Science-Series/book-series/CHDSS

An Introduction to IoT Analytics

Harry G. Perros

CRC Press
Taylor & Francis Group
Boca Raton London New York

CRC Press is an imprint of the
Taylor & Francis Group, an **informa** business

First edition published 2021
by CRC Press
6000 Broken Sound Parkway NW, Suite 300, Boca Raton, FL 33487-2742

and by CRC Press
2 Park Square, Milton Park, Abingdon, Oxon, OX14 4RN

Library of Congress Cataloging-in-Publication Data
Names: Perros, Harry G., author.
Title: An introduction to IoT analytics / Harry G. Perros.
Description: First edition. | Boca Raton : CRC Press, 2021. | Includes
    bibliographical references and index.
Identifiers: LCCN 2020038957 (print) | LCCN 2020038958 (ebook) | ISBN
    9780367687823 (hardback) | ISBN 9780367686314 (paperback) | ISBN
    9781003139041 (ebook)
Subjects: LCSH: Internet of things. | System analysis. | System
    analysis--Statistical methods. | Operations research.
Classification: LCC TK5105.8857 .P47 2021 (print) | LCC TK5105.8857
    (ebook) | DDC 004.67/8--dc23
LC record available at https://lccn.loc.gov/2020038957
LC ebook record available at https://lccn.loc.gov/2020038958

ISBN: 978-0-367-68782-3 (hbk)
ISBN: 978-0-367-68631-4 (pbk)
ISBN: 978-1-003-13904-1 (ebk)

Typeset in Minion Pro
by SPi Global, India

Access the support material: www.routledge.com/9780367686314

To those who aspire to become
better individuals

Contents

Preface

IoT IS A CLOSED LOOP SYSTEM CONSISTING OF SENSORS, SERVERS, a network connecting sensors to servers, and a data base that stores the information. Decision-making tools are used to make decisions based on the information received by the sensors which are then fed back into the system. There are numerous IoT applications in all aspects of our lives, such as, applications for smart cities, structural health, traffic congestion, smart environment, smart water, smart metering, assisted living, healthcare, security and emergency, smart retail, smart agriculture, and smart animal farming.

Analytics is a term that became popular with the advent of data mining, and it refers to the analysis of data in order to make decisions. The tools used in analytics come from the areas of Machine Learning, Statistics, and Operations Research. There is a plethora of tools from Machine Learning, including well-known tools such as, artificial neural networks, support vector machines, and hidden Markov models. Some of the commonly used tools from Statistics are multivariable linear regression, dimensionality reduction, and forecasting.

The contribution of Operation Research to analytics is less prominent because it has a different perspective. Operation Research techniques are used to study the performance of a system by developing mathematical and computer-based models of a system, which are then exercised in order to study the system's performance under different assumptions. Typically, it is applied to systems that have not been built yet, such as, new designs or new enhancements of an existing system. The applications of Operations Research to the study of the performance of computers, computer networks and IoT is known as Performance Evaluation and Modeling. Typical tools are simulation, queueing theory, and optimization methods.

In this book, I use the term "analytics" to include both data exploration techniques and techniques for evaluating the performance of a system.

There are a lot of applications of analytics to IoT-related problems, such as, security and breach detection, data assurance, smart metering, predictive maintenance, network capacity and planning, sensor management, decision-making using sensor data, fault management, and resource optimization.

This book arose out of teaching a one-semester graduate-level course on IoT Analytics in Computer Science Department, North Carolina State University. It covers the following set of analytic tools which can be seen as a minimum set of required knowledge.

- Simulation techniques

- Multivariable linear regression

- Time series forecasting techniques

- Dimensionality reduction

- Clustering

- Classification

- Artificial neural networks

- Support vector machines

- Hidden Markov models

In addition, there are two introductory chapters, one on probability theory and another on statistical hypothesis testing. These two chapters can be skipped by the knowledgeable reader. There are also two Appendices, one on some basic concepts of queueing theory and the other on the maximum likelihood estimation (MLE) technique, that supplement some of the material in this book.

An important aspect in learning these techniques is to actually use them in practice. In view of this, it is very important to do the project at the end of each chapter. For this, you can use any software, such as, Python, R, MatLab, and SAS, or a combination of functions from different software. If you are not familiar with any of this software, then this a great opportunity to learn one or two!

Finally, a set of PowerPoint presentations and a solutions manual are available at https://www.routledge.com/978036768631.

Author

Harry G. Perros is a Professor of Computer Science at North Carolina State University, an Alumni Distinguished Graduate Professor, and an IEEE Fellow. He has published extensively in the area of performance modelling of computer and communication systems, and in his free time he likes to go sailing and play the bouzouki.

CHAPTER 1

Introduction

FOLLOWING A SHORT INTRODUCTION TO INTERNET OF THINGS (IoT), THE techniques covered in this book are presented and motivated. These techniques are: simulation, multivariable linear regression, time series forecasting, dimensionality reduction, clustering, classification, artificial neural networks, support vector machines, and hidden Markov models. Simulation techniques are used to determine the design of a future IoT system so that it performs well for an anticipated load generated by its sensors. The remaining techniques are from Statistics and Machine Learning, and they are used to analyze data generated from sensors for decision-making. The set of techniques presented in this book provides a good foundation for doing IoT analytics.

1.1 THE INTERNET OF THINGS (IOT)

IoT is a closed loop system consisting of sensors, a network connecting sensors to servers, and a database that stores the information. Decision-making tools are used to make decisions that are then fed back into the system. As shown in Figure 1.1, data is obtained from sensors, analyzed, and then used to make decisions that drive the operation of the system. The feedback mechanism varies significantly depending on the system. For instance, a system may monitor temperature, humidity, satellite images, and other weather-related sensors in order to determine rainfall for the next three days. This information is then communicated to farmers so that they can decide how much to water their crop. The closed loop may not necessarily involve humans, but instead it may only involve actuators; i.e., a group of sensors sends information to a server, which in turn uses various models to determine the required action and subsequently instructs actuators to take a specific action. A good example of this is the self-driving car.

1.2 IOT APPLICATION DOMAINS

There are numerous IoT applications in all aspects of our lives. A summary of some of the application domains is given below.

Smart city: This involves a broad set of applications aiming at making cities more sustainable and enjoyable. Applications of interest include the following:

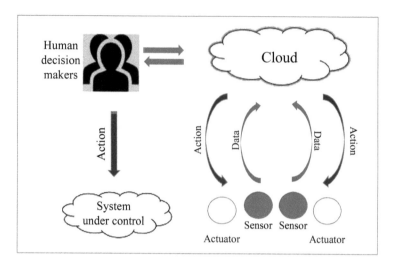

FIGURE 1.1 An IoT application is a closed loop system.

- *Structural health:* Monitoring of the status of buildings, bridges, and other infrastructures by sensing vibrations.

- *Traffic congestion:* Monitoring car and pedestrian traffic using fixed and mobile sensors, so that to determine the duration of green and red lights.

- *Smart environment:* Monitoring and preventing the occurrence of critical events in wide unpopulated areas, such as, fire in forests, landslides and avalanches, earthquakes, and exceptional air or water pollution in heavy industrial areas.

- *Smart water:* Water quality monitoring in rivers or in water distribution infrastructures, water leakage detection in pipes, river levels, dams, and reservoirs for flood or drought.

- *Smart metering:* Remote monitoring and control of large population of networked meters, such as, electricity, water, gas, oil in storage tanks, and cisterns.

Smart home: In a smart home, devices communicate with each other and with the outside environment. Applications include the following:

- *Home automation:* Applications for controlling of lights, climate, entertainment systems, and appliances.

- *Home security:* Applications for access control and alarm systems.

- *Occupancy-aware control:* Applications for sensing the presence of people in the house using sensors such as CO_2 sensors and Wi-Fi signals.

- *Pet and baby care:* Tracking the movement of pets and babies and preventing access to designated areas.

Smart retail: It includes systems to simplify stocking, storage, marketing, and selling in shops. Applications of interest include the following:

- *Smart shopping*: Tracks a customer's behavior and provides targeted information related to products of interest to the customer.
- *Stock control:* Real-time monitoring of shelves and stocks in shops to simplify the supply chain management.
- *e-payment:* Mobile-based applications for instant payments and payment transfers that will replace the wallet and debit cards.

e-health: Applications focus on monitoring patients and devices needed to assist the health of people. Applications include the following:

- *Wearable device management:* Centralized monitoring and data aggregation for various wearable medical devices and mobile e-health applications.
- *Infrastructure monitoring:* Monitoring of medical equipment, IT infrastructure, building infrastructure, and environment.
- *Patient surveillance:* Applications for ambient-assisted living including monitoring patients and fall detection.

Manufacturing: IoT can realize a seamless integration of various manufacturing devices equipped with sensing, identification, processing, communication, actuation, and networking capabilities. Applications include the following:

- *Digital control:* Process control automation, plant safety, and security.
- *Asset maintenance:* Predictive maintenance to maximize reliability.
- *Smart grid:* Applications to enable real-time energy optimization.
- *Stock control:* Real-time monitoring of inventory of raw and semi-finished material.

Smart farming: IoT can revolutionize farming operations and improve livestock monitoring. Applications include the following:

- *Smart agriculture:* Monitoring of the soil used for growing agricultural products, plants, green houses, and environmental conditions.
- *Smart animal farming:* Monitoring animal health conditions, animal tracking, and living conditions.

1.3 IOT REFERENCE MODEL

An IoT system is not a monolithic entity, but it consists of a number of interconnecting layers, as shown in Figure 1.2. Each layer has its own functionality defined in a technologically agnostic manner. It is designed and implemented separately, and it interacts only with the layer directly above and below. Breaking up the development of a complex system in layers, facilitates its development and implementation, and allows different vendors to develop compatible products for the layers. The set of layers and their functionality that describes a system is known as the *reference model*, and it is commonly used to develop systems in areas, such as, computer networks and software engineering. Below, we describe the functionality of each layer of the IoT reference model.

- *IoT devices and controllers:* This layer consists of physical devices and controllers which may control multiple devices. These are the "things" and they include a wide range of devices.

- *Networking:* This layer includes all the necessary networking devices, such as, transmission links, switches, and routers, for interconnecting IoT devices, fog computing servers, and main servers. The necessary networking protocols to enable such connectivity are also included in this layer.

- *Fog computing:* This is an intermediate level of computing, where data from sensors are processed before they are sent up to the main IoT servers. At this layer, data packets are inspected for content and protocol adherence. Then, the transmitted data may

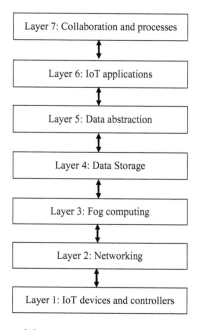

FIGURE 1.2 The IoT reference model.

be reformatted, decoded or coded, and aggregated to reduce networking overheads. Also, thresholding and event generation may take place at this layer.

- *Data storage:* Non-real-time applications expect to find the necessary data stored in a disk. Also, data may not be saved in the same disk storage, particularly when dealing with large IoT systems. Instead, they could be saved in different sites and in different formats. This layer focuses on storing multiple data formats from different sources. Data can be consolidated into a single place or into multiple sites accessed through virtualization.

- *Data abstraction:* In this layer, the data is aggregated, filtered, and reformatted so that it can be readily processed by specialized applications.

- *IoT applications:* IoT applications vary depending on the IoT system they support. In general, they use data from sensors to provide actions issued to actuators or to the users of an IoT system. Analytic tools that interpret data for decisions, the subject matter of this book, belong to this layer.

- *Collaboration and processes:* People and processes are often required to implement the actions yielded by an IoT system. Applications in this layer give business people the right data, at the right time, so they can make the right decisions.

Security of an IoT system requires security at all layers of IoT, such as, secure physical IoT devices, secure networks and protocols, secure communications, tamper resistant software, secure storage, and authentication/authorization at the application layer. These security requirements can be summarized in a separate reference model, not described here.

1.4 PERFORMANCE EVALUATION AND MODELING OF IOT SYSTEMS

This book focuses on techniques for IoT analytics. These techniques come from three different areas, namely, Performance Evaluation and Modeling, Statistics, and Machine Learning. In this section, we describe what is Performance Evaluation and Modeling and why it is important to evaluate the performance of an IoT system. In the following section, we describe the use of Statistical and Machine Learning techniques in IoT.

Performance Evaluation and Modeling is a scientific discipline that focuses on studying the performance of a system, such as, an IoT system. It is a discipline that dates back to the early 70s, and the techniques used are part of the Operations Research tool set.

Performance is measured in terms of metrics specific to the system under study. In an IoT system, the performance metrics can be grouped into (a) those related to the supporting IP network that provides connectivity between sensors, actuators, fog computing devices, and higher-layer servers executing processes in layer 4 and above, (b) those related to the computing facilities used in fog computing and high-layer server(s), and (c) those related to sensors and actuators. There are also performance metrics related to the entire IoT system. Below is a list of some of the most common metrics.

a. IP network

- Utilization of an Ethernet switch or an IP router, measured as the percent of time the device is used.

- Response time of an Ethernet switch or an IP router, measured as the time it takes to switch a packet through an Ethernet switch or an IP router.

- Packet loss measured as the percent of transmitted packets that are lost. It should be typically less than 1%.

- Throughput of the IP network, measured as the maximum number of packets per second that the network can transmit.

- Failure rate of components and time to repair a component.

- Security

b. Computing facilities

- CPU utilization, memory utilization, and disk utilization of a virtual machine (VM) that runs IoT-related processes, such as, fog computing and layer 4 and above processes.

- Execution time of a job in a VM.

- Utilization of a VM, measured as the percent of time it is busy executing jobs.

- Utilization of a computer running a virtual environment.

- Failure rate of components and time to repair a component.

- Security.

c. Sensors and actuators

- Response time of a sensor, measured as the time it takes for a sensor to sense a stimulus and transmit it out to its controller.

- Response time of an actuator, measured as the time it takes for an actuator to process and execute a command.

- Failure rate of components and time to repair a component.

- Security.

The above metrics can be used as a guidance in order to determine the capacity of an IoT system, i.e., the number of sensors and actuators, the number of fog and higher-level servers, the number of Ethernet switches and routers, and the speed of the networking links.

For non-real-time IoT application, the time it takes to receive data from the sensors, process it, and determine a response is not critical; i.e., it does not have to be executed within a few milliseconds. However, for real-time IoT applications, the time it takes for the IoT system to react to a stimulus received by a sensor is very critical. For instance, in the case of a driverless car, the breaks have to be activated immediately when the radar detects a person in front of the car. In real-time IoT applications, the end-to-end response time and jitter are very important metrics.

- *End-to-end response time:* This is the time it takes for an IoT system to react to a stimulus from a sensor. It consists of the time a sensor needs to capture and transmit the data, the time to process the data on a computing device, the time for the computing device to transmit instructions to an actuator, and the reaction time of the actuator. The propagation delay, i.e., the time it takes to propagate a data packet along a transmission link, may be negligible if the sensor, actuator, and the computing device are in close proximity. Otherwise, it has to be factored in.

- *Jitter of the end-to-end response time:* The end-to-end response time may vary over time due to congestion in the network and in the computing facilities. This variability is known as jitter, and it can be expressed in different ways. For instance, let r_i be the end-to-end response time of the ith packet sent by a sensor. Then, the jitter can be computed as the average of the differences of the successive response times over a minute, i.e., the average of $r_1 - r_2, r_2 - r_3, r_3 - r_4$, etc. Several jitter definitions have been developed for computer networks (see [1]).

The performance of an IoT system as a function of the allocated computing and networking resources and the traffic generated by the sensors is very complex and, in general, it is not known. It can be measured if the IoT system exists using existing measurement tools or new ones that can be specially designed.

However, it is not possible to measure the performance of a new design of an IoT system that has not been built yet, since it does not exist. The same problem occurs, if we are considering various alternatives to expand an existing IoT system. Again, these alternatives do not exist and therefore it is hard to measure their performance in order to determine which one is the best.

One approach to measuring the performance of a new design of an IoT system that has not been deployed yet is to develop a prototype and then experiment with it in order to establish its performance under different resources and traffic loads. This approach is feasible but it may be expensive and slow to implement.

A more typical approach is to build a model of the IoT system under study, and then use the model to carry out a what–if analysis. This is a faster and cheaper way to analyze its performance. This approach is not limited to IoT systems, but rather it is used in many sectors of the economy. For instance, a naval architect who is developing a new design for a boat carries endless experiments using simulation in order to establish the boat's hydrodynamic performance. It would be rather unthinkable to do so using a real-life prototype!

A model is a representation of a real-life system. There are different types of models, of which *symbolic* models are commonly used in Performance Evaluation and Modeling. These are models that represent the properties of real-life systems through the means of diagrams, mathematics, and computer simulation. Models based on diagrams are maps, activity diagrams, and workflow diagrams. They are useful to describe the processes involved, the tasks and sub-processes they consist of, and how these processes interact with each other. There is a variety of different diagramming software, such as the Business Process Execution Language (BPEL), the Business Process Modeling Notation (BPMN), and the Unified Modeling Language (UML) (see [1]).

Models based on mathematics can be classified into two groups: deterministic models and stochastic models. Deterministic models are models that do not contain the element of probability. These are primarily optimization models, such as: linear programming, non-linear programming, and dynamic programming. Stochastic models are models that contain the element of probability. Examples are the following: queueing theory, stochastic processes, and reliability theory.

Finally, computer simulation is an alternative to mathematical modeling and it may or may not contain probabilities, which means that it may be deterministic or stochastic. In computer simulation, the operation of a system under study is depicted in a computer program, which is then exercised for what–if analysis. Simulation techniques are very easy to use, as opposed to the mathematical models which require knowledge of the underlying mathematics. In view of this, they are commonly used to model various systems including IoT systems. In this book, we only describe simulation techniques (see Chapter 3).

The following is a partial list of problems that can be addressed using Performance Evaluation and Modeling techniques:

- Optimal replacement policies of sensors.

- Optimal number of sensors for an IoT application so that to provide a reliable system.

- Estimation of the response time to upload a message from a sensor and download the response to an actuator.

- Performance analysis of a signaling protocol for IoT.

- Dynamic dimensioning of the number of VMs needed in a cloud-based IoT service as a function of customer demand.

- Identification of bottlenecks in the networking and computing facilities of an IoT system.

- Bandwidth requirements and scheduling within a router for optimal delivery of real-time sensor data.

- Optimal allocation of data in a distributed storage environment.

1.5 MACHINE LEARNING AND STATISTICAL TECHNIQUES FOR IOT

Machine Learning grew as a scientific discipline out of the quest for artificial intelligence. It is currently driving an explosion in artificial intelligence applications that range from self-driving cars to speech translation.

Machine Learning is classified into *supervised, unsupervised*, and *reinforcement learning*. In supervised learning, the data used to train an algorithm is labeled, i.e., the data points are classified into different classes. For instance, let us assume that we have a set of temperature sensors in three different locations 1, 2, and 3. Each sensor reports the sensed temperature x every minute along with its location ℓ. The data points collected from all the sensors are of the form (x, ℓ), and we say that the data set of these data points is labeled. In this case, the label is the location ℓ, where ℓ = 1, 2, 3. On the other hand, the data set is said to be unlabeled if the location is not known; i.e., the data consists of just the temperature values.

A supervised learning algorithm uses a labeled training data set to obtain a function which can be used to infer the label of a new data point. For instance, let us assume in the above example that we obtain a data point (x, ℓ) with a corrupted location value ℓ. An appropriate supervised learning algorithm can infer the location ℓ of x, using the training labeled data set.

An unsupervised algorithm can infer the label of each data point of an unlabeled data set. In the above example, an appropriate unsupervised algorithm can infer the location of each temperature value in the unlabeled data set; i.e., assuming that we know that we have three locations, it clusters the data set into three clusters of temperatures, each corresponding to a different location.

Reinforced learning is an area of Machine Learning whereby a software agent learns by interacting with its environment in discrete time steps. The agent receives rewards by performing correctly and penalties for performing incorrectly. The agent learns without intervention from a human by maximizing its reward and minimizing its penalty.

In this book, we focus on the following supervised and unsupervised learning algorithms:

- *Clustering algorithms:* Several clustering algorithms have been proposed that deal with grouping a data set that has not been previously grouped into a number of groups or clusters of similar data points. Clustering a data set is also known as labeling the data set, since each data point is assigned a label. Clustering algorithms are unsupervised Machine Learning algorithms. In Chapter 8, we describe the following algorithms: hierarchical clustering, k-means clustering, fuzzy c-means model, Gaussian mixture decomposition, and DBSCAN.

- *Classification algorithms:* These algorithms deal with the problem of classifying an unlabeled data point given a set of labeled data. The labeled data set may come from original labeled data or it could be labeled using a clustering method. Classification algorithms are supervised Machine Learning algorithms. In Chapter 9, we describe the algorithms: k-nearest neighbor, naive Bayes classifier, decision trees, and logistic regression. Artificial neural networks, support vector machines, and Hidden Markov Models are described in separate chapters.

- *Artificial neural networks (ANNs):* An artificial neural network or simply a neural network is composed of multiple nodes, where each node imitates vaguely a biological neuron of the brain. A node can take input data from other nodes and performs a simple operation on the data. The result of this operation is passed to other nodes. The nodes are interconnected via links which are associated with a weight. A neural network is trained using a labeled data set, by altering the value of the weights of the links, and it can be used in supervised learning for classification and regression. Neural networks can also be used in an unsupervised learning for data clustering. Neural networks are described in Chapter 10.

- *Support vector machines (SVM):* This is a classification technique for two classes. For linearly separated data, SVM calculates an optimal hyperplane that separates the two classes. For non-linearly separated data, a boundary can be determined after the data is transformed to a higher-dimensional space using a kernel function, where they actually become linearly separated. Support vector machines are described in Chapter 11.

- *Hidden Markov models (HMM):* An HMM is a Markov chain which produces at each state an observable output. The parameters of an HMM are hidden from the user, and thus its name; i.e., the one-step transition probabilities of the Markov chain, the initial state probabilities, and the functions used to generate an output at each state are not known, but they can be estimated from the observable data. There are numerous applications of HMMs in classification and forecasting. Hidden Markov models are described in Chapter 12.

In addition to the above techniques, Statistical techniques, such as, linear multivariable regression, dimension reduction, and forecasting techniques are also used. Below, we briefly describe these techniques.

- *Multivariable linear regression:* This technique is applicable to the case where a random variable Y can be expressed as a linear combination of a number of input random variables X_1, X_2, ..., X_k. In this case, the multivariable linear regression technique can be used to estimate the parameters of this linear combination. The estimated linear function can then be used to determine the Y value for any given combination of values of the input variables. Multivariable linear regression is described in Chapter 5.

- *Dimension reduction:* One of the problems that arises when dealing with a large data set is that the number of variables can be very large. Each variable of the data set corresponds to a different dimension in the space of the data set. Analytic techniques can be difficult to apply to data sets with a large number of dimensions. Dimension reduction techniques transform a data set into a space with fewer dimensions than originally, which facilitates the application of analytics. Dimension reduction techniques are described in Chapter 7.

- *Forecasting techniques:* Forecasting techniques are used to determine future values of a time series, i.e., a sequence of data observed over time. These techniques are useful in IoT since a lot of the data reported by sensors are time series. Forecasting techniques are described in Chapter 6.

Machine Learning and Statistical techniques can be used in an IoT system in order to solve a variety of problems:

- Identification of anomalies in data traffic in order to trigger alarms and take corrective action through actuators.

- Identification of faulty/compromised sensors.

- Improving energy consumption, achieve carbon emission targets, and anomaly identification for preventing outages.

- Development of early warning indicators of failing components and machines in a manufacturing plant, in order to reduce unplanned downtime and scheduled maintenance, thus reducing costs and downtime.

- Forecasting the required number of sensors and gateways as a function of customer demand for an IoT application.

We note that often problems that arise in IoT systems require a solution that may involve several techniques used together from Performance Evaluation and Modeling, Machine Learning and Statistics. Finally, we note that an important problem area not related to the scope of this book is security and breach detection. This involves anomaly detection algorithms to identify potential threats and conditions, breach indicators, and automatic mitigation of in-progress attacks.

1.6 OVERVIEW OF THE BOOK

Analytics is a term that became popular with the advent of data mining, and it refers to the analysis of data in order to make decisions. In this book, we interpret the term analytics as consisting of Performance Evaluation and Modeling techniques, Machine Learning techniques, and Statistical techniques. The book covers the following techniques:

a. Performance Evaluation and Modeling

   o Simulation techniques

b. Statistical Techniques

   o Multivariable linear regression

   o Time series forecasting techniques

   o Dimensionality reduction

c. Machine Learning Techniques

- o Clustering

- o Classification

- o Artificial neural networks

- o Support vector machines

- o Hidden Markov models

In addition, two more chapters have been included, one on probability theory and another on statistical hypothesis testing, in order to provide the necessary background to the book. There are also two Appendices that supplement some of the material in the book. One is on basic queueing theory concepts and the other on the maximum likelihood estimation (MLE) technique.

The reader is assumed to have a basic knowledge of Probability Theory and Statistics. Also, knowledge of a high-level programming language is required.

An important aspect in learning these techniques is to actually use them in practice. In view of this, it is very important to do the project at the end of each chapter. For this, you can use any software, such as, Python, R, MatLab, and SAS, or a combination of functions from different software. If the reader is not familiar with any of these software, then this a great opportunity to learn one or two!

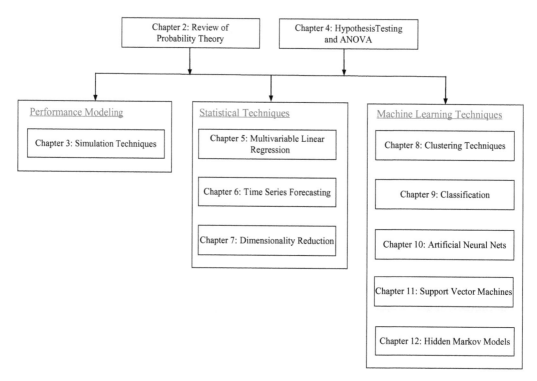

FIGURE 1.3 Layout of the book.

Figure 1.3 gives a pictorial view of the chapters in this book. Chapters 2 and 4 are the two introductory chapters, which can be skipped by the knowledgeable reader. The remaining chapters in Figure 1.3 are grouped per area, i.e., Performance Evaluation and Modeling, Statistical techniques, and Machine Learning techniques, and they can be studied in any order. The two Appendices are not shown in Figure 1.3.

EXERCISES

Select an IoT application and search the literature to identify 3–5 relevant papers. Make sure that the scope of the application is narrow, such as, "smart metering for energy consumption" as opposed to "smart cities". Based on the literature search, answer the following questions:

a. Describe the number of different solutions that have been proposed and the techniques used to obtain these solutions.

b. Compare these solutions and identify what other problems in this area need to be addressed.

REFERENCES

1. H. Perros. 2014. *Networking Services: QoS, Signaling, Processes*, https://people.engr.ncsu.edu/hp/.

CHAPTER 2

Review of Probability Theory

IN THIS CHAPTER, WE REVIEW SOME BASIC CONCEPTS OF PROBABILITY theory, such as, discrete and continuous distributions, expectations, and transformations. We assume that the reader has some rudimentary knowledge of probability theory. For an in-depth review, the interested reader should consult other books on this topic.

2.1 RANDOM VARIABLES

A *random variable* is a variable that takes different values probabilistically. They are usually notated by a capital letter. For instance, let X be a variable that indicates the number of packets transmitted by an IoT device every second, and for simplicity let us assume that 10 is the largest number of packets the device can ever transmit. Then, X takes the values 0, 1, ..., 10. However, we do not know what value X will take in the next second, but we can predict it probabilistically; i.e., we can say that $X = i$ with some probability p_i. We refer to X as a random variable because it is a variable that takes different values according to some probability distribution. These probabilities can be calculated empirically, or we can assume that they follow some known theoretical distribution. We now proceed to define a random variable more precisely.

Let us consider an experiment which may produce different outcomes. The set of all possible outcomes of the experiment is known as the *sample space*. For instance, let us consider the set of outcomes of an experiment that involves tossing a coin three times. In this case, the sample space is {HHH, HHT, HTH, THH, HTT, THT, TTH, TTT}, where H indicates head and T tail. We refer to this sample space as a discrete sample space, since it is a set of discrete values.

Now, let us consider a different experiment, where we observe the inter-departure time of successive packets transmitted by an IoT device. We define this time as the time elapsed from the moment the last bit of packet i is transmitted to the moment that the last bit of packet $i + 1$ is transmitted. Let X indicate this inter-departure time, and for simplicity let us assume that X cannot be greater than 2. Then, X takes values in the space $(0, 2]$. This is an example of a continuous sample space, since X can take any value in $(0, 2]$.

A random variable is defined as a mapping of a sample space to the real numbers $\mathfrak{R}$. For instance, in the case of the coin experiment, we can define a random variable X that gives the number of heads; i.e., if the outcome is HHT, then $X = 2$. X takes the values

{0, 1, 2, 3} with probability {0.5^3, 3×0.5^3, 3×0.5^3, 0.5^3} since heads and tails have the same probability of coming up. We can define a variety of different random variables on the same sample space. For instance, we can define a random variable X as the profit or loss from tossing the coin three times, whereby for each toss we win $10 if we get heads and we pay $5 if we get tails. Then, X takes the values {−15, 0, 15, 30} with probability {0.5^3, 3×0.5^3, 3×0.5^3, 0.5^3}.

In the case of the continuous sample space of the inter-departure time of successive packets from an IoT device, we can define a random variable X such that it is equal to 1 if the inter-departure time is less than 1 and equal to 2 otherwise.

Often, we use a random variable without referring to the actual experiment and resulting sample space. For instance, we define a random variable to indicate the inter-departure time of successive packets from an IoT device without having to describe the underlying experiment and sample space; likewise, for other random variables, such as the time it takes to transmit a packet, the inter-arrival time of packets at a server, the time-to-fail of an IoT device, etc.

Random variables are discrete or continuous, depending on whether they take discrete or continuous values. For instance, the random variable that describes the number of packets transmitted by an IoT device in each second is discrete. On the other hand, the random variable that gives the inter-departure time of successive packets transmitted by an IoT device is continuous. Below, we present various well-known theoretical discrete and continuous-time random variables and their probability distributions.

2.2 DISCRETE RANDOM VARIABLES

Let X be a discrete random variable, and let $p(X = i)$ be the probability that X takes the value i. The set of all the probabilities for all the values of X is referred to as the probability distribution, and they sum up to 1. In this section, we present a number of well-known theoretical discrete random variables and their probability distribution.

2.2.1 The Binomial Random Variable

Consider a *Bernoulli experiment*, i.e., an experiment that consists of n independent trials, where the outcome of each trial has two possible values, 1 and 2. By independent trials we mean that the outcome of one trial does not affect the outcome of another trial. Let p be the probability that 1 comes up, and $1 - p$ that 2 comes up (Obviously, these two probabilities add up to 1). Let X be a random variable defined as the number of times 1 comes up in n trials. Then,

$$p\left(X=i\right)=\binom{n}{i}p^i\left(1-p\right)^{n-i}, i=0, 1, 2, \ldots$$

This probability distribution is known as the *binomial distribution* and a random variable that follows this distribution is known as a *binomial random variable*.

2.2.2 The Geometric Random Variable

This random variable is defined as the number of times outcome 1 (or 2) occurs successively in a Bernoulli experiment, before outcome 2 (or 1) occurs for the first time. For instance, when tossing a coin n times, we want to know with what probability we will get i successive heads before tails appear for the first time. Similarly, this random variable can be seen as the number of failures that occur before we get the first success.

As above, let p and $1 - p$ be the probability that 1 or 2 comes up in a single trial, and let n be the total number of trials. Then using the coin example, tails (T) may come up at the first toss, or after several tosses, or not at all. These combinations along with the value of X are given in Table 2.1.

The probability distribution can be easily obtained since the trials have a binary outcome and they are independent from each other. For instance, the probability that $X = 2$ is the probability that we get two heads followed by tails, i.e., $p^2(1 - p)$. These probabilities are also given in Table 2.1. In general,

$$p(X = i) = p^i(1 - p), i = 0, 1, \ldots$$

This is the well-known *geometric distribution*, and a random variable that has this distribution is called a *geometric random variable*.

2.2.3 The Poisson Random Variable

Let us consider the occurrence of a particular event, such as the arrival of packets at a server. Figure 2.1 shows an example of such arrivals, indicated by arrows, over a horizontal time axis. Now, let us impose a unit time on the horizontal axis, indicated by the vertical

TABLE 2.1   An example of the geometric distribution

Outcomes	$P(X = i)$
T	$1 - p$
HT	$p(1 - p)$
HHT	$p^2(1 - p)$
HHHT	$p^3(1 - p)$
$\vdots$	$\vdots$
HHHH...H	p^n

FIGURE 2.1 Arrivals of packets at a server.

blue lines, and count the number of arrivals within each unit time, also given in Figure 2.1. We note that the unit time could be of any length of time. It could be, for instance, equal to 30 seconds, 1 minute, 5 minutes, 1 hour, etc. Let X be a random variable that gives the number of arrivals within a unit time. Of interest is the probability that X is equal to i. This probability varies depending on how the packets arrive. We say that random variable X is Poisson if

$$p(X=i)=e^{-\lambda}\frac{\lambda^i}{i!}, i=0, 1, 2, \ldots$$

where λ is the average number of arrivals in a time unit. Let N be the total number of arrivals over a period T, expressed in unit times. Then N/T is an estimate of the mean number of arrivals per unit time. From Figure 2.1, we have 15/10 = 1.5. Obviously, in order for this to be a good estimate, T has to be very large. In theory, this is expressed as follows:

$$\lambda = \lim_{T \to \infty} \frac{N}{T}.$$

The Poisson distribution is connected with the exponential distribution, as will be seen later on, in that the inter-arrival time of the packets (customers more generally) is exponentially distributed with a mean $1/\lambda$.

2.2.4 The Cumulative Distribution

Let X be a discrete random variable, and let $p(X = i)$ be the probability that X takes the value i. Without loss of generality, we assume that X takes values which are all integers. The cumulative distribution of X is denoted by $F_X(i)$ and it is given by

$$F_X(i) = p(X \le i) = \sum_{k=-\infty}^{i} p(X = k).$$

The complement of the cumulative distribution, $p(X > i)$, is known as the *residual distribution* and it is given by

$$\mathcal{F}_X(i) = p(X > i) = \sum_{k=i+1}^{+\infty} p(X = k).$$

For example, consider a random variable X that takes the values 0, 1, 2, 3 with probability 1/8, 3/8, 3/8, 1/8, respectively. The probability distribution and its cumulative distribution are shown in Figure 2.2. The following are some properties of the cumulative distribution.

$$0 \le F_X(i) \le 1$$

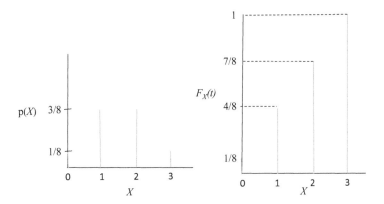

FIGURE 2.2 An example of a cumulative distribution.

$$F_X(-\infty) = 0 \text{ and } F_X(+\infty) = 1$$

$$\text{If } i \le j, \text{ then } F_X(i) \le F_X(j)$$

$$p(i \le X \le j) = F_X(j) - F_X(i-1)$$

$$p(X = i) = F_X(i) - F_X(i-1)$$

The cumulative probability distribution of the binomial distribution is

$$F_X(k) = \sum_{i=0}^{k} \binom{n}{i} p^i (1-p)^{n-i},$$

and it can be computed numerically.

The cumulative probability distribution of the geometric distribution is

$$F_X(k) = \sum_{i=0}^{k} p^i (1-p) = (1-p)\sum_{i=0}^{k} p^i = (1-p)\frac{1-p^{k+1}}{1-p}$$

$$= 1 - p^{k+1}.$$

The cumulative probability of the Poisson distribution is

$$F_X(k) = \sum_{i=0}^{k} e^{-\lambda}\frac{\lambda^i}{i!} = e^{-\lambda}\sum_{i=0}^{k}\frac{\lambda^i}{i!},$$

and it can be computed numerically.

2.3 CONTINUOUS RANDOM VARIABLES

A continuous random variable takes a continuum of values within an interval $[a, b]$, where $a < b$. Unless specified, we assume that the entire range of the real numbers is used, i.e., $a = -\infty$ and $b = +\infty$. The probability distribution of a continuous random variable X is known as the *probability density function* (pdf) and it is notated by $f_X(t)$. In general, $f_X(t)$ is a continuous and differentiable function, and its integral from $-\infty$ to $+\infty$ is 1.

The cumulative distribution $F_X(t)$ of a continuous random variable X with a pdf $f_X(t)$ is as follows:

$$F_X(t) = \int_{-\infty}^{t} f_X(s)ds.$$

Similar properties as in the discrete case presented in the previous section apply, such as: $0 \le F_X(t) \le 1$, $F_X(-\infty) = 0$, $F_X(+\infty) = 1$, and if $t_1 \le t_2$ then $F_X(t_1) \le F_X(t_2)$.

The pdf $f_X(t)$ can be obtained as the derivative of $F_X(t)$, i.e.,

$$f_X(t) = \frac{d}{dt} F_X(t),$$

and it has the following properties:

$$f_X(t) \ge 0$$

$$\int_{-\infty}^{+\infty} f_X(t)dt = 1$$

$$p(X > t) = \int_{t}^{+\infty} f_X(s)ds = 1 - F_X(t)$$

$$p(t_1 \le X \le t_2) = \int_{t_1}^{t_2} f(s)ds = F_X(t_2) - F_X(t_1)$$

Using the above property, we have

$$p(t \le X \le t + dt) = F_X(t + dt) - F_X(t) \approx f_X(t)dt.$$

$F_X(t + dt) - F_X(t)$ is the area under the curve of the probability density function from t to $t + dt$, and since dt is small, this area can be approximated by $f_X(t)dt$. As $dt \to 0$, $p(t \le X \le t + dt)$

tends to 0 since $f_X(t)dt$ tends to 0, but it also tends to $p(X = t)$, which means that $p(X = t) \to 0$. That is, the probability that the continuous random variable X is equal to t is zero, i.e., $f_X(t) = 0$. Consequently, it does not make sense to talk about the probability $f_X(t)$. Rather, we should talk about the probability $f_X(t)dt$. Despite this, for simplicity, we use $f_X(t)$ in place of $f_X(t)dt$.

We now proceed to describe some of the most well-known continuous random variables.

2.3.1 The Uniform Random Variable

This is a random variable that has the same probability of selecting any value in the interval $[a, b]$. Its pdf is

$$f_X(t) = \begin{cases} 1/(b-a), & \text{if } t \in [a, b] \\ 0, & \text{otherwise} \end{cases},$$

and its cumulative distribution is

$$F_X(t) = \int_a^t \frac{1}{b-a} dx = \frac{1}{b-a} x \Big|_a^t = \frac{t-a}{b-a}.$$

Figure 2.3 shows an example of the uniform pdf and its cumulative distribution.

2.3.2 The Exponential Random Variable

The exponential random variable is a non-negative random variable that has the following pdf.

$$f_X(t) = \lambda e^{-\lambda t}, \ t \geq 0,$$

where λ is a positive value. It is typically used to model the duration of an activity, such as, a service time. Its cumulative distribution $F_X(t)$ can be obtained as follows:

$$F_X(t) = \int_0^t f_X(s) ds = \int_0^t \lambda e^{-\lambda s} ds$$

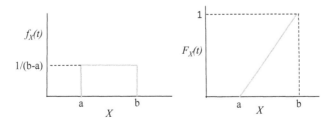

FIGURE 2.3 The uniform distribution and its cumulative.

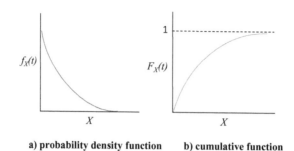

a) probability density function   b) cumulative function

FIGURE 2.4 The exponential distribution and its cumulative (a) probability density function (b) cumulative function.

$$= -\int_0^t e^{-\lambda s} d\left(-\lambda s\right)$$

$$= -e^{-\lambda s}\Big|_0^t = -e^{-\lambda t} + 1$$

$$= 1 - e^{-\lambda t}.$$

Figure 2.4 shows an example of the exponential pdf and its cumulative distribution.

As mentioned in Section 2.2.3, the exponential distribution is related with the Poisson distribution. Let X be a random variable that indicates the number of occurrences per unit time of an event, such as arrivals of packets to a server. If X is Poisson distributed, then the probability that i arrivals occur in a unit time is given by the following expression:

$$p\left(X = i\right) = e^{-\lambda} \frac{\lambda^i}{i!}, i = 1, 2, \ldots.$$

It can be shown that the inter-arrival time, i.e., the time elapsing between two successive arrivals, is exponentially distributed with the same parameter λ, i.e.,

$$f_X\left(t\right) = \lambda e^{-\lambda t}, t \geq 0.$$

The parameter λ is the *rate* of arrivals, i.e., the mean number of arrivals per unit time. The mean inter-arrival time of successive packets is $1/\lambda$. For instance, if the rate of arrivals at a server is 1000 packets/sec, then the mean inter-arrival time is 1/1000 sec, or 1 msec.

The exponential distribution has the *memoryless property*. For instance, let us assume that the inter-arrival time of a bus at a bus stop is exponentially distributed. Then, the amount of time it will take for the bus to arrive is independent of how long it has been since the previous bus came (Luckily, the inter-arrival time of a bus does not follow the exponential distribution!). More precisely, this property can be expressed as follows:

$$p(X > t + s | X > s) = p(X > t).$$

This is because

$$p(X > t + s | X > s) = \frac{p(X > t + s, X > s)}{p(X > s)}$$

$$= \frac{p(X > t + s)}{p(X > s)}$$

$$= \frac{e^{-\lambda(t+s)}}{e^{-\lambda s}} = e^{-\lambda t};$$

i.e., the probability that one has to wait more than t is independent of how long one has been waiting.

The memoryless property of the exponential distribution is not a realistic assumption when modeling real systems. However, it is popular because of its mathematical simplicity.

Another interesting property is the following:

$$p(X \leq t + \Delta t | X > t) \approx \lambda \Delta t.$$

Because of the memoryless property, we have

$$p(X \leq t + \Delta t | X > t) = p(X \leq \Delta t) = 1 - e^{-\lambda \Delta t}.$$

We recall that

$$e^x = 1 + x + \frac{x^2}{2!} + \frac{x^3}{3!} + \cdots$$

Therefore,

$$e^{-\lambda \Delta t} = 1 + (-\lambda \Delta t) + \frac{(-\lambda \Delta t)^2}{2!} + \frac{(-\lambda \Delta t)^3}{3!} + \cdots$$

The terms above that include the high orders of $(\Delta t)^n$, $n = 2, 3, \ldots$ can be ignored since they are very small, and consequently

$$1 - e^{-\lambda \Delta t} \approx \lambda \Delta t.$$

2.3.3 Mixtures of Exponential Random Variables

As mentioned above, the exponential distribution is an easy distribution to work with mathematically, but it has the memoryless property which is not a realistic assumption. Combining several exponential random variables to construct a new random variable is a common practice, because it takes advantage of the mathematical simplicity of the exponential distribution, but at the same time it gives rise to more sophisticated distributions that can model real-life situations better. A combination of exponential random variables is referred to as a *mixture of exponential random variables.*

Let us consider an example of a cloud-based application software that processes requests from users. The processing consists of n different tasks that are executed sequentially; i.e., when a request arrives, task 1 is first executed, then task 2, and so on until task n is executed and a response is returned back to the user. Now, let X_i be a random variable that describes the time to execute task i, and let us assume that it is exponentially distributed with parameter λ_i, $i = 1, 2, \ldots, n$. Then, the total time to execute a request is represented by a random variable X that is equal to the sum of the individual random variables, i.e., $X = X_1 + X_2 + \cdots + X_n$. The distribution of X is called the *Erlang* distribution.

Another simple case is where the processing of the requests involves the execution of only one task, which is selected from the n tasks probabilistically, i.e., with probability p_i task i is executed. In this case, $X = p_1 X_1 + p_2 X_2 + \cdots + p_n X_n$. The distribution of X is called the *hyper-exponential* distribution.

Finally, we note that we can generalize this idea by assuming that the execution of a request involves following a probabilistic path through these tasks. For instance, after completing task 1, it may go to task 2 with probability p_{12} or to task 3 with probability $p_{13} = 1 - p_{12}$. Likewise, from task 2, it may go to task 1 with probability p_{21} or to task 3 with probability p_{23}, or to task 4 with probability $p_{24} = 1 - p_{21} - p_{23}$, and so on. The distribution of such a mixture is called a *phase-type distribution.*

Below we examine briefly the *Erlang* and the *hyper-exponential* distributions.

The Erlang random variable represents the example where n tasks are sequentially executed. Let $X_1, X_2, \ldots, X_n$ be n identical and independent exponential random variables with parameter λ (Identical means that they have the same exponential distribution, and independent means that the outcome of one random variable does not effect the outcome of another random variable). We define a new random variable X as the sum of these n random variables, i.e., $X = X_1 + X_2 + \cdots + X_n$. An example with $n = 4$ is shown in Figure 2.5. Each circle, sometimes referred to as a *stage*, represents a different exponential random variable with the parameter shown inside the circle.

The cumulative distribution of the Erlang random variable with n stages is

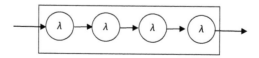

FIGURE 2.5 The Erlang random variable.

$$F_X(t) = p(X \le t).$$

We observe that the probability of having n successive exponential distributions in less than t is equal to the probability that we have at least n Poisson arrivals in less than t. This probability is equal to the complement of the probability of having 0, 1, 2, ..., $n - 1$ Poisson arrivals. Therefore, we have

$$F_X(t) = 1 - \sum_{i=0}^{n-1} p(i \text{ Poisson arrivals occur in less than } t)$$

$$= 1 - \sum_{i=0}^{n-1} e^{-\lambda t} \frac{(\lambda t)^i}{i!}.$$

The pdf of the Erlang random variable can be obtained by differentiating the above cumulative distributions. After some calculations, we obtain

$$f_X(t) = \lambda \frac{(\lambda t)^{n-1}}{(n-1)!} e^{-\lambda t}, \, t > 0.$$

We note that an extension of the Erlang random variable, known as the *generalized Erlang*, is the case where each stage has a different parameter λ.

The hyper-exponential random variable X is defined as follows. Let $X_1, X_2, ..., X_n$ be n independent exponential random variables with parameters $\lambda_1, \lambda_2, ..., \lambda_n$ respectively. Then, X is defined as follows:

$$X = \begin{cases} X_1 & \text{with prob } p_1 \\ \vdots & \\ X_n & \text{with prob } p_n \end{cases};$$

i.e., X takes the value of X_i with probability p_i where $p_1 + p_2 + \cdots + p_n = 1$. The probability distribution of X is

$$f_X(t) = \sum_{i=1}^{n} p_i \lambda_i e^{-\lambda_i t}, \, t \ge 0.$$

The cumulative distribution is equal to the probabilistic sum of the individual cumulative distributions, i.e.,

$$F_X(t) = \sum_{i=1}^{n} p_i \left(1 - \lambda_i e^{-\lambda_i t}\right) = \sum_{i=1}^{n} p_i - \sum_{i=1}^{n} p_i \lambda_i e^{-\lambda_i t}$$

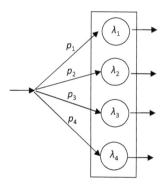

FIGURE 2.6 The hyper-exponential random variable.

$$= 1 - \sum_{i=1}^{n} p_i \lambda_1 e^{-\lambda_i t}.$$

An example of the hyper-exponential variable with $n = 4$ is shown in Figure 2.6.

2.3.4 The Normal Random Variable
The pdf of a normally distributed random variable X is given by the expression:

$$f_X(t) = \frac{1}{\sigma\sqrt{2\pi}} e^{-\frac{1}{2}\frac{(t-\mu)^2}{\sigma^2}}, \quad -\infty < t < +\infty,$$

where μ and σ^2 is the mean and variance of X, respectively. It is notated as $N(\mu, \sigma^2)$, and we use the notation $X \sim N(\mu, \sigma^2)$ to indicate that X is normally distributed with mean μ and variance σ^2. Its cumulative distribution is

$$F_X(t) = \frac{1}{\sigma\sqrt{2\pi}} \int_{-\infty}^{t} e^{-\frac{1}{2}\frac{(x-\mu)^2}{\sigma^2}} dx.$$

The normal distribution (also known as the Gaussian distribution) is very important in Statistics and in many areas of science and engineering. This is because many real-life variables are normally distributed or can be closely approximated by a normal distribution, such as, height and intelligence. Another main reason for its importance is the *central limit theorem*, which says that if a random variable X is the sum of n identical and independent random variables which may not necessarily be normally distributed, then X follows the normal distribution as n increases. More specifically, let $X_1, X_2, ..., X_n$ be n random variables that are independent and have the same distribution (any distribution) with a mean μ and a variance σ^2. Then, $X \sim N(n\mu, n\sigma^2)$ as n increases (typically, when $n \geq 30$); i.e., irregardless of the original distribution of the individual X_i random variables, the pdf of X approximates the normal distribution as n increases. This can also be stated as follows:

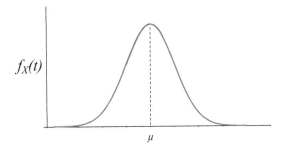

FIGURE 2.7 The normal probability density distribution.

$$\frac{1}{n}\sum_{i=1}^{n} X_i \sim N\left(\mu, \frac{\sigma^2}{n}\right);$$

i.e., the pdf of the average of $X_1, X_2, \ldots, X_n$ approximates the normal distribution with mean μ and variance σ^2/n as n increases. As a result of this expression, if we sample n points from a population which follows an arbitrary distribution with a mean μ and variance σ^2, then the sample mean is normally distributed with mean μ and variance σ^2/n, for $n \geq 30$. The population from where we sample is known as the *parent population* and the distribution that the sample mean follows is known as the *sampling distribution*.

The normal distribution has the familiar bell-shaped symmetric curve shown in Figure 2.7. Half of the points are above the mean and the other half below it, i.e., if $X \sim N(\mu, \sigma^2)$ then $p(X \leq \mu) = p(X \geq \mu) = 0.50$. In addition, about 68% of all points fall within one standard deviation of the mean, i.e., $p(\mu - \sigma \leq X \leq \mu + \sigma) = 0.6836$. Also, about 96% of the points lie within two standard deviations of the mean, i.e., $p(\mu - 2\sigma \leq X \leq \mu + 2\sigma) = 0.9644$, and finally, almost all of the points lie within three standard deviations of the mean, i.e., $p(\mu - 3\sigma \leq X \leq \mu + 3\sigma) = 0.9974$.

If $\mu = 0$ and $\sigma^2 = 1$, then the normal distribution is known as the *standard* normal distribution and its pdf is

$$f_X(t) = \frac{1}{\sqrt{2\pi}} e^{-\frac{1}{2}t^2}, \quad -\infty < t < +\infty.$$

The standard normal distribution is used to calculate the area under the curve for any normal distribution. Let $X \sim N(\mu, \sigma^2)$, and let us assume that we want to calculate the probability $p(X \geq t)$. This probability is obtained by calculating the area under the curve of the normal distribution $N(\mu, \sigma^2)$ from t to $+\infty$; i.e.,

$$p(X \geq t) = \int_{t}^{+\infty} \frac{1}{\sigma\sqrt{2\pi}} e^{-\frac{1}{2}\frac{(x-\mu)^2}{\sigma^2}} dx.$$

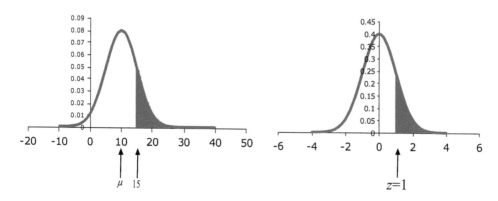

FIGURE 2.8 The normal probability density distribution.

The same probability can be obtained by calculating the area under the standardized normal distribution $N(0, 1)$ from z to $+\infty$, where z is given by the expression

$$z = \frac{t - \mu}{\sigma}.$$

This is known as the *z-transform*. The numerator $t - \mu$ in the above expression is the distance of t from its mean μ. This distance is expressed in the units of X. For instance, if X measures time in seconds, then $t - \mu$ is also in seconds. Now, if we divide $t - \mu$ by the standard deviation σ, the quantity $(t - \mu)/\sigma$ gives the distance of t from μ in standard deviations. For instance, let $\mu = 10$ sec, $\sigma = 5$ sec, and $t = 15$ sec. Then, $t - \mu = 5$ and $(t - \mu)/\sigma = 1$. If $t \leq \mu$, then $(t - \mu)/\sigma$ is negative, meaning that t lies to the left of μ. For instance, if $t = 5$ sec, then $(t - \mu)/\sigma = -1$, which means that t is one standard deviation to left of μ.

Now, in the standard normal distribution, we select point z which has the same distance from its mean 0 in standard deviations as t. For the above example of $\mu = 10$ sec, $\sigma = 5$ sec, and $t = 15$ sec, we set $z = (t - \mu)/\sigma = 1$, i.e., z is one standard deviation to the right of the mean. This is shown in Figure 2.8, where the normal distribution on the left is $N(10, 25)$ and the standard normal distribution is on the right. We recall that the area under each curve is equal to 1, since these curves are probability density functions, and the sum of all probabilities is equal to 1. In view of this, the area under the $N(0, 1)$ curve from z to $+\infty$ is the same as the area under the $N(\mu, \sigma^2)$ curve from t to $+\infty$, since both correspond to the same percentage of the total area.

Typically, when we want to calculate an area under a normal curve, we transform this to the standard normal distribution for which there is a pre-computed table that gives the area under the curve for different values of z.

2.4 THE JOINT PROBABILITY DISTRIBUTION

So far, we considered the case of a single random variable. In many cases, however, we study systems which are described by multiple random variables. For instance, packets from IoT devices that are transmitted to a server may have different priorities. In this case, they are queued by the server in different queues, one per priority. The queues are served according to a priority-based scheduler. In this case, the queueing system is depicted by a number of random variables, one per queue, that give the number of packets waiting in the queue.

The joint probability distribution of n discrete random variables is notated as $p(X_1 = i, X_2 = j, ..., X_n = k)$ and of n continuous random variables as $f_{X_1, X_2, ..., X_n}(t_1, t_2, ..., t_n)$. Their cumulative distributions are notated as $p(X_1 \geq i, X_2 \geq j, ..., X_n \geq k)$ and $F_{X_1, X_2, ..., X_n}$ $(t_1, t_2, ..., t_n)$.

Two discrete random variables X_1 and X_2 are said to be *statistically independent* if

$$p\left(X_1 = i, X_2 = j\right) = p\left(X_1 = i\right)p\left(X_2 = j\right)$$

for all (i, j) values, Otherwise, they are *statistically dependent*. More generally, n discrete random variables $X_1, X_2, ..., X_n$ are statistically independent if

$$p\left(X_1 = i, X_2 = j, ..., X_n = k\right) = p\left(X_1 = i\right)p\left(X_2 = j\right) ... p\left(X_n = k\right)$$

for all $(i, j, ..., k)$ values.

Statistical independence for continuous random variables is defined in terms of their cumulative distributions; i.e., n continuous random variables $X_1, X_2, ..., X_n$ are statistically independent if

$$F_{X_1, X_2, ..., X_n}\left(t_1, t_2, ..., t_n\right) = F_{X_1}\left(t_1\right)F_{X_2}\left(t_2\right) ... F_{X_n}\left(t_n\right)$$

for all $(t_1, t_2, ..., t_n)$ values.

For example, let us consider the experiment of tossing a pair of dice, where one is red and the other green. Let X_1 indicate the number on the green die and X_2 the number on the red die. These two random variables are statistically independent, since $p(X_1 = i, X_2 = j) = 1/36$, and $p(X_1 = i)p(X_2 = j) = 1/6 \times 1/6 = 1/36$. This conclusion of course is not surprising since the outcome of one die does not affect the outcome of another. However, if we define a new random variable X_3 equal to the sum of the numbers shown on both dice, i.e., $X_3 = X_1 + X_2$, then X_1 and X_3, and X_2 and X_3 are not independent. This can be easily verified by applying the above rule of statistical independence. For instance, $p(X_1 = 2, X_3 = 5) = 1/36$. On the other hand, $p(X_1 = 2) = 1/6$ and $p(X_3 = 5) = 2/6$, and of course $1/36 \neq (1/6) \times (2/16)$.

2.4.1 The Marginal Probability Distribution

When studying a problem, it is possible that we may be able to calculate the joint probability distribution of a number of random variables, but we may be only interested in the probability distribution of one of these random variables. A probability distribution calculated from a joint distribution is called the *marginal* distribution.

Let X_1 and X_2 be two discrete random variables, and let $p(X_1 = i, X_3 = j)$ be their joint probability distribution. Then, the marginal distribution of X_1 is as follows:

$$p(X_1 = i) = \sum_{j=-\infty}^{+\infty} p(X_1 = i, X_2 = j).$$

Likewise, the marginal distribution of X_2 is as follows:

$$p(X_2 = j) = \sum_{i=-\infty}^{+\infty} p(X_1 = i, X_j = j).$$

If X_1 and X_2 are continuous random variables, and $f_{X_1,X_2}(t_1, t_2)$ is their joint probability distribution, then, the marginal for X_1 is

$$f_{X_1}(t) = \int_{-\infty}^{+\infty} f_{X_1,X_2}(t,s)\,ds.$$

The above expressions generalize easily to more than two random variables.

2.4.2 The Conditional Probability

The conditional probability is the probability that a random variable X_1 takes some value i given that another variable X_2 has taken a value j. This is notated as $p(X_1 = i \mid X_2 = j)$. The following relation holds:

$$p(X_1 = i \mid X_2 = j) = \frac{p(X_1 = i, X_2 = j)}{p(X_2 = j)}.$$

If X_1 and X_2 are statistically independent, then $p(X_1 = i, X_2 = j) = p(X_1 = i)p(X_2 = j)$ and therefore $p(X_1 = i \mid X_2 = j) = p(X_1 = i)$. The last expression is another definition of statistical independence provided that it holds for all i and j values.

If X_1 and X_2 are continuous random variables, then

$$f_{X_1|X_2=s}(t) = \frac{f_{X_1,X_2}(t,s)}{f_{X_2}(s)}.$$

With continuous random variable, more often we work with their cumulative distributions, i.e.,

$$p(X_1 \leq t \mid X_2 \leq s) = \frac{p(X_1 \leq t, X_2 \leq s)}{p(X_2 \leq s)}.$$

2.5 EXPECTATION AND VARIANCE

The expected value of a random variable is the average of all the values it takes. The expected value of a discrete random variable X with a probability distribution $p(X_1 = i)$, $-\infty < i < +\infty$ is

$$E(X) = \sum_{i=-\infty}^{+\infty} ip(X = i)$$

and the expected value of a continuous random variable X with a pdf $f_X(t)$ is

$$E(X) = \int_{-\infty}^{+\infty} tf_X(t)dt.$$

Below, we describe some of the properties of the expectation.

Let X be a random variable, then the following two relations hold:

$$E(aX) = aE(X)$$

$$E(aX + b) = aE(X) + b,$$

where a and b are constants (not functions of X).

Let X be a random variable, and let Y be a new random variable which is obtained by transforming X through a function $g(X)$, i.e., $Y = g(X)$. Then, the expected value of Y is as follows:

$$E(g(X)) = \begin{cases} \displaystyle\sum_{i=-\infty}^{+\infty} g(x_i)p(X = x_i), & \text{if } X \text{ is discrete} \\[2em] \displaystyle\int_{-\infty}^{+\infty} g(x)f_x(x)dx & \text{if } X \text{ is continuous} \end{cases}$$

Let $X_1, X_2, \ldots, X_n$ be n random variables (they may be mutually dependent or independent). Then,

$$E(X_1 + X_2 + \cdots + X_n) = E(X_1) + E(X_2) + \cdots + E(X_n).$$

Let $X_1, X_2, \ldots, X_n$ be n random variables (which are mutually independent). Then,

$$E(X_1 X_2 \ldots X_n) = E(X_1)E(X_2)\ldots E(X_n).$$

The variance of a random variable is a metric that describes the variability of the values around the mean. If it is small, then most of the values are clustered around the mean. If it is large, then the values are scattered away from the mean. The variance of a discrete random variable X with a probability distribution $p(X = i)$, $-\infty < i < +\infty$, is

$$Var(X) = \sum_{i=-\infty}^{+\infty} (i - E(X))^2 \, p(X = i)$$

and the variance of a continuous random variable X with a pdf $f_X(t)$ is

$$Var(X) = \int_{-\infty}^{+\infty} (t - E(X))^2 \, f_X(t) dt.$$

Below, we describe some of the properties of the variance.

Let X be a random variable, then the following two relations hold:

$$Var(aX) = a^2 Var(X)$$

$$Var(X + b) = E(X),$$

where a and b are constants (not functions of X).

Let $X_1, X_2, \ldots, X_n$ be n random variables. Then,

$$Var(X_1 + X_2 + \cdots + X_n)$$

$$= Var(X_1) + Var(X_2) + \cdots + Var(X_n) + 2\sum_{i=1}^{n-1}\sum_{j=i+1}^{n} Cov(X_i, X_j),$$

where $Cov(X_i, X_j)$ is the covariance between the two random variables X_i and X_j. If $X_1, X_2, \ldots, X_n$ are mutually independent, then the covariance terms $Cov(X_i, Y_j)$ are all zero and we have the following expression similar to the one obtained for the expectation:

$$Var(X_1 + X_2 + \cdots + X_n) = Var(X_1) + Var(X_2) + \cdots + Var(X_n).$$

The covariance $Cov(X, Y)$ of two random variables X and Y measures the degree of their dependency, and it is defined as follows:

$$Cov(X, Y) = E\left[(X - E(X))(Y - E(Y)) \right]$$

$$= E\left[XY - YE(X) - XE(Y) + E(X)E(Y) \right]$$

$$= E(XY) - E(X)E(Y).$$

If X and Y are independent, then $E(XY) = E(X)E(Y)$, and therefore $Cov(X, Y) = 0$. The covariance takes values in $(-\infty, +\infty)$.

An alternative way to measure the dependency between two random variables X and Y is to use the correlation $Corr(X, Y)$, which is the normalized $Cov(X, Y)$, defined as follows:

$$Corr(X, Y) = \frac{Cov(X, Y)}{\sqrt{Var(X)}\sqrt{Var(Y)}},$$

where $\sqrt{Var(\cdot)}$ is the standard deviation of X and Y. The correlation takes values in $(-1, 1)$. In practice, it is used more than the covariance.

We conclude this section by giving the expressions for the *nth moment about the origin* $E(X^n)$ and the *nth central moment* $E[(X - E(X))^n]$ of a random variable X. If X is discrete, we have

$$E(X^n) = \sum_{i=-\infty}^{+\infty} i^n p(X = i)$$

$$E\left[(X - E(X))^n\right] = \sum_{i=-\infty}^{+\infty} (i - E(X))^n p(X = i).$$

If X is continuous, we have

$$E(X^n) = \int_{-\infty}^{+\infty} t^n f_X(t) dx$$

$$E\left[(X - E(X))^n\right] = \int_{-\infty}^{+\infty} (t - E(X))^n f_X(t) dt.$$

2.5.1 The Expectation and Variance of Some Random Variables

In Sections 2.2 and 2.3, we examined several discrete and continuous random variables. Tables 2.2 and 2.3 give their expectation and variance. Below, we derive the expected value of the binomial, geometric, Poisson, and exponential random variables.

The expected value of a binomial random variable X can be easily obtained by adding up the expected value of all tosses. For the ith toss, let $X_i = 1$ if heads come up and $X_i = 0$ if tails come up. Then, the expected value of X_i is $E(X_i) = 0 \times q + 1 \times p = p$, and consequently,

$$E(X) = E(X_1 + X_2 + \cdots + X_n) = \sum_{i=1}^{n} E(X_i) = np.$$

TABLE 2.2   Discrete random variables

X	Parameters	E(X)	Var(X)
Binomial	n, p	np	npq
Geometric	n, p	p/q	p/q^2
Poisson	λ	λ	λ

TABLE 2.3   Continuous random variables

X	Parameters	E(X)	Var(X)
Uniform	a, b	$(a + b)/2$	$(b - a)^2/12$
Exponential	λ	$1/\lambda$	$1/\lambda^2$
Erlang	n, λ	n/λ	n/λ^2
Hyper-exponential	p, λ_1, λ_2	$p/\lambda_1 + (1 - p)/\lambda_2$	$2(p/\lambda_1^2 + (1-p)/\lambda_2^2) - (p/\lambda_1 + (1-p)/\lambda_2)^2$

The geometric random variable has the probability distribution

$$p(X = k) = p^k q, \ k \geq 0,$$

and its expected value is p/q. This can be obtained as follows:

$$E(X) = \sum_{k=0}^{+\infty} kp^k q = pq \sum_{k=0}^{+\infty} kp^{k-1} = pq \sum_{k=0}^{+\infty} \frac{d}{dk} p^k$$

$$= pq \frac{d}{dk} \left(\sum_{k=0}^{+\infty} p^k \right) = pq \frac{d}{dk} \frac{1}{1-p} = pq \frac{1}{(1-p)^2}$$

$$= pq \frac{1}{q^2} = \frac{p}{q}.$$

In the above derivation, we make use of the expressions

$$\frac{1}{1-p} = 1 + p + p^2 + \cdots$$

and

$$\frac{d}{dx} \frac{1}{f(x)} = -\frac{1}{(f(x))^2} \frac{d}{dx} f(x).$$

The Poisson random variable has the probability distribution

$$p(X = i) = e^{-\lambda} \frac{\lambda^i}{i!}, \, i = 0, 1, 2, \ldots$$

and its expected value is λ. This can be obtained as follows:

$$E(X) = \sum_{k-0}^{+\infty} kp(X = k) = \sum_{k-0}^{+\infty} k e^{-\lambda} \frac{\lambda^\kappa}{\kappa!}$$

$$= e^{-\lambda} \lambda \sum_{k-0}^{+\infty} k \frac{\lambda^{\kappa-1}}{\kappa!} = e^{-\lambda} \lambda \sum_{k-0}^{+\infty} \frac{d}{d\lambda} \frac{\lambda^\kappa}{\kappa!} = e^{-\lambda} \lambda \frac{d}{d\lambda} \left(\sum_{\kappa=0}^{+\infty} \frac{\lambda^\kappa}{\kappa!} \right)$$

$$= e^{-\lambda} \lambda \frac{d}{d\lambda} e^\lambda = \lambda.$$

In the above derivation, we make use of the expressions

$$e^x = 1 + \frac{\lambda^1}{1!} + \frac{\lambda^2}{2!} + \cdots + \frac{\lambda^n}{n!} + \cdots$$

and

$$\frac{d}{dx} e^x = e^x.$$

Finally, the exponential random variable has the pdf

$$f_X(t) = \lambda e^{-\lambda t}, \, t \geq 0$$

and its expected value is $1/\lambda$. This can be shown as follows:

$$E(X) = \int_0^{+\infty} \lambda t e^{-\lambda t} dt = \frac{1}{\lambda} \int_0^{+\infty} (\lambda t) e^{-\lambda t} d(\lambda t).$$

Setting $u = \lambda t$, we have

$$E(X) = \frac{1}{\lambda} \int_0^{+\infty} u e^{-u} du = \frac{1}{\lambda} \left[u(-e^{-u}) \Big|_0^{+\infty} - \int_0^{+\infty} e^{-u} d(-u) \right]$$

$$= \frac{1}{\lambda}\left[0 + \int\limits_0^{+\infty} e^{-u} d(-u)\right] = \frac{1}{\lambda}.$$

In this derivation, we make use of the following integration rule:

$$\int\limits_{-\infty}^{+\infty} f(x)g'(x)dx = f(x)g(x)\Big|_{-\infty}^{+\infty} - \int\limits_{-\infty}^{+\infty} f'(x)g(x)dx.$$

EXERCISES

1. An unbiased die is thrown once. Compute the probability of the following events: (a) The number shown is odd, (b) The number shown is less than 3, and (c) The number shown is a prime number.

2. A prisoner in a prison is put in the following situation. A regular deck of 52 cards is placed in front of him. He must choose cards one at a time, and once a card is chosen, the card is replaced in the deck and the deck is shuffled. If the prisoner happens to select three consecutive red cards, he is executed. If he happens to select six cards before three consecutive red cards appear, he is granted freedom. What is the probability that the prisoner is executed?

3. Three marksmen fire simultaneously and independently at a target. What is the probability of the target being hit at least once, given that marksman one hits a target 9 times out of 10, marksman two hits a target 8 times out of 10 while marksman three only hits a target 1 out of every 2 times.

4. IoT devices transmit packets to a server according to the Poisson distribution with a rate of 100 per hour. Suppose that the machine breaks down from time to time and it takes 1 hour to be repaired. What is the probability that (a) zero, (b) two, and (c) five new packets will arrive during this time?

5. Assume that IoT devices fail in accordance with a Poisson distribution at the rate of one device per five weeks. If there are two spare devices on hand and if a new supply will arrive in eight weeks, what is the probability that during the next eight weeks the IoT system will not be fully operational (i.e., not all IoT devices will be working) for a week or more owing to lack of devices?

6. IoT requests processed by a server require a CPU time that can be modeled by an exponential distribution with mean 140 msec. The CPU scheduling algorithm is based on time-slicing; i.e., if an IoT request does not complete within a time slice of 100 msec, it is routed back to the end of the queue where it waits its turn again for another 100 msec slice, and so on until it is fully processed and departs from the server.

a. What is the probability that an arriving request will be forced to wait for a second time-slice?

b. Of the 500 requests coming during an hour, how many are expected to finish within the first time-slice?

REFERENCES

1. W.J. Stewart. *Probability, Markov Chains, Queues, and Simulation*, July 2009, Princeton University Press.
2. S. Ross. *Introduction to Probability Models*, 2019 12 edn, Academic Press.

In addition, you will find a lot of material in the Internet.

CHAPTER 3

Simulation Techniques

COMPUTER SIMULATION IS A MODELING TOOL THAT CAN BE USED to evaluate the performance of different systems, such as an IoT system. Simulation techniques are easy to use and in view of this they are very popular. When simulating a system, its operation is depicted in a computer program, which is then exercised for what–if analysis. This chapter starts by describing how to develop a simple hand simulation model using an example that deals with the recertification of IoT devices. Subsequently, a list of relevant topics that one must know in order to develop a complete simulation model are presented. These are as follows: pseudo-random number generation, random variates generation, simulation designs, estimation techniques, and validation of a simulation. A set of exercises and a simulation project is given at the end of the chapter.

3.1 INTRODUCTION

There are two major categories of simulation depending upon the system that is being modeled: *discrete-event* and *continuous event*. Discrete-event simulation is used to model systems whose state takes discrete values. For instance, let us assume that we want to simulate the queueing of IoT requests in a server. The state of this system is discrete, because it is depicted by the number n of requests queueing for service, $n = 0, 1, 2, ...$, which is a discrete variable. Discrete-event simulation is used to model systems such as IoT systems, computer networks, production systems, etc. Continuous-event simulation, on the other hand, is totally different to discrete-event simulation and it is based on differential equations. It is used for systems whose state is a continuous variable, such as the position in the trajectory of a rocket. In this chapter, we deal with discrete-event simulation techniques.

Discrete-event simulation is easy to use, and therefore it is a very popular modeling technique. This is in contrast to other modeling techniques, such as queueing theory and optimization techniques, that require in-depth mathematical knowledge.

A discrete-event simulation model is a computer program that depicts a system under study. It uses random numbers to represent the occurrence of events and it tracks the evolution of the various events that occur in the system. Statistical techniques are used to obtain estimates of performance measures generated by a simulation model.

There are several discrete-event simulation languages that enable a rapid development of a simulation model, such as, ns-3, SLAM II, ARENA, MATLAB-Simulink, OPNET, and JMT. In this chapter, we describe the basic simulation techniques for building a discrete-event simulation model from scratch without using a simulation language.

The best way to understand the material in this chapter is to do the simulation project also given at the end of the chapter.

3.2 THE DISCRETE-EVENT SIMULATION TECHNIQUE

We present the discrete-event approach for building a simulation model by means of an example. The example has to do with IoT devices sending recertification requests to a server. In the following section, we present a simple model of this problem, and in Section 3.2.2 we extend this model to a more complex one.

3.2.1 Recertification of IoT Devices: A Simple Model

We consider the case where a large number of IoT devices send recertification requests to an authentication server. The requests are placed into a single queue, as shown in Figure 3.1 and they are processed by the server in the order in which they arrive, i.e., first-in-first-out (FIFO). For simplicity, the transmission time from an IoT device to the server is not modeled. Also, we are not interested what happens to the requests after they are processed by the server.

The first and most important step in building a simulation model is to identify the *events* whose occurrence alter the *state* of the system. The state of the system is described by a number of variables which are related to the system under study. Furthermore, the selection of the state variables depends also on what performance measures we want to obtain.

In this problem, the state variable of interest is n: the number of requests waiting in the queue plus the one being served. If $n = 0$, then the queue is empty and the server is idle. If $n = 1$, then the queue is empty but the server is busy. If $n > 1$, then the server is busy and there are $n - 1$ requests in the queue.

Having identified the state variables, the next step is to identify the events whose occurrence will cause them to change. In our example, the following two events will cause n to change:

1. Arrival of a request

2. Completion of the service time of a request

Having identified the events, the next step is to identify the logic that has to be executed when an event occurs.

1. Arrival of a request

    a. If there are other requests waiting in the queue or the queue is empty but the server is busy at the moment of arrival, then the request joins the end of the queue.

    b. If the server is idle at the moment of arrival, then the server begins processing the request; i.e., a new service begins.

2. Completion of the service time of a request

    a. If there are other requests waiting in the queue, then the server starts processing the request at the top of the queue; i.e., a new service begins.

    b. If there are no requests waiting in the queue, then the server becomes idle.

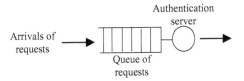

FIGURE 3.1 Queueing of requests at the authentication server.

In order to incorporate the events in a simulation model, we use a set of variables, known as *clocks*, which keep track of the time instants at which events occur. In our case, we need two clocks, an *arrival clock* (*AC*) and a *departure clock* (*DC*). The arrival clock gives the time instant at which the next arrival will occur, and the departure clock gives the time instant at which a service completion will occur. In addition, we use a clock that gives the current time, known as the *master clock* (*MC*). We note that these clocks are just variables and have nothing to do with the real clock on the wall or the computer's clock.

The heart of the simulation model centers around the management of these two events. In particular, using the arrival and departure clocks, the model decides which of the two events will occur next; i.e., it locates the event which has the smallest clock value, say *t*. Then, the master clock is advanced to time *t*, and the model takes action according to the type of event as described above. It then goes back to compare the two clocks, and so on. The event management approach is depicted in the flow chart in Figure 3.2.

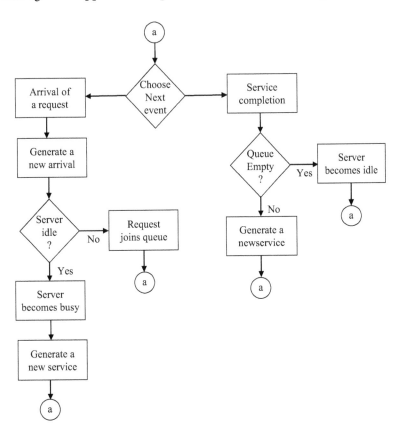

FIGURE 3.2 Event management.

If we assume that a request is in service, then we have an arrival event and a departure event pending. In this case, choosing the next event involves simply the comparison of *AC* and *DC*.

1. If $DC < AC$, then the next event to occur is the service completion. In this case, we move the master clock to that time, by setting $MC = DC$, branch out on the right-hand side of the flow chart and follow the logic described above for a service completion; i.e., if there is a request waiting, then it starts its service. In this case, we change the departure clock to point to the time when this new service will be completed. For simplicity, we assume that the service time is constant equal to 10 units of time, and therefore $DC = DC + 10$. If the queue is empty, then we set the state of the server to idle and go back to the beginning to look for the next event.

2. If $AC < DC$, then the next event is an arrival. In this case, we move the master clock to that time, by setting $MC = AC$, and branch to the left of the flow chart. Before we proceed to execute the logic associated with the arrival event described above, we schedule the time that the next arrival will occur. Again, for simplicity we assume that the inter-arrival time is constant equal to 6 units of time, and therefore, $AC = AC + 6$. Subsequently, we check the state of the server. If it is busy, then there is one request being served and possible one or more waiting in the queue. In this case, the new arrival is queued and we go back to the beginning to look for the next event. If the server is idle, then the new arrival starts its service immediately. For this, we set the state of the server to busy and schedule the time when the new service will be completed, i.e., $DC = MC + 10$. Then go back to the beginning to look for the next event.

Above, we examined the case where there are two events pending, i.e., an arrival and a departure. If the server is idle, then there is only one event pending, that of an arrival. In this case, we move the master clock to the time of arrival, by setting $MC = AC$, and branch to the left of the flow chart. Before we proceed to execute the logic associated with the arrival event, we schedule the time of the next arrival. Since the server is idle, the new arrival starts its service immediately. We set the state of the server to busy and schedule the time when this service will be completed, i.e., $DC = MC + 10$. Then, we go back to the beginning to look for the next event.

We are now ready to do the hand simulation given in Table 3.1. In order to start the simulation, we need to specify the initial conditions; i.e., we need to decide in what state the system is at time $MC = 0$. As will be seen later on, the initial conditions do not affect the metric we want to estimate. So, we assume that there is one request in service that is due to complete its service at time 2, the queue is empty, and the first arrival is due to occur at time 6. In Table 3.1, the column "queue" gives the number of customers waiting to receive service, and the column "server" gives the state of the server, busy or idle.

When the clocks are integers, it is possible that multiple events may occur at the same time. In this case, we need to decide in advance the order in which they will be executed. Often the order of execution is immaterial to the calculation of the final estimate, but sometimes it does have an impact, and therefore the order of execution needs to be determined so

TABLE 3.1   Hand simulation

MC	AC	DC	Queue	Server
0	6	2	0	busy
2	6	–	0	idle
6	12	16	0	busy
12	18	16	1	busy
16	18	26	0	busy
18	24	26	1	busy
24	30	26	2	busy
26	30	36	1	busy
30	36	36	2	busy
36	36	46	1	busy
36	42	46	2	busy
42	48	46	3	busy
46	48	56	2	busy
48	54	56	3	busy
54	60	56	4	busy
56	60	66	3	busy
60	66	66	4	busy

that to imitate the system as closely as possible. In our hand simulation in Table 3.1, we see that often $AC = DC$ (In fact, it is repeated every 10 units of time, but this is of no significance.). In this case, the order of execution of the events is to first do the departure event and then the arrival event. However, if we did it in the other way, it will not affect the estimation of the final estimate, which in this case is the distribution of the waiting time in the queue, and more specifically, the mean and 99th percentile of the waiting time in the queue.

The observant reader will realize that the number in the system keeps increasing. This is because the mean arrival time $1/\lambda$ is 6 and the mean service time $1/\mu$ is 10; i.e., the arrival rate (the average number of arrivals per unit time) $\lambda = 1/6 = 0.167$, and the service rate (the average number of departures if the server is continuously busy) is $1/10 = 0.1$, and therefore $\lambda > \mu$. In view of this, requests arrive faster than they are served, and as a result the queue of waiting requests keeps increasing. In this case, the queue is *unstable*. Typically, we want to use values of λ and μ such that $\lambda < \mu$, so that the queue is *stable*, i.e., it does not grow infinitely large. For more details, see Appendix A.

3.2.2 Recertification of IoT Devices: A More Complex Model

In this section, we expand the simple model examined above to a slightly more complex model. As before, we assume that a large number of IoT devices are recertified by a server. In order to avoid overflows of the buffer allocated to the certification process that runs on a server, we impose an upper bound B on the total number of requests waiting to be processed, including the one in service, as shown in Figure 3.3. A request that arrives at the server when there are less than B requests in the system is allowed to join the queue.

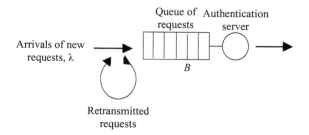

Queue of Authentication
requests server

Arrivals of new
requests, λ

B

Retransmitted
requests

FIGURE 3.3 A queueing representation of the model with retransmissions.

Otherwise it is rejected and a message is sent back to the IoT device. In this case, the IoT device goes through a delay d and retransmits its request.

We assume that requests arrive at a fixed rate. We will refer to these requests as *new requests*. A blocked request is retransmitted after a delay d and it is referred to as a *retransmitted request* (In queueing theory, it is also referred to as an *orbiting customer.*).

The state variables of interest are the number of requests that have succeeded to get into the buffer, and the number of retransmitted requests. The following events will cause these two state variables to change:

1. Arrival of a new request

2. Arrival of a retransmitted request

3. Completion of a service time

The occurrence of one event may trigger the occurrence of one or more events, as will be seen below. Having identified the events, the next step is to identify the logic that has to be executed when an event occurs. We assume that the inter-arrival time, retransmission time, and service time are all constant, equal to 6, 5, and 10, respectively, and $B = 2$. Also, we define the following clocks: CLA, CLS, and CLR, where CLA is the clock that indicates the time of the next new arrival, CLS indicates the service completion time, and CLR indicates the time a retransmitted request will arrive. There may be multiple CLR values, one for each request that is being retransmitted. MC indicates the master clock and shows the current simulation time.

1. Arrival of a new request

   a. Once a new request arrives, the next arrival of a new request is determined, i.e., $CLA = MC + 6$. If the total number of requests (i.e., those in the queue plus the one in service) is less than B, then the new arrival enters the buffer. If the buffer is empty and the server is idle, then a new service completion is scheduled, i.e., $CLS = MC + 10$.

   b. If the total number of requests is equal to B, then the new arrival is rejected. In this case, the rejected request will be retransmitted after a constant retransmission delay 5. For this, a new event for the arrival of the retransmitted request at time $CLR = MC + 5$ is scheduled.

2. Arrival of a retransmitted request

    a. If the total number of requests is less than B, the arrival enters the buffer. If the buffer is empty and the server is idle, a new service completion is scheduled, i.e., $CLS = MC + 10$.

    b. If the total number of requests is equal to B, then the arrival is rejected again, and a new arrival time for the retransmitted request is scheduled at time $CLR = MC + 5$.

3. Service completion

    a. Once a service is completed, the next service completion time is generated if the queue is not empty. This will occur at $MC + 10$, where 10 is the constant service time. If the queue is empty, then the server becomes idle.

We are now ready to do the hand simulation given in Table 3.2. For the initial conditions at master clock time 0, we assume that the queue is empty, the server is idle, there are no retransmitted requests, and the first new request will arrive at $CLA = 2$.

TABLE 3.2   Hand simulation

MC	CLA	CLS	Total number of requests	CLR
0	2	–	0	–
2	8	12	1	–
8	14	12	2	–
12	14	22	1	–
14	20	22	2	–
20	26	22	2	25
22	26	32	1	25
25	26	32	2	–
26	32	32	2	31
31	32	32	2	36
32	38	32	2	36, 37
32	38	42	1	36, 37
36	38	42	2	37
37	38	42	2	42
38	44	42	2	42, 43
42	44	42	2	43, 47
42	44	52	1	43, 47
43	44	52	2	47
44	50	52	2	47, 49
47	50	52	2	49, 52
49	50	52	2	52, 54
50	56	52	2	52, 54, 55

Due to the fact that the clocks are all integers, it is possible that two or more events may be scheduled to occur at the same time. In the previous example, the order in which multiple events occurring at the same time are executed did not matter. However, in this case it does matter. We will use the following order: first execute a retransmitted arrival, then a new arrival, and finally a service completion.

We note that column "Total number of requests" in Table 3.2 gives the number of requests waiting in the queue plus the one in service. Also, column "*CLR*" may contain multiple clock values, one per retransmitted request.

3.3 GENERATING RANDOM NUMBERS

In the section, we will discuss methods for generating random numbers. Numbers chosen at random are used in a variety of applications. In simulation, they are used to introduce randomness in a model. For instance, in the example discussed in Section 3.2.2, we assumed that the inter-arrival, retransmission time, and service time are all constant. It is possible that the service time is constant, since it may execute a fixed number of instructions, but it is highly unlikely that the requests arrive at constant intervals and also that the retransmission time is also constant. The variability of the inter-arrival and retransmission times, and in general of any variable used in a simulation, is typically represented by an empirical or a theoretical probability distribution. In view of this, in order to make a simulation model more realistic, one should be able to use numbers from a given theoretical or empirical distribution generated in a random way.

We distinguish between generating random numbers which are uniformly distributed in [0, 1], known as *pseudo-random numbers*, and random numbers which follow an empirical or theoretical distribution, known as *random variates or stochastic variates*. Pseudo-random numbers are used in the generation of random variates. In view of this, we first discuss how to generate pseudo-random numbers and then discuss how to generate random variates.

3.3.1 Generating Pseudo-Random Numbers

Random numbers uniformly distributed in [0, 1] are generated using various mathematical algorithms. These algorithms produce numbers in a deterministic fashion; i.e., given a starting value, known as the *seed*, the same sequence of random numbers can be produced each time as long as the seed remains the same. Despite the deterministic way in which random numbers are created, these numbers appear to be random since they pass a number of statistical tests designed to test various properties of random numbers. In view of this, these random numbers are referred to as *pseudo-random numbers*.

An advantage of generating pseudo-random numbers in a deterministic fashion is that they are reproducible, since the same sequence of random numbers is produced each time we run a pseudo-random generator with the same starting value. This is helpful when debugging a simulation program, as we typically want to reproduce the same sequence of events in order to verify the correctness of the simulation.

In general, an acceptable method for generating random numbers must yield sequences of numbers or bits that are uniformly distributed, statistically independent, reproducible, and non-repeating for any desired length. Several algorithms have been proposed for pseudo-random number generation, such as, the *congruential* method, the *Tausworthe* generators, the *lagged Fibonacci* generators, and the *Mersenne twister*. Below, we examine the congruential method.

The congruential method is a very popular method and most of the pseudo-random number generators in programming languages are based on some variation of this method. The advantage of the congruential method is that it is very simple, fast, and it produces pseudo-random numbers that are statistically acceptable for computer simulation. The congruential method uses the following recursive relationship to generate random numbers:

$$x_{i+1} = \left(ax_i + b\right)\left(mod\ m\right),$$

where x_i, a, b, and m are all non-negative numbers. Given that the previous random number was x_i, the next random number x_{i+1} can be generated as follows. Multiply x_i by a and then add b. Then, compute the modulus m of the result; i.e., divide the result by m and set x_{i+1} equal to the remainder of the division. For example, if $x_0 = 0$, $a = b = 7$, and $m = 10$, then we can obtain the following sequence of numbers: 7, 6, 9, 0, 7, 6, 9, 0, ... The initial value x_0 is the seed and it is required to get the generator started. The seed value does not affect the long-term behavior of a simulation model after a small set of random numbers has been generated.

The method using the above expression is known as the *mixed congruential method*. A simpler version of this method is the *multiplicative congruential method* given by the expression:

$$x_{i+1} = ax_i \left(mod\ m\right).$$

The numbers generated by a congruential method are between 0 and $m - 1$. Uniformly distributed random numbers between 0 and 1 can be obtained by simply dividing the resulting x_i by m.

The number of successively generated pseudo-random numbers after which the sequence starts repeating itself is called the *period*. If the period is equal to m, then the generator is said to have a *full period*. It is important to note that one should not use any arbitrary values for a, b, and m. Systematic testing of various values for these parameters has led to generators which have a full period and which are statistically satisfactory. Such generators can be found in all programming languages and in statistical and mathematical software packages.

The mixed congruential method is a special case of a following general case: $x_{i+1} = f(x_i, x_{i-}, ...)$ $(mod\ m)$, where $f(.)$ is a function of previously generated pseudo-random numbers. Another well-known special case is the *quadratic* congruential generator: $x_{i+1} = (ax_i^2 + bx_{i-1} + c)(mod\ m)$.

Congruential generators can be combined together to construct composite generators, which may have better statistical behavior than that of the individual generators. One good example involves two congruential generators. The first generator is used to fill a vector of size n with random numbers. The second generator is then used to generate a random integer k uniformly distributed over the numbers 1, 2, ..., n. The random number stored in the kth position of the vector is the random number returned by the composite generator. The first generator then replaces the random number in the kth position with a new random number. The procedure repeats itself in this fashion. It has been demonstrated that this composite generator has good statistical properties.

3.3.2 Generating Random Variates

There are many techniques for generating random variates, of which the *inverse transformation* method is one of the most commonly used. This method is applicable only to cases where the cumulative distribution is invertible. Let X be a random variable and let $f_X(t)$ and $F_X(t)$ be its pdf and cumulative distribution, respectively. Now, let us say that we want to generate random variates from $f_X(t)$. We note that $F_X(t)$ is defined in the region [0, 1]. In view of this, we can select a random value of $F_X(t)$ by generating a pseudo-random number r and set it equal to $F_X(t)$, i.e., $F_X(t) = r$. Inverting $F_X(t)$ gives

$$t = F_X^{-1}(r),$$

where $F_X^{-1}(.)$ indicates the inverse of $F_X(.)$.

For example, let us assume that we want to generate random variates with probability density function $f_X(t) = 2t$, $0 \le t \le 1$. The cumulative distribution $F_X(t)$ is

$$F_X(t) = \int_0^t 2x\,dx = 2\left(\frac{1}{2}x^2\Big|_0^t\right) = t^2,\ 0 \le t \le 1.$$

Now, let r be a pseudo-random number, then $r = t^2$, and consequently $t = \sqrt{r}$. This inversion is shown graphically in Figure 3.4. Each time we want to generate a random variate t from the above distribution, we generate a pseudo-random number r and then obtain t by taking the square root of r.

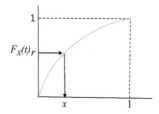

FIGURE 3.4 The inverse transformation method.

Below, we employ the inverse transformation method to generate random variates from some well-known continuous and discrete probability distributions and also from empirical distributions (A review of continuous and discrete distributions is given in Sections 2.2 and 2.3.).

a. Generating Uniformly Distributed Random Variates
The uniform random variable X defined in the interval $[a, b]$ has the following pdf:

$$f_X(t) = \begin{cases} 1/(b-a), & \text{if } t \in [a, b] \\ 0, & \text{otherwise} \end{cases},$$

and its cumulative distribution is

$$F_X(t) = \frac{t-a}{b-a}.$$

Let r be a pseudo-random number. Then, using the inverse transformation method we have

$$r = \frac{t-a}{b-a}$$

or

$$t = a + r(b-a).$$

Using the above expression, we can generate random variates from the uniform distribution. Each time, we generate a pseudo-random number r and then obtain a sample t using the above expression.

b. Generating Exponentially Distributed Random Variates
The exponential random variable X has the following pdf:

$$f_X(t) = \lambda e^{-\lambda t}, t \geq 0,$$

and its cumulative distribution $F_X(t)$ is

$$F_X(t) = 1 - e^{-\lambda t}.$$

Let r be a pseudo-random number. Then, using the inverse transformation method we have

$$r = 1 - e^{-\lambda t}.$$

Solving for t, we have: $e^{-\lambda t} = 1 - r$, or $-\lambda t = \log(1 - r)$, or

$$t = -\frac{1}{\lambda}\log_e(1-r).$$

Since $1 - r$ is also a random number, we can simplify the above expression to

$$t = -\frac{1}{\lambda}\log_e(r),$$

where $1/\lambda$ is the mean value of X. We note that the right-hand side of the above expression is always positive, since $\log_e(1 - r)$ is negative.

c. Generating Normally Distributed Random Variates

The normal random variable X has the following probability density function:

$$f_X(t) = \frac{1}{\sigma\sqrt{2\pi}}e^{-\frac{1}{2}\frac{(t-\mu)^2}{\sigma^2}}, \quad -\infty < t < +\infty.$$

Random variates can be obtained by inverting the cumulative function:

$$F_X(t) = \frac{1}{\sigma\sqrt{2\pi}}\int_{-\infty}^{t}e^{-\frac{1}{2}\frac{(x-\mu)^2}{\sigma^2}}\,dx.$$

For this, there are functions available in programming languages and statistical and mathematical packages.

An alternative method is to use the Box–Muller algorithm that generates two normal variates at a time. Let r_1 and r_2 be two pseudo-random numbers. Then, we can obtain two normal variates t_1 and t_2 as follows:

$$t_1 = \sqrt{-2\log_e r_1}\,\cos(2\pi r_2)$$

and

$$t_2 = \sqrt{-2\log_e r_1}\,\sin(2\pi r_2).$$

d. Generating Binomial Distributed Random Variates

The binomial random variable X gives the number of times outcome 1 occurs in a Bernoulli experiment. Let p and $1 - p$ be the probability that 1 or 2 comes up in a single trial, and let n be the total number of trials. Then, the probability distribution of X is

$$p\left(X=i\right)=\binom{n}{i}p^{i}\left(1-p\right)^{n-i}, i=0, 1, 2, \ldots$$

The easiest way to generate random variates is to imitate the Bernoulli experiment. Let i be a counter set to zero. We generate a pseudo-random number r_1 and if $r_1 \leq p$, then outcome 1 occurs in the first trial and i is increased by one, otherwise outcome 2 occurs and i is not modified. Next, we generate a second random number r_2, and if $r_2 \leq p$, then outcome 1 occurs in the second trial and i is increased by one, otherwise outcome 2 occurs and i is not modified. We continue in this way until we generate n pseudo-random numbers. The final value of i is a random variate from the binomial distribution.

e. Generating Geometrically Distributed Random Variates

The geometric random variable X gives the number of times outcome 1 occurs in a Bernoulli experiment, before outcome 2 occurs. Let p and $1 - p$ be the probability that 1 or 2 comes up in a single trial, and let n be the total number of trials. Then, the probability distribution of X is

$$p\left(X=i\right)=p^{i}\left(1-p\right), i=0, 1, \ldots, n.$$

Its cumulative probability function is

$$F_X\left(k\right)=1-p^{k+1}.$$

We observe that $1 - F_X(k)$ also varies between 0 and 1. Therefore, if r is a pseudo-random number, then

$$r = p^{k+1},$$

or

$$\log r = \left(k+1\right)\log p,$$

or

$$k = \frac{\log r}{\log q} - 1.$$

The above expression can be further simplified since the quantity $(1 - F_X(k))/p = p^k$ also varies between 0 and 1. We have

$$r = p^{k},$$

or

$$k = \frac{\log r}{\log p}.$$

f. Generating Poisson Distributed Random Variates

The Poisson random variable gives the number of times an event occurs in a unit time and has the following probability distribution:

$$p(X = i) = e^{-\lambda} \frac{\lambda^i}{i!}, \, i = 0, 1, 2, \dots$$

where λ is the average number of times the event occurs in a unit time.

We recall that for Poisson distributed arrivals with a mean λ, the inter-arrival time between successive arrivals is exponentially distributed with a mean $1/\lambda$. Therefore, the easiest way to generate Poisson variates is to generate exponentially distributed random variates until their sum exceeds the unit time. Let us assume that the unit time is equal to 1 (it could be any value for that matter). Then, we generate exponentially distributed random variates $t_1, t_2, \dots, t_n, t_{n+1}$, so that

$$\sum_{i=1}^{n} t_i \leq 1 < \sum_{i=1}^{n+1} t_i.$$

The value n is the Poisson random variate. We recall that an exponential variate t is obtained using the expression: $t = -(1/\lambda)\log_e(r)$, where $1/\lambda$ is the mean and r is a pseudo-random number. Consequently, the above expression can be re-written as

$$\sum_{i=1}^{n} -\frac{1}{\lambda} \log_e r_i \leq 1 < \sum_{i=1}^{n} -\frac{1}{\lambda} \log_e r_i,$$

or

$$\sum_{i=1}^{n} \log_e r_i \geq -\lambda > \sum_{i=1}^{n} \log_e r_i,$$

or

$$\log_e \left(\prod_{i=1}^{n} r_i \right) \geq -\lambda > \log_e \left(\prod_{i=1}^{n+1} r_i \right),$$

or

$$\prod_{i=1}^{n} r_i \geq e^{-\lambda} > \prod_{i=1}^{n+1} r_i.$$

The value n for which the above expression holds is a Poisson random variate.

g. Generating Random Variates from an Empirical Distribution

Empirical probability distributions are used often in simulation, and they can be either approximated by a known theoretical distribution or used as is. Generating random variates directly from an empirical distribution is quite easy and it is done using the inverse transformation method.

Consider a discrete random variable X that takes the values 0, 1, 2, 3 with probability 1/8, 3/8, 3/8, 1/8, respectively. Its cumulative distribution takes the values 1/8, 4/8, 7/8, 1, as shown in Figure 3.5. Let r be a pseudo-random number. Then we select a random variate as follows:

1. If $r \leq 1/8$, then $X = 0$

2. If $1/8 \leq r \leq 4/8$, then $X = 1$

3. If $4/8 \leq r \leq 7/8$, then $X = 2$

4. If $7/8 \leq r \leq 1$, then $X = 3$

In Figure 3.5, r falls in between 4/8 and 7/8, and therefore we chose 2.

If the random variable X is continuous, then it is described by a histogram, as shown in Figure 3.6. Let x_i and $f_X(x_i)$, $i = 1, 2, ..., 7$ be the mid-point and height of the ith rectangular. The cumulative distribution of X can be approximated by the points $F_X(x_i) = f_X(x_1) + f_X(x_2) + \cdots + f_X(x_i)$, $i = 1, 2, ..., 7$, as shown in Figure 3.7.

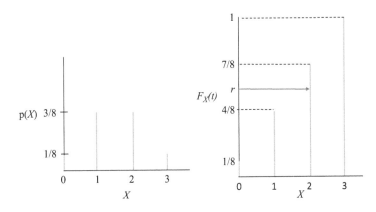

FIGURE 3.5 Generating a random variate from a discrete distribution.

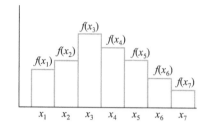

FIGURE 3.6 A histogram of a continuous random variable.

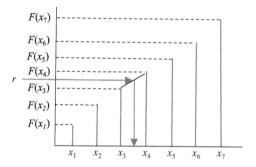

FIGURE 3.7 Generating a random variate from a continuous distribution.

Generating a random variate then is straightforward. Let r be a pseudo-random number, and let $F_X(x_{i-1}) \leq r \leq F_X(x_i)$. Then, we select a random variate x by interpolating between the values x_{i-1} and x_i, i.e.,

$$x = x_{i-1} + (x_i - x_{i-1}) \frac{r - F_X(x_{i-1})}{F_X(x_i) - F_X(x_{i-1})}.$$

3.4 SIMULATION DESIGNS

There are two different designs for implementing event-based simulation models, namely, *event-advance* and *unit-time advance*. These two designs use a different approach to advancing time in a simulation. There is also an alternative design to the event-based design, known as the *activity-based design*, but it has limited applicability. A description of the activity-based design can be found in [1].

The event-advance design was used in the two examples described in Section 3.2. In this design, each pending event is associated with a clock which gives the time instance in the future that this event will occur. The simulation model, upon completion of processing an event E, say at time t, regroups all the possible events that will occur in the future and finds the one with the smallest clock value, say t_1. It then advances to time t_1, by setting the master clock equal to t_1. It takes appropriate action as dictated by the event, and then repeats the process of finding the next event, and so on. The simulation model, therefore, moves through time by simply jumping to the time instances at which events occur, as depicted in Figure 3.8.

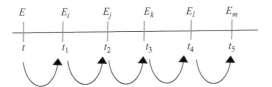

FIGURE 3.8 Advancing from event to event.

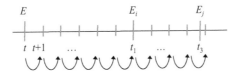

FIGURE 3.9 Unit-time advancing from event to event.

The unit-time advance design offers an alternative method to finding the next event, by advancing the master clock in fixed small time increments, referred to as unit times. Each time the master clock is advanced by a unit time, all future event clocks are compared with the current value of the master clock. If any of these clocks is equal to the master clock, then the associated event has just occurred and appropriate action takes place. If no clock is equal to the current value of the master clock, then no event has occurred and the master clock is again increased by a unit time and the cycle is repeated.

This mode of advancing the simulation through time is depicted in Figure 3.9. Let us say that at time t the simulation has just completed processing event E, and that the next event E_i will occur at time t_1. We increase the master clock by a unit time and compare to all the event clocks. Since none of the clocks is equal to the master clock, the master clock is increased by another unit time, and so on, until after some iterations the master clock becomes equal to t_1. Event E_i is then processed, the set of future events is modified accordingly, and then the master clock is increased by one unit time repeatedly until it hits the next event, and so on.

As can be seen in the example in Figure 3.9, the clock value of event E_j falls between two successive values of the master clock. This happens when a clock is not an integer multiple of the unit time. In view of this, in the unit-time advance design we select an event when its clock value becomes less than or equal to the master clock value.

An alternative way to implement the unit-time advance design is to use the actual durations of the events instead of their clock times. Typically, we schedule a new event, say the time of the next arrival, by generating an inter-arrival time which is then added to the master clock. This new clock time is an actual time instant in the future when the event will occur. For instance, if the master clock is 100 and the inter-arrival time is 20, then the arrival clock will be set to 120. An alternative way is to store only the durations of these events, i.e., in the case of the inter-arrival time we will store the value 20. Then, and each time we advance the master clock by a unit time, we reduce each event duration by a unit time. The next event to occur is the event whose duration becomes zero or less than zero when the master clock is increased by a unit time.

Below, we will use event clocks and not event durations, but the same applies to event durations.

3.4.1 The Event List

Let us assume that a simulation model has just processed an event and the master clock is currently at time t. The set of all the events scheduled to occur in the future is known as the *pending event list* or just the *event list*. For each event scheduled to occur in the future, the list contains the following information:

1. Time of occurrence, i.e., the value of the event's clock

2. Type of event

3. Additional parameters

The event type is used in order to determine what action should be taken when the event occurs. For instance, using the event type the program can determine which procedure to call. Additional parameters regarding the event may be also kept in the event list.

In the event-advance design, finding the next event requires locating the event in the list with the smallest clock value. In the simulation example in Section 3.2.1, there were only two events, an arrival event and a departure event. Consequently, the next event is determined by a simple comparison between the two clocks. In the example in Section 3.2.2, we have an arrival event of a new request, a departure event, and possibly one or more arrival events of retransmitted requests. If B is small and the new requests arrive fast, we may end up with a large number of arrival events of retransmitted requests. In this case, finding the next event might require more than a few comparisons.

In general, when simulating complex systems, the number of events may be very large. Therefore, it is important to have an efficient algorithm for finding the next event since this operation may well account for a large percentage of the total computation involved in a simulation program. The efficiency of this algorithm depends on how the event list is stored in the computer. An event list should be stored in such a way so as to lend itself to an efficient execution of the following operations.

1. Locating the next event

2. Deleting an event from the list after it has occurred

3. Inserting newly scheduled events in the event list

The simplest way to store an event list is to store all clocks in an array, as shown in Figure 3.10. Each event is associated with an integer number i which is also the position in the array where its clock is stored. Finding the next event is reduced to the problem of locating

$CL1$	$CL2$	$CL3$	...	CLn

FIGURE 3.10 Event list stored in an array.

the clock in the array with the smallest value. The location of the clock then permits us to access the code associated with the event.

An event is not deleted from the array after it has occurred. If an event is not valid at a particular time, then its clock can be set to a very large value so that it will never be selected. A newly scheduled event i is inserted in the list by simply updating its clock in the ith location of the array.

The advantage of storing an event list in an array is that insertions of new events and deletions of caused events can be done very easily, and in constant time. The time it takes to find the smallest number in the array depends on the length of the array, and its complexity is linear in time. Locating the smallest number in an array does not take much time if the array is small. However, if the array is large, then it becomes time consuming. To overcome this problem, one can use a linked list to store the event list.

A linked list provides a mechanism whereby each data element of a list of data elements is stored in a different part of the memory; i.e., they are not stored in contiguous memory locations as in an array. This permits data elements to be dynamically added or removed at runtime. In order to access the data elements in the list, we store along with a data element the address of the next data element. This pointer is referred to as a *link*. The data element and the link is referred to as a *node*. In general, a node may consist of a number of data elements and links. Linked lists are drawn graphically as shown in Figure 3.11. Each node is represented by a box consisting of as many compartments as the number of data elements and links stored in the node. In the example given in Figure 3.11, each node consists of three compartments. The first one has the clock value, the second the type of event, and the third the pointer to the next node. The pointer *head* points to the first node in the list. The pointer of the last node is set to NULL, a special value indicating that this is the last node in the linked list. Due to the fact that two successive nodes are connected by a single pointer, this data structure is known as a *singly linked list*.

The data elements in Figure 3.11 are linked together so that their clock values are in an ascending order; i.e., the first node contains the event with the smallest clock value, the second node contains the event with the next larger clock value, and so on. In view of this, the next event to occur is given by the node at the top of the linked list, which can be accessed directly using the head pointer. When a new event is scheduled, the new node has to be appropriately inserted in the linked list so that to continue to maintain the ascending order of the clock values. In order to insert a node in the linked list, we have to traverse the linked list and compare each node until we find the correct insertion position. The maximum number of nodes compared in the worst case is the total number of nodes in the list. Thus, the complexity of the insert operation is linear on n.

Searching a linked list might be time consuming if n is very large. In this case, one can employ better searching procedures. For example, a simple solution is to maintain a pointer

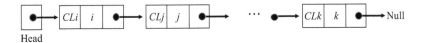

Head

FIGURE 3.11 Event list stored in a linked list.

to a node which is in the middle of the linked list. This node separates logically the list into two sublists. By comparing the clock value of the node to be inserted with the clock value stored in this node, we can establish easily in which sublist the insertion is to take place. The actual insertion can then be located by sequentially searching the nodes of the sublist.

There are many other data structures that can be used, and the reader should consult a book on data structures. Reference [1] also contains relevant material. Finally, we note that programming languages provide commands for creating and manipulating linked lists.

3.4.2 Selecting the Unit Time

The unit time is readily deduced when all event clocks are integer variables as each event clock is simply a multiple of the unit time. However, quite frequently event clocks are represented by real variables. In this case, it is quite likely that an event may occur in between two successive time instants of the master clock, as shown in Figure 3.9 for event E_j.

In the simulation model, new events are scheduled based on the current value of the master clock. So, when we have a case such as that of event E_j, we can assume that it actually occurs at the master clock time; i.e., we introduce an error as to when exactly event E_j occurs. This error may not be significant if the time unit is very small. Alternatively, we can reset the master clock value to the value of the clock associated with E_j. Resetting the master clock should not affect the results of the simulation.

In general, a unit time should be small enough so that at most one event occurs during the period of a unit of time. This simplifies the simulation logic. Multiple events occurring during the period of a unit of time can also be handled, but it may require some additional logic. However, if the unit time is too small, the simulation program will spend most of its time in non-productive mode, i.e., advancing the master clock and checking whether an event has occurred or not. One simple heuristic rule is to generate a large number of variates for all the random variables used in the simulation model, and then set the unit time equal to one-half of the smallest generated variate.

3.5 ESTIMATION TECHNIQUES

So far, we have examined techniques for building a simulation model. These techniques were centered around the topics of random number generation and the event-based simulation design. The reason why we develop a simulation model is because we want to estimate various performance metrics related to the system under study. These metrics are obtained by collecting and analyzing data captured during the simulation. This data is known as *endogenous* data, i.e., data generated from within the simulation. The input data to a simulation fixed before the execution of the simulation are referred to as *exogenous* data.

3.5.1 Collecting Endogenously Created Data

Endogenous data can be collected sometimes by simply introducing in the simulation program single variables which are used as counters. However, quite often we are interested in the duration of an activity. For instance, in the problem of the recertification of the IoT devices described in Section 3.2.2, we are interested in estimating the amount of time, T, it takes to process a request. This is the amount of time elapsed from the moment a new request arrives to the moment it is processed by the server and departs from the system. Also, of interest may be the amount of time D, it takes for a new request to enter the server's queue if it is rejected when it first arrived. This is, the amount of time it spends being retransmitted.

In order to capture endogenously created data, we have to enhance the simulation model once more with additional data structures. For instance, in the above example, the simplest way to measure T and D is to maintain a two-dimensional array A where each row is associated with a specific request. The row number is used as the request's id. When a new request arrives, we record its arrival time in $A(i, 1)$, where i is the next available row. If the new request cannot enter the queue, we record the time it begins its retransmission delay, which is the same as its arrival time, in $A(i, 2)$. The time it enters the queue after one or more retransmissions is stored in $A(i, 3)$, and the time it departs from the system is recorded in $A(i, 4)$. The values of T and D for each request can then be easily computed. $A(i, 4) - A(i, 1)$ gives us the value for T, and $A(i, 3) - A(i, 2)$ gives us the value for D for the ith request.

In order to populate this array, we need to keep track of the id of each request. If a request is being retransmitted or being served, then the corresponding event in the event list should be associated with the request's id number. This can be easily done by enhancing the event list. If the event list is maintained in an array, then this array should be redefined as a two-dimensional array, where the first column contains the clock value and the second column contains the id. If the event list is maintained in a linked list, then each node should have an additional field for the id. A new request is given an id at the time of its arrival.

In addition, requests waiting in the queue should be identified by their id. This can be done by representing the queue by a linked list, as shown in Figure 3.12, where each node simply corresponds to a request waiting in the queue. When a request joins the queue, a new node is created with its id and added at the end of the linked list. When a request completes its service and departs from the system, the next request at the top of the linked list begins service, and its node is deleted from the linked list.

Since the total number of requests in the server's queue plus the one in service cannot exceed B, we can also represent the queue by an array Q of size B. In addition, we use two pointers P and E to point to the beginning of the queue and to the end of the queue, respectively. When an arrival enters the queue, we record its id in position $E + 1$. When a service is completed, the next request to go into service is given by P, which is then increased by one if the queue does not become empty. The pointers wrap around as requests enter and leave the queue. An example of this is shown in Figure 3.13.

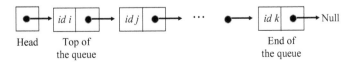

FIGURE 3.12 Queue represented by a linked list.

FIGURE 3.13 Queue represented by an array.

The endogenously created data collected during the simulation is processed after the simulation is over in order to obtain performance metrics. For instance, for the above case, we can use all the generated values of T to construct an empirical probability distribution of T. We can do the same for D. Most often, we are not interested in the entire probability distribution. Rather, we are interested in the mean and percentiles, such as the 95th percentile, and their confidence intervals. The 95th percentile is a useful metric as it gives us some idea of the length of the tail of the distribution.

3.5.2 Transient-State versus Steady-State Simulation

In order to start a simulation, we have to make an assumption about the state of the system at time zero. For instance, in the hand simulation in Section 3.2.2, we assumed that the system at time zero is empty and that the first request arrives at time 2. The state of the system we assume at time zero, including the seed for the pseudo-random generator, is known as the *initial condition*.

An initial condition affects the behavior of the system for an initial period of time. Thereafter, the simulation will behave statistically in the same way whatever the initial condition. During this initial period, the simulation is said to be in a *transient state*. After that period, the simulation is said to be in a *steady state*.

Transient analysis is of interest when we want to analyze the behavior of a system for a short period of time. For instance, in the recertification example of IoT devices, we may be interested in knowing the total time to recertify the first few requests. In this case, the initial condition will have an impact on the results. Transient analysis is also used for systems which constantly change, and therefore do not have a steady state.

Transient analysis is not very common when modeling systems, such as, IoT systems, packet-switched networks, and production lines. In such cases, steady-state simulation is the norm. For this, the simulation model has to run long enough so that to get away from the transient state, also known as the warm-up period.

There are two basic methods for choosing the initial condition. The first is to begin with an empty system; i.e., we assume that there are no activities going on in the system at the beginning of the simulation. The second method is to make the initial condition to be as representative as possible of a typical state that the system might find itself in. This method reduces the duration of the transient period, but it requires an a priori knowledge of the system.

When collecting endogenously created data we should be careful of the effects of the transient period. For, the data created during the transient period are dependent on the initial condition. Two methods are commonly used to remove the effects of the transient period. The first one requires a very long simulation run, so that the amount of data collected during the transient period is insignificant relative to the amount of data collected during the steady-state period. In the second method, we discard the data collected during the transient state. This brings us to the problem of how to determine when the simulation system has reached its steady state. There are various methods to do that, but in practice, we just discard the first few hundreds of the collected data. For instance, in the example of the recertification of IoT devices, it suffices to ignore the data collected for the first 100 requests. If in doubt, run your simulation multiple times, and each time increase the transient period. Select the transient period past which no change in the performance metrics collected during the steady state is observed.

Below, we describe various estimation procedures, and we assume that the observations have been collected when the simulation is in steady state.

3.5.3 Estimation of the Confidence Interval of the Mean

Let us assume that we have collected a sample of n endogenous data for a performance metric, such as the amount of time it takes to recertify an IoT device, when the simulation is in steady state. Let $x_1, x_2, \ldots, x_n$ be the n observations. Then, the sample mean is

$$\bar{x} = \frac{1}{n}\sum_{i=1}^{n} x_i.$$

Typically, when we estimate the mean of a performance metric, we also provide an error of the estimation. For instance, in opinion polls, we typically hear that x% of people prefer candidate X and that the error is 3%. This means that the true percentage of the population who prefer candidate X is within $x \pm 3$%. This interval is known as the *confidence interval*, because the true population percentage lies in the interval with some probability, such as 95%. This probability is known as the *confidence level*. The confidence interval of the sample mean $\bar{x}$ is given by the expression:

$$\left(\bar{x} - 1.96\frac{\sigma}{\sqrt{n}}, \bar{x} + 1.96\frac{\sigma}{\sqrt{n}}\right)$$

at 95% confidence level. The confidence interval provides an indication of the error associated with the sample mean. It is a very useful statistic and it should always be computed. Unfortunately, quite frequently it is ignored. The confidence interval tells us that the true population mean lies within the interval 95% of the time; i.e., if we repeat the above experiment 100 times, 95% of these times, on the average, the true population mean will be within the interval.

The derivation of the confidence interval is very simple. We recall from Section 2.3.4 that if we sample n observations from a population which follows an arbitrary distribution with a mean μ and variance σ^2, then the sample mean $\bar{x}$ is normally distributed with mean μ and variance σ^2/n, for $n \geq 30$. Now, let us fix points a and b in this distribution so that

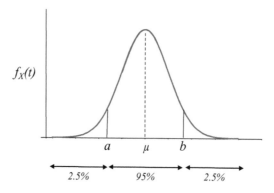

FIGURE 3.14 Points a and b in the normal distribution.

95% of the observations (i.e., of the sample means $\bar{x}$) fall in-between the two points. Points a and b are symmetrical around μ, and the two tails $(-\infty, a)$ and $(b, +\infty)$ account for 5% of the total distribution, as shown in Figure 3.14. Using the table of the standard normal distribution, we have that a is 1.96 standard deviations below μ, i.e., $a = \mu - 1.96\sigma / \sqrt{n}$, and b is 1.96 standard deviations above μ, i.e., $b = \mu + 1.96\sigma / \sqrt{n}$. Now, if we consider an arbitrary observation $\bar{x}$, then this observation will lie in the interval $[a, b]$ 95% of the time; i.e., its distance from μ will be less than $1.96\sigma / \sqrt{n}$ 95% of the time. Or, otherwise stated, 95% of the time μ will be less than $1.96\sigma / \sqrt{n}$ from $\bar{x}$; i.e., μ will lie in the confidence interval $\left(\bar{x} - 1.96\sigma / \sqrt{n}, \bar{x} + 1.96\sigma / \sqrt{n} \right)$ 95% of the time.

In general, a confidence interval can be calculated for any confidence level. Most typical confidence levels are 99%, 95%, and 90%. For each value, points a and b can be calculated from the table of the standard normal distribution. The z value at 90% and 99% confidence is 1.645 and 2.56, respectively. We observe that a small confidence interval comes at the expense of having lower confidence about the result!

In order to construct a confidence interval of the sample mean $\bar{x}$, we need to know the variance σ^2. If it is not known, then we use the sample variance s^2 given by the familiar expression:

$$s^2 = \frac{1}{n-1} \sum_{i=1}^{n} \left(x_i - \bar{x} \right)^2.$$

Note that dividing the above summation $\sum_{i=1}^{n} \left(x_i - \bar{x} \right)^2$ by $n - 1$ gives an unbiased estimator of σ^2, whereas dividing it by n gives the maximum likelihood estimator of σ^2 (see Appendix B for an explantion of the maximum likelihood estimation method). Either denominators can be used.

The above expression for s^2 is correct when the observations are independent of each other. To compute the above expression, we need to store all the observations first and then at the end of the simulation compute the sample variance. The following equivalent expression allows the calculation on the fly without having to store all the observations:

$$s^2 = \frac{1}{n-1}\left[\sum_{i=1}^{n} x_i^2 - \frac{1}{n}\left(\sum_{i=1}^{n} x_i\right)^2\right];$$

i.e., when the kth observation is obtained, we update the cumulative sums $\sum_{i=1}^{k} x_i^2$ and $\sum_{i=1}^{k} x_i$, and total count of observations. At the end of the simulation, the variance can be easily computed using the above expression.

In the case where the standard deviation σ is not known and has to be estimated from the sample data, the confidence interval is computed using the *t-student distribution* with $n - 1$ degrees of freedom; i.e.

$$\left(\bar{x} - t_{n-1}\frac{s}{\sqrt{n}}, \bar{x} + t_{n-1}\frac{s}{\sqrt{n}}\right).$$

The t-student distribution is symmetric and bell-shaped, like the normal distribution, but has heavier tails, meaning that it is more prone to producing values that fall far from its mean. It is commonly used instead of the z value, which is calculated from the normal distribution, since the standard deviation σ is typically not known. The t-student distribution is also used if the sample size is small, less than 30. As the degrees of freedom increase, the t-distrubtion tends to the normal distribution. It is calculated for different confidence levels in the same way as the z values.

Quite frequently, the observations x_1, x_2, ..., x_n that we obtain endogenously from a simulation model are correlated. For instance, if we have several recertification requests waiting in the server's queue, the waiting time of a request i depends on how long request j in front of it has to wait. If j waits for a long time, then i has to wait for a long time as well. On the other hand, if j does not wait long, then i will not wait for long either. Therefore, successive processing times of requests are correlated. In general, two random variables are uncorrelated, or positively correlated or negatively correlated, as shown in Figure 3.15.

Correlation between observations does not affect the calculation of the sample mean $\bar{x}$, but the expression of the sample standard deviation given above no longer holds. The correct procedure, therefore, for obtaining the confidence interval of the sample mean is to first check if the observations x_1, x_2, ..., x_n are correlated. If they are not, then we can proceed as described above. On the other hand, if the observations are correlated, then we have to use a different method to estimate the variance. Below, we examine the following three methods: (a) estimation of the autocorrelation function (ACF), (b) batch means, and (c) replications. For further information, see [1].

a. *Estimating the Autocorrelation Function (ACF)*
Let X and Y be two random variables. We recall from Section 2.5 that the $Cov(X, Y)$ is given by the expression:

$$Cov(X,Y) = E(XY) - E(X)E(Y).$$

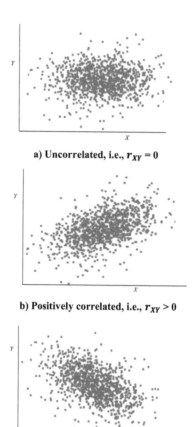

a) Uncorrelated, i.e., $r_{XY} = 0$

b) Positively correlated, i.e., $r_{XY} > 0$

c) Negatively correlated, i.e., $r_{XY} < 0$

FIGURE 3.15 Three cases of correlation (a) Uncorrelated, i.e., $r_{XY} = 0$; (b) Positively correlated, i.e., $r_{XY} > 0$; (c) Negatively correlated, i.e., $r_{XY} < 0$.

The covariance takes values in the space $(-\infty, +\infty)$, and it is not dimensionless, which may make its interpretation difficult. In view of this, we use the correlation $Corr(X, Y)$, which is the normalized $Cov(X, Y)$. This is a dimensionless metric, defined as follows:

$$Corr(X,Y) = \frac{Cov(X,Y)}{\sqrt{Var(X)}\sqrt{Var(Y)}}.$$

Let $(x_1, y_1), (x_2, y_2), \ldots, (x_n, y_n)$ be n pairs of observations of (X, Y). Then, the sample covariance s_{XY} is given by the expression:

$$s_{XY} = \frac{\sum_{i=1}^{n}(x_i - \bar{x})(y_i - \bar{y})}{n-1},$$

and the sample correlation r_{XY}, obtained by normalizing the sample covariance, is

$$r_{XY} = \frac{\sum_{i=1}^{n}(x_i - \bar{x})(y_i - \bar{y})}{\sqrt{\sum_{t=1}^{n}(x_i - \bar{x})^2}\sqrt{\sum_{t=1}^{n}(y_i - \bar{y})^2}}.$$

It can be shown that $-1 \leq r_{XY} \leq 1$. X and Y are positively correlated if $r_{XY} > 0$, and they are strongly positively correlated if $r_{XY} = 1$. X and Y are negatively correlated if $r_{XY} < 0$, and they are strongly negatively correlated if $r_{XY} = -1$. Finally, they are uncorrelated, i.e., independent from each other, if $r_{XY} = 0$. An example of two uncorrelated, positively correlated, and negatively correlated random variables X and Y is shown in Figure 3.15.

Now, let us consider n observations $x_1, x_2, \ldots, x_n$ from a random variable X. We form the following $n-1$ pairs of observations: $(x_1, x_2), (x_2, x_3), \ldots, (x_{n-1}, x_n)$. We regard the first observation in each pair as coming from X and the second observation as coming from a random variable Y. Then, r_{XY} is

$$r_{XY} = \frac{\sum_{i=1}^{n-1}(x_i - \bar{x})(x_{i+1} - \bar{x})}{\sqrt{\sum_{t=1}^{n-1}(x_i - \bar{x})^2}\sqrt{\sum_{t=1}^{n-1}(x_{i+1} - \bar{x})^2}},$$

where

$$\bar{x} = \frac{1}{n-1}\sum_{i=1}^{n-1}x_i \text{ and } \bar{y} = \frac{1}{n-1}\sum_{i=2}^{n}x_i.$$

The above correlation r_{XY} is called the *lag 1 autocorrelation* of a random variable, and we refer to it as r_1, where the subscript stands for lag 1. If n is large, it can be approximated as follows:

$$r_1 = \frac{\sum_{i=1}^{n-1}(x_i - \bar{x})(x_{i+1} - \bar{x})}{\sum_{t=1}^{n}(x_i - \bar{x})^2},$$

where $\bar{x}$ is the sample mean of the n observations. In a similar fashion, we can obtain the *lag k autocorrelation* that is the correlation between observations which are k apart. This is given by the expression:

$$r_k = \frac{\sum_{i=1}^{n-k}(x_i - \bar{x})(x_{i+k} - \bar{x})}{\sum_{t=1}^{n}(x_i - \bar{x})^2}, \quad k \geq 1,$$

where $\bar{x}$ is the overall sample mean. The lag k autocorrelations, $k \geq 1$, are also referred to as the *autocorrelation coefficients*.

The autocorrelation can also be computed by first computing the auto-covariances s_1, s_2, ..., s_n where s_k is given by the expression:

$$s_k = \frac{1}{n} \sum_{i=1}^{n-k} (x_{i} - \bar{x})(x_{i+k} - \bar{x}).$$

The lag k autocorrelation is computed as the ratio:

$$r_k = \frac{s_k}{s^2},$$

where s^2 is the sample variance. We note that $r_0 = 1$, since $s_0 = s^2$. The set of sample autocorrelations computed from a given sample are typically presented in a graph, known as the *correlogram*. Two examples of a correlogram are shown in Figures 3.16 and 3.17.

Let us now return to our estimation problem. Having obtained a sample of n observations $x_1, x_2, ..., x_n$, we calculate the autocorrelation coefficients, and then the variance can be estimated using the expression:

$$s^2 = s_X^2 \left[1 + 2 \sum_{k=1}^{n/4} \left(1 - \frac{k}{n} r_k \right) \right],$$

where s_X^2 is the sample variance given by

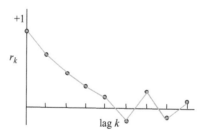

FIGURE 3.16 A correlogram with short-term correlation.

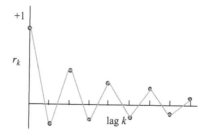

FIGURE 3.17 A alternating correlogram.

$$\left| x_1, x_2, \ldots, x_b, \right| x_{b+1}, x_{b+2}, \ldots, x_{2b}, \left| \quad \cdots \quad \right| x_{(m-1)b+1}, x_{(m-1)b+2}, \ldots, x_{mb} \right|$$

FIGURE 3.18 The batch means method.

$$s^2 = \frac{1}{n-1} \sum_{i=1}^{n} \left(x_i - \bar{x} \right)^2, \text{ where } \bar{x} = \frac{1}{n} \sum_{i=1}^{n} x_i;$$

i.e., in the case of correlated observations, the sample variance s^2 calculated under the assumption that the observations are uncorrelated is modified by the above term in the square brackets.

b. *The Batch Means Method*

This is a popular technique and very easy to implement. It involves dividing successive observations into batches as shown in Figure 3.18. Each batch contains the same number of observations. Let the batch size be equal to b. Then, batch 1 contains observations x_1, x_2, ..., x_b, batch 2 observations x_{b+1}, x_{b+2}, ..., x_{2b}, and so on. The total number of observations n should be selected so that $n = mb$. The observations at the end of a batch may be correlated with those at the beginning of the next batch. Let $\bar{x}_i$ be the sample mean of the ith batch. If we chose the batch size b to be large enough, then the sequence of the m sample means $\bar{x}_1$, $\bar{x}_2$, ..., $\bar{x}_m$, are approximately uncorrelated. Consequently, we can calculate the standard deviation using the standard expression. We have

$$\bar{\bar{x}} = \frac{1}{m} \sum_{i=1}^{m} \bar{x}_i$$

$$s^2 = \frac{1}{m-1} \sum_{i=1}^{m} \left(\bar{x}_i - \bar{\bar{x}} \right)^2,$$

and the confidence interval of $\bar{\bar{x}}$ is

$$\left(\bar{\bar{x}} - t_{0.95} \frac{s}{\sqrt{m}}, \bar{\bar{x}} + t_{0.95} \frac{s}{\sqrt{m}} \right).$$

An estimate of b can be obtained by plotting out the correlogram of all the mb observations, which can be obtained from a preliminary simulation run. Statistical packages that plot a correlogram also plot two symmetric lines about the x-axis, i.e., one above the x-axis and the other below it. These two lines give the confidence interval bounds for the null hypothesis that the lag k autocorrelation is zero (see Chapter 4 for an explanation of the null hypothesis). In simple terms, if the lag k autocorrelation is above the upper line or below the lower line, then we assume that it is not zero. If these confidence interval lines are not given, then the rule of thumb is to use a cutoff of around 0.15; i.e., if the absolute value of lag k is less than 0.15, then we assume that lag k is zero; i.e., the pairs (x_i, x_{i+k}), $i \geq 1$, are not correlated. Going back to estimating b, we select the first lag k which falls within the confidence interval bounds, and we fix b so that it is at least 5 times k. The number of batches m is typically more than 30.

The batch size and the number of batches simulated may be a lot bigger than what was described above. In general, they are fixed so that the confidence interval is very small. This is an important issue and it is discussed further in Section 3.5.6.

c. The Replication Method

Another approach to constructing a confidence interval of a mean is to replicate the simulation several times. For each replication, we use a different seed, which gives rise to a different sequence of pseudo-random numbers, and consequently a different sequence of events. As a result, we get a different set of observations for each replication. The difference between two successive seeds should be large.

Suppose we make m replications each with l observations, and we get the following sets of observations:

Replication 1: $x_{11}, x_{12}, \ldots, x_{1l}$
Replication 2: $x_{21}, x_{22}, \ldots, x_{2l}$
$\vdots$

Replication m: $x_{m1}, x_{m2}, \ldots, x_{ml}$

Let $\bar{x}_i$ be the sample mean of the ith replication, i.e.,

$$\bar{x}_i = \frac{1}{l} \sum_{j=1}^{l} x_{ij}, \, i = 1, 2, \ldots, m.$$

The sample means $\bar{x}_1, \bar{x}_2, \ldots, \bar{x}_m$ are independent from each other and therefore the confidence interval is

$$\left(\bar{\bar{x}} - t_{0.95} \frac{s}{\sqrt{m}}, \, \bar{\bar{x}} + t_{0.95} \frac{s}{\sqrt{m}} \right),$$

where

$$\bar{\bar{x}} = \frac{1}{m} \sum_{i=1}^{m} \bar{x}_i \text{ and } s^2 = \frac{1}{m-1} \sum_{i=1}^{m} \left(\bar{x}_i - \bar{\bar{x}} \right)^2.$$

For each replication, we allow the simulation to go through the warm-up period and then run it to collect the l observations. This procedure is repeated m times.

The replication method appears to be similar to the batch means method. However, in the batch means method, the batch size is relatively small and the number of batches is large. In the replications method, it is the other way around; i.e., each replication is very large and the number of replications can be small, as little as 5.

3.5.4 Estimation of the Confidence Interval of a Percentile

Percentiles are important performance metrics used in *service level agreements* (SLAs). They are often ignored in favor of the mean of a random variable, which in general is not a good metric. We are primarily interested in the 90th, 95th, and 99th percentile. Given a random variable X, its 95th percentile (or any other percentile) is a value $x_{0.95}$ such that 95% of the values of the random variable X are less than $x_{0.95}$, i.e., $Prob[X \leq x_{0.95}] = 0.95$. Correspondingly, 5% of the values of the random variable X are greater than $x_{0.95}$, i.e., $Prob[X \geq x_{0.95}] = 0.05$.

These high percentiles give us an idea of the shape of the right-hand side tail of the probability distribution. For, if $x_{0.95}$ is close to the mean, then the tail is very short. On the other hand, if it is far away from the mean, then the distribution has a long tail. We are interested in the shape of the right-hand tail of the distribution as it provides a probabilistic upper bound of a random variable. For instance, let us say that from the simulation we determine that the 95th percentile of the time to process recertification requests is 500 msec. This can be interpreted as an upper bound, in the sense that 95% of the time a recertification request will be processed within 500 msec.

Percentiles are calculated as follows. Let $x_1, x_2, \ldots, x_n$ be a sample of X values generated by the simulation. The sample is sorted out in an ascending order, and let $y_1 \leq y_2 \leq \ldots \leq y_n$ be the sorted observations. Then, the 95th percentile of X (or any other percentile) is the value y_k where $k = \lceil 0.95 \times n \rceil$. (The function $\lceil y \rceil$ is the ceiling function that maps a real number x to the smallest integer not less than x.) For instance, for a sample of 950 observations, $k = \lceil 0.95 \times 950 \rceil = 903$, and the 95th percentile is the value y_{903}.

The confidence interval on a percentile can be calculated easily using the replication method or the batch means method. Let us assume that we are interested in the 95th percentile. After m replications or batches, we obtain a sample of m independent estimates of the 95th percentile, i.e., $x_{0.95,1}, x_{0.95,2}, \ldots, x_{0.95,m}$. Then, we have

$$\left(\bar{x}_{0.95} - t_{0.95} \frac{s}{\sqrt{m}}, \bar{x}_{0.95} + t_{0.95} \frac{s}{\sqrt{m}} \right)$$

where

$$\bar{x}_{0.95} = \frac{1}{m} \sum_{i=1}^{m} x_{0.95,i} \text{ and } s^2 = \frac{1}{m-1} \sum_{i=1}^{m} \left(x_{0.95,i} - \bar{x}_{0.95} \right)^2.$$

The estimation of high percentiles requires very long replication runs, or very long batches. For, if the runs are not long, then the estimates will be biased. The calculation of a percentile requires that we store the entire sample of observations until the end of the simulation and that we order the sample of observations in an ascending order. These two operations can be avoided by constructing a frequency histogram of the random variable on the fly. When an observation becomes available, it is immediately classified into the appropriate bin of the histogram. Thus, it suffices to keep track of how many observations fall within each bin. At the end of each replication, the percentile can be easily picked out from the histogram. Obviously, the accuracy of this implementation depends on the chosen width of the bins of the histogram.

3.5.5 Estimation of the Confidence Interval of a Probability

The estimation of the confidence interval of the probability p that a particular event E occurs can be handled in the same way as the estimation of the confidence interval of a mean or a percentile using the batch means method or the replication method. For instance, in the case of the recertification of IoT devices, let us say that we want to calculate the probability p that a new request will not get into the queue upon arrival at the server. We introduce the variables N and K in the simulation; the first one keeps count of the total number of new requests that arrived, and the second one keeps count of the total number of those requests that were denied entrance upon arrival. Then, $p = K/N$. Using the replication method or the batch means method, we calculate m independent estimates of p, i.e., p_1, $p_2, ..., p_m$. Then, we have

$$\left(\bar{p} - t_{0.95} \frac{s}{\sqrt{m}}, \bar{p} + t_{0.95} \frac{s}{\sqrt{m}} \right)$$

where

$$\bar{p} = \frac{1}{m} \sum_{i=1}^{m} p_i \text{ and } s^2 = \frac{1}{m-1} \sum_{i=1}^{m} (p_i - \bar{p})^2.$$

3.5.6 Achieving a Required Accuracy

In general, the accuracy of an estimate of a statistic depends on the width of the confidence interval $t_{0.95} \left(s / \sqrt{n} \right)$, where n is the number of observations used. The width of the confidence interval is also referred to as the error. The smaller the width, the higher is the accuracy. For instance, if it is concluded in an opinion poll that 30% of the population will vote for a candidate with an error of ±5 %, then that means that the true population percent of people who will vote for the candidate lies in interval (25%, 35%). However, this is not the same as concluding that 30% will vote for the candidate with an error of ±1%, i.e., the confidence interval is (29%, 31%). Obviously, in the second case the accuracy is much higher.

We note that the width $t_{0.95} \left(s / \sqrt{n} \right)$ is proportional to $1/\sqrt{n}$. Therefore, the width can be reduced if we increase n. For instance, in order to halve the width, n has to be increased four times since $1/\sqrt{4n} = (1/2) \left(1/\sqrt{n} \right)$; i.e., we need to run the simulation long enough so that to collect a total of $4n$ observations. A rule of thumb is that n should be fixed so that the width is about 10% or less of the estimated statistic. The typical approach is to run the simulation with increasing values of n, until we get the desired result. Then, we can proceed with the simulation experiments with the chosen value of n.

An alternative method to reducing the width $t_{0.95}\left(s/\sqrt{n}\right)$ is to reduce the variance s^2. For that, there are special *variance reduction techniques*, some of which can be found in [1].

3.6 VALIDATION OF A SIMULATION MODEL

Validation of a simulation model is very important, but it is often overlooked. How accurately does a simulation model (or, for that matter, any kind of model) reflect the operations of the system under study? How confident can we be that the obtained simulation results are accurate and meaningful? Below, we list a number of checks that can be carried out in order to validate a simulation model.

a. *Check the pseudo-random number generator.* A battery of statistical tests can be carried out to make sure that the pseudo-random number generator used in the simulation program creates numbers that are uniformly distributed in [0,1] and are statistically independent.

b. *Check the stochastic variate generators.* Similar statistical tests can be carried out for each random variate generator used in a simulation model.

c. *Check the logic of the simulation program.* This is a rather difficult task. One way of going about it is to print out the status variables, the future event list, and other relevant data structures each time an event takes place. Then, check by hand whether the data structures are updated appropriately. This is a rather tedious task. However, using this method one can discover logical errors and also get a good feel about the simulation model.

d. *Relationship validity.* Verify that the assumptions you made when building the simulation model are the same as in the system under study.

e. *Output validity.* There are various ways that the validity of the output can be confirmed. A simple method is to consider special cases for which you can predict qualitatively the behavior of the results. For instance, let us consider the example of the recertification of IoT devices. If we assume that B is very large and that the service time is extremely small, then the time to process a request is almost zero. This can be confirmed by setting B to a large value and then run the simulation with different decreasing service times and observe the processing time of requests go to zero. Other validity tests can be carried out by obtaining graphs of a particular measure of interest for different values of the input parameters and then trying to interpret these graphs intuitively.

3.7 SIMULATION LANGUAGES

A discrete-event simulation language takes away the burden of having to develop your own simulation. It provides basic house-keeping functions to manipulate the event list, store and retrieve files, and collect and analyze statistically endogenously created data. In addition, it provides a library of pre-defined building blocks that can be used to model a variety of different problems. Many languages have a graphical tool to create a simulation model, and in addition they may also provide an animation while the simulation is running. Simulation languages allow the user to develop a simulation model very quickly, unlike the case where a simulation program has to be built from scratch. The main problem with simulation languages is that the user has to develop a model using the provided building blocks, which may not necessarily reflect the logic of the system under study. Such problems may be circumvented through good working knowledge of the simulation language.

At this point, the interested reader is encouraged to learn and use a simulation language, such as, ns-3, SLAM II, ARENA, MATLAB-Simulink, OPNET, and JMT.

EXERCISES

1. Redo the hand simulation in Table 3.2, assuming that retransmitted requests have a higher priority of being processed by the server than new requests; i.e., when the server completes a request, it searches the queue for a retransmitted request. The first such request waiting in the queue is the next request that will be processed. If there are no retransmitted requests, then it processes the first new request waiting in the queue. For this set $B = 5$ and run the simulation sufficiently long so that to be able to apply this priority scheme a few times.

2. Use the inverse transformation method to generate random variates with a probability density function $f_X(t) = 3t^2, 0 \leq t \leq 1$.

3. Apply the inverse transformation method to generate random variates with the following probability density function:

$$f_X(t) = \begin{cases} 5t, & 0 \leq t \leq 4 \\ t-2, & 4 < t \leq 10 \end{cases}$$

(Note that $f_X(t)$ needs to be normalized.)

4. Set up a procedure to generate random variates from

$$f_X(t) = \begin{cases} t, & 0 \leq t \leq 0.5 \\ 1-t, & 0.5 < t \leq 1 \end{cases}$$

5. A server processes two types of requests issued by IoT devices. The CPU time required by the first type of requests has an Erlang distribution with three stages, where each stage is exponentially distributed with a mean 10. The CPU required by the second type of requests has an Erlang distribution with four stages, where each stage is exponentially distributed with a mean 6. A request has a probability of 0.4 or 0.6 being the first type or the second type, respectively. Set up a procedure to generate random variates of the CPU time required by a request.

SIMULATION PROJECT

Task 1 (to be completed after you read Section 3.2)

Do the hand simulation for the model described in Section 3.2.2 until $MC = 100$. Make sure you understand how the events are created and how the master clock moves from event to event.

Subsequently, write a program in a high-level language of your choice to do the same thing. Run your simulation using the same parameters as the hand simulation until $MC = 200$; i.e., the stopping rule of the simulation is the value of the master clock. (We will change this stopping rule in the next task.) Use the same initial conditions. Print a line of output each time you handle an event, i.e., you advance the master clock, with the same information as in Table 3.2. Go over the results by hand to make sure that the simulation advances properly from event to event. This is an important validation check to make sure that your simulation works correctly.

Task 2 (to be completed after you read Section 3.3)

Use the same program you developed in task 1 but assume that the inter-arrival times and retransmission times are exponentially distributed. The mean values of these two exponential distributions are the same as in task 1, i.e., 6 and 5, respectively. The service time remains constant equal to 10 as in task 1, since the execution time of a request is more or less constant. Each time you want to generate an inter-arrival time or a retransmission time, draw a pseudo-random number r and then use it to obtain an exponential variate as described in Section 3.2.2. Print a line of output each time you handle an event, i.e., you advance the master clock, with the same information as in task 1. Run your simulation until MC goes over 200. Go over the results by hand to make sure that the simulation advances properly from event to event.

Task 3 (to be completed after you read Section 3.5)

The objective of this task is to use the simulation in order to calculate:

a. The mean and 95th percentile and their confidence interval of the time T it takes to certify an IoT device.

b. The mean and 95th percentile and their confidence interval of the time D it takes a new request to enter the queue if it is rejected the first time it arrived.

For this, you have to modify your program as follows: stop printing a line of output each time you handle an event, add additional data structures to collect data on T and D, implement the batch means method to construct the confidence intervals, and calculate the mean, 95th percentile and their confidence interval, after the simulation has run.

The length of the simulation is controlled by m, the number of batches and the number of observations b within a batch. Set $m = 51$, and $b = 1,000$. (Recall that we need a large batch size in order to estimate the percentile accurately.) In order to allow for the initial conditions, ignore the results from the first batch and calculate all your performance metrics using the remaining 50 batches. As before, the inter-arrival times and retransmission times are exponentially distributed with mean 6 and 5, respectively. The buffer size B including the one being processed is the same as above, i.e., $B = 5$. The service time s is constant.

Obtain the following results:

1. Vary the service time as follows: 1, 2, 3, 4, 5, 6. For each value obtain the statistics:

    a. Mean and 95th percentile and confidence intervals of T.

    b. Mean and 95th percentile and confidence intervals of D.

    Graph your results and comment on your results.

2. Investigate the effect of varying B on all the above statistics for $s = 5$. Graph your results and comments on your findings.

Increase m and b if the confidence intervals are not sufficiently small.

REFERENCES

1. H. Perros, *Computer Simulation Techniques: The Definitive Introduction!*, https://people.engr.ncsu.edu/hp/files/simulation.pdf.

In addition, there is a lot of relevant material on the Internet.

CHAPTER 4

Hypothesis Testing

$\mathbf{I}$N THIS CHAPTER, WE DESCRIBE STATISTICAL HYPOTHESIS TESTING AND ANALYSIS of variance (ANOVA). These are well-known tests used in many techniques in Statistics, such as, regression and forecasting, to test the validity of various estimated parameters.

4.1 STATISTICAL HYPOTHESIS TESTING FOR A MEAN

Statistical hypothesis is used in Statistics to test whether a hypothesis, of which the truth is not known, is correct or not. The hypothesis, notated as H_0, is known as the *null hypothesis*, and the opposite to the hypothesis, notated as H_a, is known as the *alternative hypothesis*.

Let us consider the population of the weights of all men aged 20–29 in the USA . According to the 1970 census, the population mean and standard deviation were 170 and 40 lbs, respectively. Now, let us assume that we want to test if the population mean has changed in 2017. We formulate the null hypothesis that the population mean is still 170 versus that it has changed (increased or decreased); i.e.,

$$H_0: \mu = 170$$

$$H_a: \mu \neq 170$$

This is known as a *two-tailed test* or a *two-sided test*. Depending upon the nature of the hypothesis, we may also formulate a *one-tailed test* (or *one-sided test*) *to the right*. For instance, if we are only concerned whether the average population weight has increased, then we formulate the hypothesis:

$$H_0: \mu = 170$$

$$H_a: \mu > 170$$

On the other hand, if we want to test whether the average population weight has decreased, then we formulate a *one-tailed test* (or *one-sided test*) *to the left*, as follows:

$$H_0: \mu = 170$$

$$H_a: \mu < 170$$

The way we test the two-tailed test hypothesis is to draw a random sample of weights from the 2017 population of all men aged 20–29 in the USA and calculate the sample mean $\bar{x}$. We recall from Section 2.3.4 (Chapter 2) that if we sample n observations from a population which follows an arbitrary distribution with a mean μ and variance σ^2, then the sample mean $\bar{x}$ is normally distributed with mean μ and variance σ^2/n for $n \geq 30$. Accordingly, if the null hypothesis is true, i.e., the population mean has not changed, then the sample mean $\bar{x}$ is normally distributed with mean 170 and standard deviation $40/\sqrt{n}$, where n is the sample size. Assuming that $n = 64$, we have that $40/\sqrt{64} = 5$.

Now, we compare $\bar{x}$ to the population mean $\mu = 170$. If $\bar{x} = 170$, then we can argue that the null hypothesis is true, but what can we say if $\bar{x} = 165$ or 180? In general, if $\bar{x}$ is very close to μ, then we accept the null hypothesis that the population mean has not changed. On the other hand, if it is far away from μ, then we reject the null hypothesis and accept the alternative hypothesis that the population mean has changed (increased or decreased).

In view of this, we need a cut-off point which will act as a threshold. In fact, since we are doing a two-tailed test, we need two points a and b. We note that this is the same concept used in Section 3.5.5 (Chapter 3) to construct confidence intervals. We choose the two symmetrical points so that 95% of the population (or any other percentage, such as 90% and 99%) lies between them, as shown in Figure 4.1. This means that 2.5% of the population lies within each tail $(-\infty, a)$ or $(b, +\infty)$. If the population mean has not changed, then 95% of the sample means $\bar{x}$ are between a and b. Consequently, if $\bar{x}$ lies in-between a and b, we have a 95% probability that the null hypothesis is correct. Otherwise, we reject it and accept the alternative hypothesis. This probability is known as the *confidence level*.

We note that the correct way to say that "we accept the null hypothesis" is to say that "we fail to reject the null hypothesis". However, for simplicity, the former expression is often used. In the opposite direction, if we reject the null hypothesis, then it is correct to say that "we accept the alternative hypothesis".

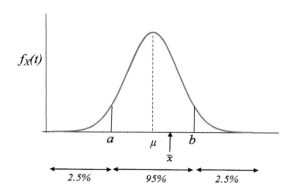

FIGURE 4.1 Accept H_0 if $\bar{x}$ is between a and b.

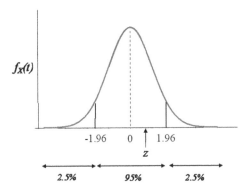

FIGURE 4.2 Two-tailed test at 95% confidence.

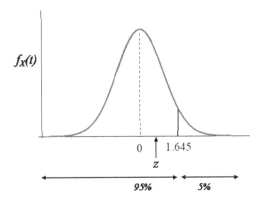

FIGURE 4.3 One-tailed test on the right at 95% confidence.

To simplify the test, we transform points a, b, and $\bar{x}$ to the standard normal distribution $N(0, 1)$, as described in Section 2.3.4 (Chapter 2). At 95% confidence, a corresponds to -1.96, b to 1.96, and $\bar{x}$ to $z = (\bar{x} - 170)/5$, as shown in Figure 4.2. For the two-tailed test, we accept the null hypothesis if $|z| \leq 1.96$. Otherwise, we reject it. For instance, let $\bar{x} = 173$. Then, $z = 3/5 = 0.6$ which is less than 1.96, and therefore we accept the null hypothesis that the population mean of weights has not changed; likewise, if $\bar{x} = 165$, since $z = -5/5 = -1 > -1.96$. However, if $\bar{x} = 180$, then we reject the null hypothesis since $z = 10/5 = 2 > 1.96$.

In the case of the one-tailed test on the right, we are only concerned with how far $\bar{x}$ is to the right of the population mean μ; i.e., we choose point b so that the tail $(b, +\infty)$ accounts for 5%, for a 95% confidence. This point corresponds to 1.645 in the standard normal distribution, as shown in Figure 4.3. Now, given a sample mean $\bar{x}$, we calculate its corresponding z value, i.e., $z = (\bar{x} - 170)/5$, and we accept the null hypothesis if $z < 1.645$. Otherwise, we reject it and accept the alternative hypothesis. For instance, if $\bar{x} = 175$, then $z = 1 < 1.645$, and we accept the null hypothesis; likewise, if $\bar{x} = 155$, for $z = -3 < 1.645$. However, if $\bar{x} = 180$, then $z = 2 > 1.645$, and therefore we reject the null hypothesis and accept the alternative hypothesis.

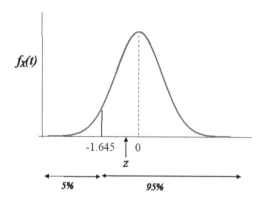

FIGURE 4.4 One-tailed test on the left at 95% confidence.

Finally, in the case of the one-tailed test on the left, we are only concerned with how far $\bar{x}$ is to the left of the population mean μ; i.e., we choose point a so that the tail $(-\infty, a)$ accounts for 5%, for a 95% confidence. This point corresponds to −1.645 in the standard normal distribution, as shown in Figure 4.4. Now, given a sample mean $\bar{x}$, we calculate its corresponding z value, i.e., $z = (\bar{x} - 170)/5$, and we accept the null hypothesis if $z > -1.645$. Otherwise, we reject it and accept the alternative hypothesis. For instance, if $\bar{x} = 165$, then $z = -1 > -1.645$, and we accept the null hypothesis; likewise, if $\bar{x} = 180$, for $z = 2 > -1.645$. However, if $\bar{x} = 155$, then $z = -3 < -1.645$, and therefore we reject the null hypothesis and accept the alternative hypothesis.

We remind the reader that we use the standard normal distribution if the population variance σ^2 in known and/or the sample size $n \geq 30$. Otherwise, we use the t-student distribution. Typically, the t-student distribution is used.

4.1.1 The p-Value

In order to carry out the hypothesis test, we need to compare the z value that corresponds to $\bar{x}$ against the cut-off points a and b or one of them depending on whether we do a two-tailed test or a one-tailed test. These points are determined using the standard normal distribution or the t-student distribution for a given level of confidence.

An easier way to do this is to calculate the p-value of $\bar{x}$. As shown in Figure 4.5, the p-value is the area under the standard normal curve (or the t-student distribution curve) from z to $+\infty$, where z is the value that corresponds to $\bar{x}$.

For example, let us say that for a one-tailed test on the right, we compute a p-value of 0.3. At 95% confidence, the area (b, ∞) of the right tail is 0.05. Since 0.3 > 0.05, we conclude that $z < 1.645$ and accept the null hypothesis. On the other hand, if the p-value is 0.001, then that means that z falls on the right side of 1.645, and in this case, we reject the null hypothesis.

Hypothesis testing is often carried out to test whether statistically estimated coefficients, such as those of a regression model, are zero or not. In this case, statistical packages automatically report the p-value. For a one-sided test to the left or to the right, the p-value is reported as the area under the t-student distribution from $|t|$ to $+\infty$, where t is the

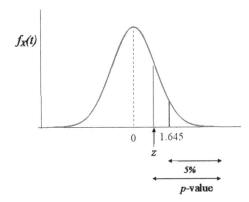

FIGURE 4.5 The *p*-value.

calculated *t*-student statistic. For a two-tailed test, the reported *p*-value is two times the area of the tail |*t*| to +∞.

The *p*-value allows us to decide quickly whether to accept or reject a null hypothesis. Typically, we test a hypothesis at the confidence levels of 90%, 95%, and 99%. For a one-tailed test, the area of the tail corresponding to these three confidence levels is 0.10, 0.05, and 0.01, respectively. For a two-tailed test, the area of both tails is also 0.10, 0.05, and 0.01, respectively. So, irrespectively of whether we do a one-tailed or a two-tailed test, we simply compare the *p*-value to these three thresholds, 0.10, 0.05, and 0.01, and accordingly we accept or reject the null hypothesis.

In general, it is safer to make sure that the conclusion holds for all three confidence levels; i.e., if the reported *p*-value is greater than 0.10, then we accept the null hypothesis. If it is less than 0.01, then we reject the null hypothesis. However, if it falls in-between these values, say it is 0.07, then the conclusion is questionable. For, we reject the null hypothesis at 90% confidence but accept it at 95%and 99% confidence.

4.1.2 Hypothesis Testing for the Difference between Two Population Means

We apply the same ideas as above to test the difference of two population means from two different populations.

Two-tailed test:

$$H_0: \mu_1 - \mu_2 = 0$$

$$H_a: \mu_1 - \mu_2 \neq 0$$

One-tailed test on the right:

$$H_0: \mu_1 - \mu_2 \geq 0$$

$$H_a: \mu_1 - \mu_2 < 0$$

One-tailed test on the left:

$$H_0: \mu_1 - \mu_2 \leq 0$$

$$H_a: \mu_1 - \mu_2 > 0$$

We draw a sample from each population, and let $\bar{x}_1$, $\bar{x}_2$, be the sample means and n_1 and n_2 the sample sizes. If the population variances σ_1^2 and σ_2^2 are known, then

$$z = \frac{\bar{x}_1 - \bar{x}_2}{\sqrt{\dfrac{\sigma_1^2}{n_1} + \dfrac{\sigma_2^2}{n_2}}}.$$

If σ_1^2 and σ_2^2 are not known, then we use the following t-student statistic with $n_1 + n_2 - 2$ degrees of freedom:

$$z = \frac{\bar{x}_1 - \bar{x}_2}{\sqrt{\dfrac{(n_1 - 1)s_1^2 + (n_2 - 1)s_2^2}{n_1 + n_2 - 2} \times \dfrac{n_1 + n_2}{n_1 n_2}}}.$$

4.1.3 Hypothesis Testing for a Proportion

Let us assume that in a large IoT deployment the number of functioning IoT sensors should be 95%. We want to test whether this is the case. As above, we take a sample from the population of the sensors and calculate the proportion $\hat{p}$ of them that are functioning. We note that the sample proportion $\hat{p}$ can be seen as a mean if we code the population of sensors in 1's and 0's. A sensor is coded 1 if it is functioning and 0 if it is not. Consequently, a sample from the population of the IoT sensors will consist of 1's and 0's. Let us say that we obtain n_1 1's and n_2 0's. Then, the sample mean is the sum of all observations, which is n_1 divided by the total number of observations $n_1 + n_2$, i.e., $n_1/(n_1 + n_2)$, which is also the proportion $\hat{p}$ of functioning sensors.

Consequently, we can treat the hypothesis testing for a proportion as a hypothesis testing for a mean. A two-tailed test will allow us to test whether a population proportion p is equal to p_0 or it has changed (increased or decreased). We have

$$H_0: p = p_0$$

$$H_a: p \neq p_0$$

If we only want to test the hypothesis whether the proportion p has increased, then we formulate the one-tailed test on the right:

$$H_0: p = p_0$$

$$H_a: p > p_0$$

On the other hand, if we want to test whether the proportion p has decreased, then we formulate the one-tailed test to the left:

$$H_0: p = p_0$$

$$H_a: p < p_0$$

Let $\hat{p}$ be the sample proportion calculated from a sample of size n. Then,

$$z = \frac{\hat{p} - p_0}{\sqrt{\dfrac{p_0(1 - p_0)}{n}}}.$$

Since we know the population variance of the proportion, we can use the normal distribution to calculate the p-value.

4.1.4 Type I and Type II Errors

Hypothesis testing is not a precise science! It is possible that we may accept the null hypothesis when in fact it is wrong, or reject the null hypothesis when in fact it is true. Associated with hypothesis testing are two errors:

Type 1 error: Erroneous rejection of the null hypothesis
Type II error: Erroneous retention of the null hypothesis

Table 4.1 summarizes the four possible decisions in hypothesis testing. Columns 1 and 2 correspond to the case where the null hypothesis is true or false, and rows 1 and 2 show the decision made, retain, or reject the null hypothesis. The terms in the parenthesis are primarily used in the medical domain. We note that the familiar terms *false positive* and *false negative* correspond to Type I and II errors. The second term, positive or negative, corresponds to the case where the hypothesis is true or false, and the first term, true or false, corresponds to whether we have made the right decision or not. For instance, in true

TABLE 4.1   Four possible decisions in hypothesis testing

	H_0 is true	H_0 is false
Retain H_0	Correct retention (True positive) Probability = $1 - \alpha$	Type II error (False negative) Probability = β
Reject H_0	Type I error (False positive) Probability = α	Correct rejection (True negative) Probability = $1 - \beta$

positive, the hypothesis is correct (positive) and we concluded that the hypothesis is correct (true). In false positive, the hypothesis is correct (positive) but we concluded that the null hypothesis is not correct (false).

We notate the compliment of the level of confidence by α; i.e., if the confidence level is 95%, then $\alpha = 0.05$ and $1 - \alpha = 0.95$. Then, the probability that we accept the null hypothesis when it is true is $1 - \alpha$, and the probability that we make a Type I error is α. Once we set the confidence level, the probability of making a Type 1 error is fixed, and there is nothing more we can do to lower it.

Now, let β be the probability of committing a Type II error. Then, $1 - \beta$ is the probability of rejecting the null hypothesis when it is wrong. Obviously, we want β to be as low as possible. A high value of $1 - \beta$ means that the test is good, and a low value means that the test is bad. Therefore, $1 - \beta$ is a measure of how good the test is, and it is called the *power of a test*. A test with a power of 0.8 is considered good. There are several factors that affect the power of a test, such as, the sample size, the standard deviation, the difference between hypothesized and true mean, the significance level, and whether it is a two-tailed or one-tailed test. The power of a test can be calculated by a statistical package or by a power calculator available on the Internet.

4.2 ANALYSIS OF VARIANCE (ANOVA)

Let us consider an experiment whereby we use wearable sensors to measure the head pressure caused by a simulated accident for three different car sizes: small, medium, and large. We want to test whether car size makes a difference. The ANOVA technique allows us to compare samples taken for each car size in order to determine statistically whether there is a difference or not. The null hypothesis is that there is no difference. The alternative hypothesis is that car size makes a difference.

The hypothesis is formulated as follows. We assume that each sample comes from a different population with mean μ_i, $i = 1, 2, 3$. If car sizes make no difference, then the population means are all the same, or otherwise stated, the three samples come from the same population. We have

$$H_0: \mu_1 = \mu_2 = \mu_3$$

$$H_a: \text{At least one mean is not statistically equal to the other two}$$

We note that we can use the t-test described in Section 4.1 to compare two means at a time, but this is not statistically the same as comparing all three means at the same time. In addition, comparing two means at a time will lead to three different decisions. Also, for a given confidence level, say 95%, our confidence in the overall process is 0.95x0.95x0.95, a lot less than 0.95. On the other hand, ANOVA allows us to test all three means at the same time at 95% confidence.

In the above example, we carry out a *one-way ANOVA* to compare the groups, since we use a single variable for comparison, i.e., head pressure. A *two-way ANOVA* is carried out

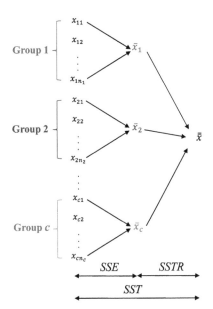

FIGURE 4.6 *SSTR, SSE, SST*.

when we have two variables, such as, head pressure and chest pressure. In general, *N-way ANOVA* is used for *N* variables. Below, we describe the one-way ANOVA.

Let us consider *c* groups of samples, as shown in Figure 4.6, and let n_i be the sample size of group *i*, and x_{ij} be the *j*th observation in group *i*, $i = 1, 2, ..., c$. The samples come from *c* different populations, which we assume that they are normally distributed with the same variance σ^2. We want to test whether the population means are equal or not.

For each group *i*, we calculate the sample mean $\bar{x}_i$ and then we calculate the grand mean $\bar{\bar{x}}$ as the average of all the sample means, i.e., $\bar{\bar{x}} = (1/c)\sum_{i=1}^{c} \bar{x}_i$. We define the following sum of squares *SST, SSTR,* and *SSE*. Figure 4.6 gives a pictorial view of these terms.

Total sum of squares (SST):

$$SST = \sum_{i=1}^{c}\sum_{j=1}^{n_i}\left(x_{ij} - \bar{\bar{x}}\right)^2$$

Viewed all groups as one sample, the above sum measures the total variability among observations by summing the squares of the differences between each observation x_{ij} and the grand mean $\bar{\bar{x}}$.

Treatment sum of squares (SSTR):

$$SSTR = \sum_{i=1}^{c} n_i \left(\bar{x}_i - \bar{\bar{x}}\right)^2$$

This is the weighted sum of the squares of the differences between each sample mean $\bar{x}_i$ and the grand mean $\bar{x}$, and it measures the variability *between groups* (or *treatments* as it is known in the design of experiments.)

Error sum of squares (SSE):

$$SSE = \sum_{i=1}^{c}\sum_{j=1}^{n_i}\left(x_{ij}-\bar{x}_i\right)^2$$

An error is the difference of an individual observation from its sample mean. *SSE* measures the variability *within each group.*

The total variability *SST* can be partitioned to the variability between groups and the variability within groups. We have

$$SST = SSTR + SSE.$$

Using the above sum of squared terms, we define the following terms:

$$\text{Mean square of treatments } \left(MSTR\right): MSTR = SSTR/\left(c-1\right)$$

$$\text{Mean square of error } \left(MSE\right): MSE = SSE/\left(n-c\right),$$

where n is the total number of samples, i.e., $n = \sum_{i=1}^{c} n_i$. The denominators in the above expressions are known as *degrees of freedom* (df), see explanatory note in Section 4.2.1.

In the ANOVA method, we test the difference between population means by analyzing the variance. *MSE* estimates the population variance σ^2 independent of whether the null hypothesis is true or not. *MSTR* estimates the population variance σ^2 if the population means are equal. If the population means are not equal, then it estimates a quantity larger than σ^2. Therefore, if *MSTR* is much larger than *MSE*, then the population means are unlikely to be equal. On the other hand, if *MSTR* is about the same as *MSE*, then the data are consistent with the null hypothesis that the population means are equal.

As in the previous hypothesis tests, we need to decide by how much larger should *MSTR* be than *MSE* before we conclude that the null hypothesis is not correct. This is determined statistically using the *F* statistic due to Fisher, which is the ratio of *MSTR* and *MSE*, i.e.,

$$F = \frac{MSTR}{MSE}.$$

The *F* statistic is a function of the degrees of freedom of the numerator and the degrees of freedom of the denominator, which in our case are $c - 1$ and $n - c$, respectively. The *F*-test is a one-tailed test to the right, because we are concerned with deviations to the right. Consequently, we compare the *p*-value reported by the statistical package, to the

TABLE 4.2 Values for each car size

Small size	Medium size	Large size
643	469	484
655	427	456
702	525	402
$\bar{x}_1 = 666.67$	$\bar{x}_2 = 473.67$	$\bar{x}_3 = 447.33$
$s_1 = 31.18$	$s_2 = 49.17$	$s_3 = 41.68$

TABLE 4.3 One-way ANOVA

Source	SS	df	MS	F	p-value
Between groups	86049.55	2	43024.78	25.17	0.001
Within groups	10254	6	1709		
Total	96303.55	8			

theoretical values 0.1 (90% confidence), 0.05 (95% confidence), and 0.01 (99% confidence). If p-value < 0.01, then we reject the null hypothesis, and if p-value > 0.10 then we accept the null hypothesis.

We now proceed with the example of determining whether car size has an effect on head pressure. Table 4.2 gives three observations per group, and the computed sample mean and standard deviation per group. The intermediate steps leading to the calculation of F and the p-value are summarized in an ANOVA table shown in Table 4.3. The column "Source" gives the source of the variation of the data: between groups ($SSTR$), within groups (SSE), and total (SST). "SS" gives the sum of squares for between groups, within groups, and total. "df" gives the degree of freedom, "MS" gives the mean square values for $MSTR$ and MSE, and "F" gives the F statistic. The table, calculated by a statistical package, also gives the p-value. Based on the calculated p-value of 0.001, we reject the null hypothesis that all sample means come from the same population and we accept the alternative hypothesis that car size does matter, which makes sense intuitively.

If we calculate the F table by hand, it is easier to determine the critical F^{cv} value for the given degrees of freedom and confidence level from the F table (available in Statistics books and in the Internet). We accept the null hypothesis if $F < F^{cv}$, otherwise we reject it.

4.2.1 Degrees of Freedom

The number of degrees of freedom (df) is the number of independent pieces of information available to estimate another piece of information. For instance, let $x_1, x_2, \ldots, x_n$ be n observations drawn from the same population, and let $\bar{x}$ be their mean. The differences $x_i - \bar{x}$ are known as residuals, and they sum up to zero since

$$\sum_{i=1}^{n}\left(x_i - \bar{x}\right) = \sum_{i=1}^{n} x_i - \sum_{i=1}^{n} \bar{x} = n\bar{x} - n\bar{x} = 0.$$

Consequently, if we know the $n - 1$ residuals, we can determine the last one. In this case, the degrees of freedom is $n - 1$.

EXERCISES

1. A manufacturer of sensors claims that the mean time to failure of a sensor is 30 months. 100 sensors were randomly monitored until they failed. The sample mean is 28.8 months with a standard deviation of 2.8. Do the data present sufficient evidence to indicate that the manufacturer's claim is correct?

2. An IoT user suspects that a particular brand of sensors is affected by humidity. The user places a sensor in a humid environment and another one in a dry environment and collects data for a month. The sensors produce one observation every hour. Describe how you can test statistically whether humidity affects the two sensors.

3. Four different sensor manufacturers make a sensor that senses a rare odor. A potential buyer wants to make sure that there is no difference in the observations that the different sensors report.

   a. State the null and alternative hypothesis.

   b. Use the information in Table 4.4 to show that $SSTR = 1186$ and $SSE = 4840.934$.

   c. Construct the ANOVA table and calculate the F value. Then use the F table (available in Statistics books and in the Internet) to determine whether to accept or reject the hypothesis.

4. Fill in the ANOVA Table 4.5 given that we have four different groups, each of 20 observations.

TABLE 4.4   Data for Exercise 3

Group	n_i	Mean	St. deviation
1	10	100.40	11.68
2	10	103.00	11.58
3	10	107.10	10.05
4	10	114.80	10.61

TABLE 4.5   ANOVA table for Exercise 4

Source	SS	df	MS	F
Between groups	136			
Within groups	432			
Total				

5. A field of wheat is divided into five areas, and a sensor that measures humidity is placed in each area. Describe how ANOVA can be used to test if the five areas of the field require different amount of watering.

REFERENCES

The material in this chapter can be found in any introductory book on Statistics.

CHAPTER 5

Multivariable Linear Regression

M ULTIVARIABLE OR MULTIPLE LINEAR REGRESSION IS USED IN THE CASE where a random variable Y can be expressed as a linear combination of a number of random variables $X_1, X_2, ..., X_k$. That is, Y is given by the expression:

$$Y = a_0 + a_1 X_1 + a_2 X_2 + \cdots + a_k X_k + \varepsilon,$$

where a_0 is the intercept, a_i is the coefficient of the ith random variable X_i, and ε is a random error (noise), which is assumed to be normally distributed with mean 0 and a variance σ^2. Multivariable linear regression can be used to estimate the parameters $a_0, a_1, ..., a_k$. The random variables $X_1, X_2, ... , X_k$ are known as *independent* variables or *predictors*, and Y is known as the *dependent* variable.

For example, let us assume that we want to estimate the amount of time it takes for a process to run on a server as a function of the CPU time consumed, the number of disk I/Os issued, the number of pages paged in, and the number of Ethernet packets transmitted and received. We define the following random variables:

X_1: total CPU used by a process

X_2: number of disk I/Os issued by a process

X_3: number of pages paged in during the execution of the process

X_4: number of Ethernet packets transmitted and received

Y: elapsed time, i.e., the time elapsed from the moment the process was loaded into the main memory to the moment it was fully executed.

We assume that the dependent variable Y is a linear function of the independent variables X_1, X_2, X_3, and X_4; i.e.,

$$Y = a_0 + a_1 X_1 + a_2 X_2 + a_3 X_3 + a_4 X_4 + \varepsilon.$$

We collect data in the form of $(y_i, x_{i1}, x_{i2}, x_{i3}, x_{i4})$, where the subscript i refers to the ith process, y_i is the observed elapsed time and $x_{i1}, x_{i2}, x_{i3}, x_{i4}$ are the observed values for the random variables X_1, X_2, X_3, and X_4. Subsequently, we use multivariable linear regression to estimate the coefficients and the variance of the error. The regression model can then be used for predictive purposes. That is, given a new tuple (x_1, x_2, x_3, x_4) we can estimate y.

In this chapter, the simple linear regression model that has only one independent random variable is first described followed by the general case of multivariable regression. A discussion of residual analysis and other measures of goodness of fit used to determine the accuracy of a regression model is then given. Subsequently, polynomial regression, confidence and prediction intervals, and the ridge, lasso, and elastic net regression techniques are presented. A set of exercises and a regression project are given at the end of the chapter.

5.1 SIMPLE LINEAR REGRESSION

A simple linear regression is a multivariable linear regression model with only one independent variable, i.e.,

$$Y = a_0 + a_1 X_1 + \varepsilon.$$

Given a set of data $(x_1, y_1), (x_2, y_2), ..., (x_n, y_n)$, we fit a straight line through the data given by the expression:

$$\hat{y}_i = \hat{a}_0 + \hat{a}_1 x_i,$$

where $i = 1, 2, ..., n$, $\hat{a}_0$ and $\hat{a}_1$ are the estimates of the intercept a_0 and coefficient a_1, and $\hat{y}_i$ is the estimate of y_i for a given value of x_i, see Figure 5.1. For a value x_i, the difference between its corresponding original value y_i and its estimate $\hat{y}_i$ is known as the error. The estimates of the coefficients a_0 and a_1 are obtained by minimizing the sum of squared errors over all the data points (SSE), i.e,

$$F(a_0, a_1) = \sum_{i=1}^{n} (\hat{y}_i - y_i)^2.$$

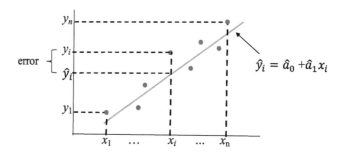

FIGURE 5.1 A simple regression.

The better the fit, the lower is the SSE. In view of this, we obtain the estimates $\hat{a}_0$ and $\hat{a}_1$ by minimizing $F(a_0, a_1)$. For this, the partial derivatives with respect to a_0 and a_1 should be zero. We have

$$\frac{\partial F}{\partial a_0} = \sum_{i=1}^{n} \left[\left(a_0 + a_1 x_i - y_i \right)^2 \right]' = \sum_{i=1}^{n} 2 \left(a_0 + a_1 x_i - y_i \right) = 0.$$

$$\frac{\partial F}{\partial a_1} = \sum_{i=1}^{n} \left[\left(a_0 + a_1 x_i - y_i \right)^2 \right]' = \sum_{i=1}^{n} 2 \left(a_0 + a_1 x_i - y_i \right) x_i = 0.$$

From the above two equations, we obtain the system of equations:

$$\begin{cases} n a_0 + a_1 \sum_{i=1}^{n} x_i = \sum_{i=1}^{n} y_i & (1) \\[2em] a_0 \sum_{i=1}^{n} x_i + a_1 \sum_{i=1}^{n} x_i^2 = \sum_{i=1}^{n} x_i y_i & (2) \end{cases}$$

It can be shown that the above system of equations has a unique solution unless all x values are identical. The solution specifies the values of a_0 and a_1 for which $F(a_0, a_1)$ is minimum. The straight line determined in this fashion is called the *regression line*.

We note that equation (1) of the above system can be obtained by observing that the best fit line always passes through the point $(\bar{x}, \bar{y})$, where $\bar{x}$ is mean of all x values, and $\bar{y}$ is the mean of all y values; i.e.,

$$\bar{y} = a_0 + a_1 \bar{x}$$

or

$$a_0 = \bar{y} - a_1 \bar{x}$$

$$n a_0 = n \bar{y} - n a_1 \bar{x}$$

$$n a_0 + a_1 \sum_{i=1}^{n} x_i = \sum_{i=1}^{n} y_i. \qquad (3)$$

The solution to the above system of equations can be obtained by substituting (3) into (2). We have

$$\frac{1}{n} \left(\sum_{i=1}^{n} y_i - a_1 \sum_{i=1}^{n} x_i \right) \sum_{i=1}^{n} x_i + a_1 \sum_{i=1}^{n} x_i^2 = \sum_{i=1}^{n} x_i y_i$$

or

$$\frac{1}{n}\left(\sum_{i=1}^{n} y_i\right)\left(\sum_{i=1}^{n} x_i\right) - \frac{a_1}{n}\left(\sum_{i=1}^{n} x_i\right)^2 + a_1\sum_{i=1}^{n} x_i^2 = \sum_{i=1}^{n} x_i y_i$$

or

$$a_1 = \frac{\sum_{i=1}^{n} x_i y_i - \frac{1}{n}\left(\sum_{i=1}^{n} y_i\right)\left(\sum_{i=1}^{n} x_i\right)}{\sum_{i=1}^{n} x_i^2 - \frac{1}{n}\left(\sum_{i=1}^{n} x_i\right)^2}$$

which can be shown to be

$$\hat{a}_1 = \frac{\sum_{i=1}^{n}(x_i - \bar{x})(y_i - \bar{y})}{\sum_{i=1}^{n}(x_i - \bar{x})^2}$$

and

$$\hat{a}_0 = \bar{y} - \hat{a}_1 \bar{x}.$$

We use the notation $\hat{a}_0$ and $\hat{a}_1$ to indicate that these are estimates of the true values of a_0 and a_1, since they are calculated using a sample of observations. $\hat{a}_0$ is the intercept and $\hat{a}_1$ is the slope of the line.

This method of estimating the coefficients a_0 and a_1 by minimizing the SSE is known as the *ordinary least squares* method. The same method is used for the general case of multi-variable linear regression.

Since $\hat{a}_1$ is estimated using a sample of observations, we test the hypothesis that it may be zero. That is, we test the hypothesis:

$$H_0: \hat{a}_1 = 0$$

$$H_a: \hat{a}_1 \neq 0$$

If the null assumption holds, then there is no statistical relation between the independent variable X and the dependent variable Y. Likewise, we test for the intercept. These two hypothesis tests are automatically carried out by a statistical package, which calculates the p-value for each coefficient. Upon examination of the p-values we determine whether the corresponding null hypothesis is true or not.

The variance σ^2 of the error ε can be easily estimated once we have estimated the two coefficients a_0 and a_1. For a given x_i, the error is equal to the difference between the actual value y_i and the predicted value $\hat{y}_i$, i.e., $\hat{\varepsilon}_i = y_i - \hat{y}_i$, where $\hat{y}_i = \hat{a}_0 + \hat{a}_1 x_i$. The sample variance of these errors is an estimate of σ^2.

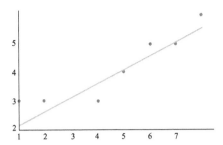

FIGURE 5.2 An example of a simple regression.

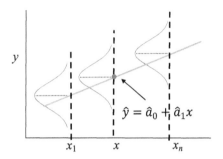

FIGURE 5.3 The prediction values are normally distributed.

For example, let us consider the following data set: $(1,3)$, $(2,3)$, $(3,2)$, $(4,3)$, $(5,4)$, $(6,5)$, $(7,5)$, $(8,6)$. Using a statistical package we obtain the regression line $\hat{y}_i = 1.6786 + 0.4881 x_i$ shown in Figure 5.2. The reported p-values are 0.0206 and 0.004 for the intercept $\hat{a}_0$ and the coefficient $\hat{a}_1$ estimates, respectively. Consequently, we reject the null hypothesis that $\hat{a}_1$ is zero at 90%, 95%, and 99% confidence. Also, we reject the null hypothesis that $\hat{a}_0$ is zero at 90%, 95% but not at 99%. Overall, we can treat these estimates as being non-zero.

For a given x, the regression line $\hat{y} = \hat{a}_0 + \hat{a}_1 x$ gives the mean prediction value. Other prediction values for x can be obtained by adding a sample of the error from the normal distribution $N(0, \sigma^2)$. Since ε is normally distributed, the set of all predictions for a given x follows the normal distribution $N(\hat{a}_0 + \hat{a}_1 x, \sigma^2)$ as shown in Figure 5.3.

5.2 MULTIVARIABLE LINEAR REGRESSION

In a multivariable linear regression, we predict the value of the dependent variable Y as a linear combination of one or more independent variables, i.e.,

$$Y = a_0 + a_1 X_1 + a_2 X_2 + \cdots + a_k X_k + \varepsilon,$$

where a_0 is the intercept, a_i is the coefficient of the ith random variable X_i, and ε is a random error (noise) normally distributed with mean 0 and a variance σ^2. The parameters a_0, a_1, ..., a_k are estimated using the least squares method, that is, by minimizing the SSE.

An important step when developing a regression model is to verify its goodness of fit, that is, to establish how well it models the data. For this, there is a number of checks, as described below, that need to be carried out. Typically, regression packages produce enough additional information so that the user can carry out these checks. If this is not the case, the user may have to use functions from other software packages.

5.2.1 Significance of the Regression Coefficients

It is possible that an estimated coefficient may have a very small value, and consequently we may be inclined to assume wrongly that it is zero. In view of this, we carry out a hypothesis testing similar to the one described in Section 5.1 for the simple linear regression, for each of the estimated regression coefficients $\hat{a}_0, \hat{a}_1, \ldots, \hat{a}_k$; That is, we test the hypothesis:

$$H_0: \hat{a}_i = 0$$

$$H_a: \hat{a}_i \neq 0$$

for $i = 1, 2, \ldots, k$. If the null assumption holds, then there is no statistical relation between the dependent variable X_i and the independent variable Y. That is, $\hat{a}_i$ can be treated as being zero. If the null hypothesis is rejected, then $\hat{a}_i$ is not zero, irrespective of how small is its value. These hypothesis tests are automatically carried out by a regression package and the p-value for each coefficient is reported.

The above procedure tests one coefficient at a time. In addition, we can test all the coefficients at the same time using one-way ANOVA. As discussed in Section 4.2, ANOVA is a technique that allows us to compare samples taken from different groups in order to determine statistically whether there is a difference in their mean or not. In a regression model, a group consists of all the y values that correspond to the same x_i. ANOVA is used to test whether the regression relation is significant. That is, there is one or more independent variables in the model that can predict Y. Specifically, the null hypothesis is that there is no relation between the independent variables and the dependent variable, that is, $a_1 = \cdots = a_k = 0$, and the alternative hypothesis is that at least one coefficient is not zero. The hypothesis test is as follows:

$$H_0: a_1 = \cdots = a_k = 0.$$

$$H_a: \text{at least one } a_i \neq 0, \ i = 1, 2, \ldots, k$$

The statistical package calculates the F value and its corresponding p-value. We recall that this is a one-tailed test on the right, and we can determine whether the null hypothesis is true depending on the p-value.

5.2.2 Residual Analysis

Given that we have estimated the coefficients a_0, a_1, ..., a_k, and the variance σ^2 of the error ε, we can predict y_i of the ith data point $(y_i, x_{i1}, x_{i2}, ..., x_{ik})$ used in the regression. We have

$$\hat{y}_i = \hat{a}_0 + \hat{a}_1 x_{i1} + \cdots + \hat{a}_k x_{ik},$$

where $\hat{a}_0$, $\hat{a}_1$, ..., $\hat{a}_k$ are the estimates of the regression coefficients and $\hat{y}_i$ is the estimate of y_i. The difference between the predicted value $\hat{y}_i$ and the actual value y_i is an estimate of the error ε, that is, $\hat{\varepsilon}_i = y_i - \hat{y}_i$, and it is also referred to as the residual. In order for a regression model to be a good fit, several assumptions have to hold for the residuals as discussed below.

a. Residuals Should be Normally Distributed

Since we assumed that the error ε follows the normal distribution $N(0, \sigma^2)$, the residuals should also follow the same distribution. This assumption can be verified by testing that the histogram of the residuals is normally distributed with mean 0 and variance σ^2 by carrying out a *chi-squared* (χ^2) test and a *quantile-quantile* (Q-Q) plot.

The χ^2 test is used to check whether an empirical distribution follows a specific theoretical distribution. In particular, it tests whether differences between an empirical and a theoretical distribution are due to random fluctuations or because the empirical distribution does not follow the theoretical distribution.

We first divide the entire range of the residual values into m bins (i.e., m successive and non-overlapping intervals). Then, we construct the histogram for the given set of bins, and let O_i be the observed frequency for bin i, i.e., the number of residuals that fall within the ith bin. Subsequently, we construct the expected frequency E_i for each bin i using the normal distribution $N(0, \sigma^2)$. Let $F(x)$ be the cumulative distribution of $N(0, \sigma^2)$. Then, $E_i = n(F(x_{i, u}) - F(x_{i, l}))$, where n is the total number of residuals, and $x_{i, u}$ and $x_{i, l}$ are the upper and lower limits of bin i. The quantity $F(x_{i, u}) - F(x_{i, l})$ gives the theoretical probability that an observation falls in bin i, and multiplying it by n gives us the theoretical expected number of residuals that belong to the bin. The χ^2 is calculated as follows:

$$\chi^2 = \sum_{i=1}^{m} \frac{(O_i - E_i)^2}{E_i}.$$

The null hypothesis is that the residuals follow the normal distribution $N(0, \sigma^2)$, and the alternative hypothesis is that they do not. This is a one-tailed test on the right, and for given level of confidence, say 95%, we can obtain the $\chi^2_{0.95, l-c}$ value from a χ^2 table with $l - c$ degrees of freedom, where l is the number of non-empty bins and c is the number of the estimated parameters, i.e., $c = 2$. The null hypothesis is accepted if $\chi^2 < \chi^2_{0.95, l-c}$.

The test is sensitive to the choice of the bin width. Most reasonable choices should produce similar, but not identical, results. For the test to be valid, the expected number of observation in each bin should be at least 5. Bins that do not fulfill this requirement, typically toward the tails, can be combined. This test is not valid for small number of residuals.

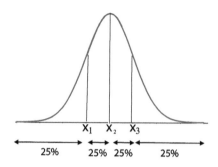

FIGURE 5.4 4-quantiles for a normal distribution.

The Q-Q plot compares two probability distributions by plotting their *quantiles* against each other. The comparison is subjective since it relies on visual inspection of the plot, but the advantage is that it allows us to see at a glance if the two distributions are similar or not.

The *q*-quantiles are $q - 1$ points that divide the area under the curve of a probability distribution into *q* equal parts. For instance, for a normal distribution with $q = 4$, we have three points x_1, x_2, and x_3 such that the area under the curve is divided into four equal parts $(-\infty, x_1)$, (x_1, x_2), (x_2, x_3), and $(x_3 + \infty)$, as shown in Figure 5.4. Let $F(x)$ be the cumulative distribution, then we have that $F(x_1) = 1/4$, $F(x_2) = 2/4$, and $F(x_3) = 3/4$. Consequently, the three points x_1, x_2, and x_3 can be obtained by inverting $F(x)$ at the points 1/4, 2/4, and 3/4, respectively. Quantiles can be calculated in the same way as percentiles (see Section 3.5.4, Chapter 3) in the case where we only have a sample of observations, rather than a theoretical distribution. Let $y_1 \le y_2 \le \dots \le y_n$ be the sample sorted out in an ascending order. Then, for $q = 4$, the three quantiles x_1, x_2, and x_3 are given by the value y_k where $k = \lceil 0.25 \times n \rceil$, $\lceil 0.50 \times n \rceil$, $\lceil 0.75 \times n \rceil$, respectively.

To do a Q-Q plot, we first have to decide how many quantiles to use. Then, we compute the quantiles for both distributions and plot their scatter diagram. A point (x_i, y_i) on the plot corresponds to the *i*th quantile of the first distribution and second distribution, respectively. If the two distributions being compared are similar, the points in the Q-Q plot will approximately lie on a straight line, as shown in Figure 5.5. If they differ, then the plot will deviate from the straight line as shown in Figure 5.6. In this case, we observe that the lower quantiles are similar, but they start to deviate as we move to higher quantiles. Specifically, the quantiles of the distribution on the *x*-axis are larger than their counterparts of the distribution on the *y*-axis. This means that the distribution on the *x*-axis has a longer tail than the other distribution.

b. Residuals Should Have No Trends

The shape of the scatter diagram of the residuals $\hat{\varepsilon}_i = y_i - \hat{y}_i$ versus the predicted value $\hat{y}_i$ provides an indication of the goodness of fit of the regression model. Ideally, the scatter diagram should show no trends as shown in Figure 5.7. That is, the residuals should be randomly distributed and not correlated. Examples of positively and negatively correlated residuals are shown in Figures 5.8 and 5.9, respectively. The randomness of the residuals can be also verified using statistical tests, such as, the frequency test and the monobit test (see [1]). The correlation of the residuals can be also checked by constructing their

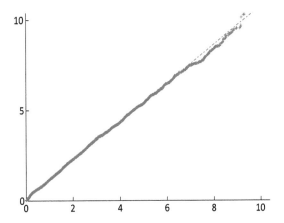

FIGURE 5.5 An example of a Q-Q plot of two similar distributions.

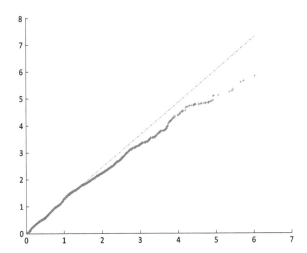

FIGURE 5.6 An example of Q-Q plot of two dissimilar distributions.

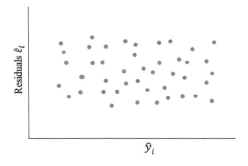

FIGURE 5.7 Scatter diagram of residuals with no trends.

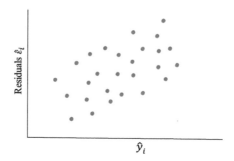

FIGURE 5.8 Positively correlated residuals.

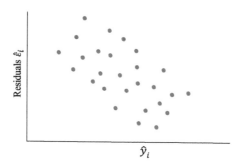

FIGURE 5.9 Negatively correlated residuals.

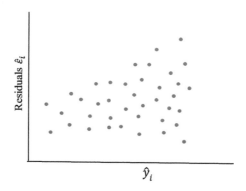

FIGURE 5.10 An example where residuals do not have the same variance.

correlogram (see Section 3.5.5). This is a graph in which the autocorrelation r_k is plotted against lag k. If the residuals are not correlated, then the lag 1 autocorrelation and all subsequent autocorrelations should be zero, that is, they fall within the upper and lower bounds, or in the absence of these bounds, their absolute value is less than 0.15.

In addition, the residuals should have the same variance for all the data points. That is, they should not display a shape as the one shown in Figure 5.10. This happens for instance when the fitted line is such that the actual observations are getting further away from the predicted values as $\hat{y}_i$ increases. Another example where the variance is not constant is the case where the actual observations are getting closer to the fitted value as $\hat{y}_i$ increases.

5.2.3 *R*-Squared

Another indicator of the goodness of fit of a regression model is *R*-squared, also known as the coefficient of determination, or the coefficient of multiple determination in the case of multivariable regression. In the case of the simple linear regression model *R*-squared is calculated as follows. Define the following sum of squares, similar to those defined in ANOVA:

$$SS_{reg} = \sum_{i=1}^{n}(\hat{y}_i - \bar{y})^2$$

$$SS_{error} = \sum_{i=1}^{n}(\hat{y}_i - y_i)^2$$

$$SS_{total} = \sum_{i=1}^{n}(y_i - \bar{y})^2,$$

where $y_i, \hat{y}_i, \bar{y}$ are the original value and the estimated value for x_i, and the mean of all the y_i values, respectively, $i = 1, 2, ..., n$. SS_{reg} is the same as *SSTR* in ANOVA, i.e., the sum of squares between sample means and the grand total (In regression, a sample is the set of all the y values that correspond to the same x_i value.). Likewise, SS_{error} and SS_{total} correspond to *SST* and *SSE* in ANOVA. *R*-squared is calculated as follows:

$$R\text{-squared} = \frac{SS_{reg}}{SS_{total}} = 1 - \frac{SS_{error}}{SS_{total}}$$

since $SS_{total} = SS_{reg} + SS_{error}$.

R-squared varies between 0% and 100%. If the fit is good, then SS_{error} is small and *R*-squared is close to 100%. If the fit is not good, then SS_{error} is large, and *R*-squared is close to 0%. In general, the higher the *R*-squared, the better the model fits the data. However, one needs to exercise caution is using this indicator, as it is possible to have a good model with a low *R*-squared, or a model that does not fit the data well but has a high *R*-squared. One should use the *R*-squared value together with examining the residuals and checking the significance of the regression coefficients as discussed above, in order to determine if a regression model is good or not.

R-squared can be used for simple linear regressions. For multivariable regressions, it is recommended to use the *adjusted R-squared*. This is because *R*-squared tends to increase as more independent variables are added in the regression model, irrespective of how well they are correlated to the dependent variable. The adjusted *R*-squared only increases if adding a new independent variable improves the goodness of fit of the model. It is given by the expression:

$$\text{Adjusted } R\text{-squared} = 1 - \frac{n-1}{n-k-1}(1 - R\text{-squared}),$$

where n is the sample size and k the number of independent variables. The adjusted R-squared may be negative if the model is too complex for the sample size and/or the independent variables have too little predictive value.

5.2.4 Multicollinearity

Multicollinearity is a term used to indicate the presence of high correlation among the independent variables. In a regression model signs of multicollinearity are

- High correlation between the independent variables.

- Coefficients whose sign and magnitudes do not make sense.

- Statistically non-significant coefficients of important independent variables.

- Extreme sensitivity on the sign and magnitude of coefficients to removal or addition of independent variables.

The *variance inflation factor* (*VIF*) is used to find out if there is multicollinearity present. It is given by the expression:

$$VIF_i = \frac{1}{1 - R_i^2},$$

where R_i^2 is the R-squared of the regression of the ith independent variable on all the remaining independent variables. That is, the R-squared is calculated by constructing a regression model where X_i is the dependent variable and X_1, ..., X_{i-1}, X_{i+1}, ...X_k are the independent variables. The better we can predict X_i from the remaining independent variables, the more R-squared tends to 1, and consequently VIF_i tends to infinity. On the other hand, if the goodness of fit of the regression is not good, then R-squared tends to 0, and VIF_i tends to 1. If VIF_i is greater than 5, then we have a multicollinearity problem.

In general, high correlation between two independent variables may not affect the goodness of fit of the regression model as long as the *VIF* values are less than 5, and the residual analysis and adjusted R-squared are satisfactory.

A simple way to avoid multicollinearity is to compute the correlation of all pairs of independent variables and remove an independent variable if it is highly correlated with another one. However, we may lose information because of the removed variable.

In general, the presence of multicollinearity does not affect the efficiency of extrapolating the fitted model to new data provided that the predictor variables follow the same pattern of multicollinearity in the new data as in the data on which the regression model is based.

5.2.5 Data Transformations

In linear regression, the relation between the independent variables and the dependent variable is assumed to be linear. In the case of the simple linear regression, this is easy to verify by plotting the scatter diagram of X and Y. In the case of linear multivariable

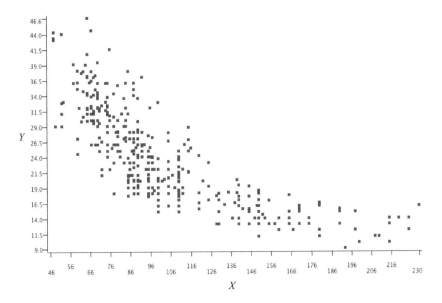

FIGURE 5.11 Original scatter diagram.

regression, this is not possible, but we can deduce linearity by plotting the scatter diagram of the residuals (see Section 5.2.4). In addition to the non-linearity issue, the residuals may not be normally distributed and/or they may not have equal variance, i.e., the scatter diagram of the residuals displays a funneling effect.

In case where we only have non-linearity while the normal distribution and equal variance apply, we need only to transform X, by taking the $log_e X$. Transforming Y by taking the $log_e Y$, can be useful when we have non-normality and/or unequal variances. This transformation may also help to *linearize* the relationship between X and Y. Finally, transforming both X and Y by taking the $log_e X$ and the $log_e Y$ may be helpful when we have non-linearity, and the residuals are not normal and have unequal variances.

An example of the effect of transforming the data is shown in Figures 5.11 and 5.12. We observe non-linearity and unequal variance in the scatter plot in Figure 5.10. These two issues are resolved after taking $log_e X$ and $log_e Y$, as shown in Figure 5.11.

5.3 AN EXAMPLE

In this section, we give an example of fitting a regression model to a data set, reported in http://archive.ics.uci.edu/ml/index.html, that contains house power consumption data. The original data is given as averages over a minute, but for this example the data was averaged over a day. We use the following four variables from the data set:

- *Total*: Total energy used, averaged every day in watt-hours.

- *Lost*: Total energy lost due to power dissipation, averaged every day in watt-hours.

- *Volts*: Voltage measured at the house, averaged every day.

- *Amps*: Total current, averaged every day in ampere.

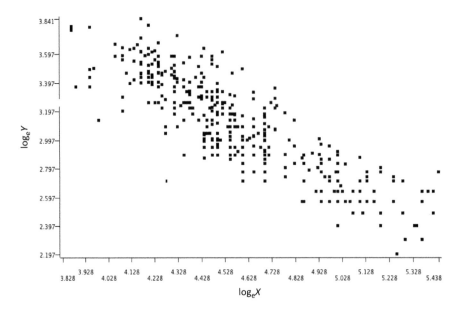

FIGURE 5.12 Scatter diagram after log–log transformation.

We want to construct a linear multivariable regression model so that to predict *Total* from the remaining variables. That is, *Total* is the predicted variable and *Lost*, *Volts*, and *Amps* are the independent variables.

We first calculate the correlation matrix in order to get a sense of how these variables are related. This matrix gives the correlations between all possible pairs of the variables, and it is given in Table 5.1. We note that there is no significant correlation between the independent variables. Next, we fit a multivariable regression model, and the results are summarized in Table 5.2. We note that all the coefficients are significant, since their *p*-value is zero. The residuals standard deviation is 0.01, and the adjusted *R*-squared is 0.9994.

TABLE 5.1   The correlation matrix

	Total	**Lost**	**Volts**	**Amps**
Total	1	0.019	0.153	0.999
Lost		1	−0.054	0.042
Volts			1	0.126
Amps				1

TABLE 5.2   Estimated coefficients for the multivariable regression

	Coefficient	**St. error**	**t value**	**$p > |t|$**
Lost	−0.2453	0.0076	−32.3433	0
Volts	0.0049	0.0001	38.9144	0
Amps	0.2414	0.0002	1,472.03	0
Intercept	−1.1862	0.0306	−38.771	0

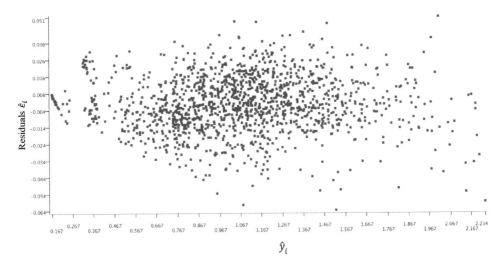

FIGURE 5.13 Residuals scatter plot for the multiple regression model.

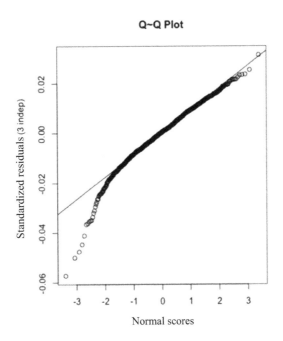

FIGURE 5.14 Q-Q plot of residuals for the multiple regression model.

Figures 5.13 and 5.14 give the residuals scatter plot and the Q-Q plot. We note that there are no discernible trends (correlation and unequal variance) in the residual scatter plot. From the Q-Q plot, we see that the left tail of the distribution of the residuals does not match that of the normal distribution, as it extends longer than that of the normal distribution. (This can be also seen by plotting the residuals histogram.) However, overall, this regression model appears to have a good fit.

TABLE 5.3   Estimated coefficients for the simple regression

	Coefficient	St. error	t value	p > |t||
Amps	0.242	0.0003	870.212	0
Intercept	−0.0279	0.0013	−20.88	0

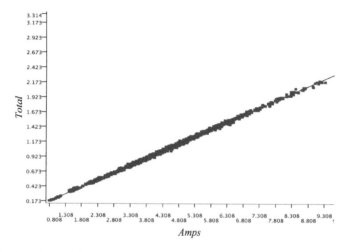

FIGURE 5.15   The fitted simple linear regression model.

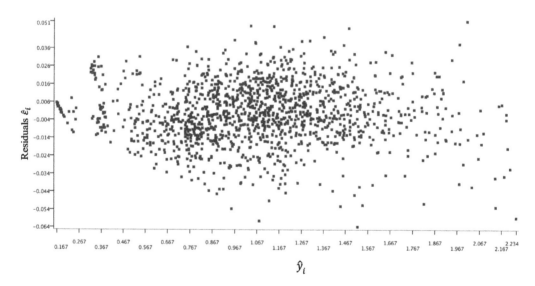

FIGURE 5.16   Residuals scatter plot for the multiple regression model.

We observe from Table 5.1 that the correlation between *Total* and *Amps* is practically 1. The obvious question, therefore, is whether we can do better by fitting a simple linear regression with only *Amps* as the independent variable. Table 5.3 gives the estimates of the coefficients and their *p*-values of the simple linear regression. The adjusted *R*-squared is 0.9981, and the residuals standard deviation is 0.017. Figure 5.15 shows the fitted regression line, and Figures 5.16 and 5.17 give the residuals scatter plot and the Q-Q plot, respectively. From

Q~Q Plot

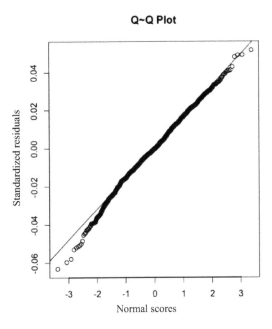

FIGURE 5.17 *Q-Q plot of the residuals for the simple regression model.*

the scatter plot we see that there is some inequality in the variance for the lower fitted values, but the Q-Q plot seems to be slightly better than the one for the multivariable regression.

It would seem that it is sufficient to just use a simple linear regression with *Amps* as the independent variable, and the addition of the other independent variables do not change the adjusted *R*-squared in any significant way.

5.4 POLYNOMIAL REGRESSION

An alternative method to data transformations when the relation between *X* and *Y* is not linear is to use a polynomial regression. This is a polynomial in *X* given by the expression:

$$Y = a_0 + a_1 X + a_2 X^2 + \cdots + a_k X^k + \varepsilon.$$

It is a special case of the multivariable regression model where $X_i = X^i, i = 1, 2, ..., k$. In view of this, its parameters $a_0, a_1, ..., a_k$ are estimated in the same way as in multivariable regression with *k* independent variables.

For example, Figure 5.18 gives a scatter diagram where *X* and *Y* are not linearly related. A simple linear regression model was first fitted, the plot of which is given in the same figure. The estimated model is *Y* = 5.043 + 0.225*X*, and the adjusted *R*-squared is 0.1363

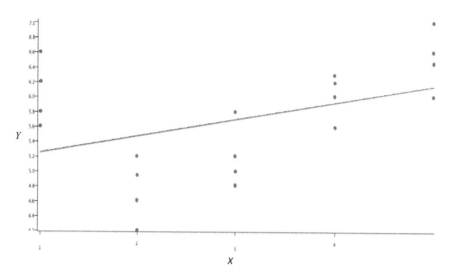

FIGURE 5.18 Simple linear regression.

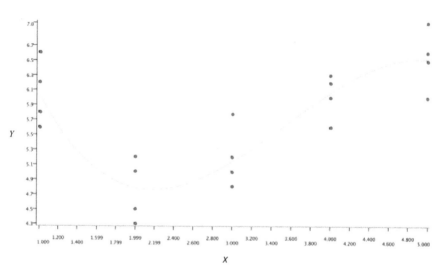

FIGURE 5.19 Second-order polynomial regression.

which confirms our visual observation that the model is a bad fit. The p-value for the coefficient of X is 0.057, which means that we reject the null hypothesis that it is zero at 90% but not at 95% and 99%. The p-value for the intercept is zero.

Subsequently, we fitted various higher-order polynomials, of which the third-order polynomial shown in Figure 5.19 gave the best fit. The fitted polynomial is: $Y = 9.935 - 5.5619X + 1.840X^2 - 0.1729X^3$. All p-values are less than 0.01 and the adjusted R-squared is 0.723. As can be seen, in this case we were able to obtain a good fit using polynomial regression.

5.5 CONFIDENCE AND PREDICTION INTERVALS

In this section, we discuss how to predict a value y, for a given input $(x_1, \ldots, x_k)$. We first note that different predictions can be obtained for a given input, of which the regression line $\hat{y} = \hat{a}_0 + \hat{a}_1 x_1 + \cdots + \hat{a}_k x_k$ gives the mean of these predictions. Other predictions can be obtained by adding a randomly sampled error ε from the distribution $N(0, \sigma^2)$ to $\hat{y}$, that is, $\hat{y} + \varepsilon$. Since $\varepsilon \sim N(0, \sigma^2)$, the predictions for a given input $(x_1, \ldots, x_k)$ also follow a normal distribution with mean $\hat{a}_0 + \hat{a}_1 x_1 + \cdots + \hat{a}_k x_k$ and variance σ^2. Note that the mean of this distribution is conditioned on the input value.

A point estimate, such as the mean $\hat{y} = \hat{a}_0 + \hat{a}_1 x_1 + \cdots + \hat{a}_k x_k$ or any other estimate $\hat{y} + \varepsilon$ is not very useful by itself since it does not convey a sense of the error involved in making the prediction. A better prediction can be obtained using a confidence interval or a *prediction interval*.

A confidence interval is built around the mean prediction $\hat{y}$ and it is an estimate of an interval within which the true population mean μ_x lies with a certain level of confidence, such as 95%. The true population mean μ_x is the population mean of all the predictions for a given input $x = (x_1, \ldots, x_k)$. Confidence intervals were described in detail in Section 3.5.3 and the subsequent sections (Chapter 3).

A prediction interval is an estimate of an interval within which a future observation corresponding to the same input x will lie with a given level of confidence, such as 95%. Otherwise stated, if we happen to obtain an observation in the future for an input x, then this observation will lie in the prediction interval with probability 0.95. Other levels of confidence can be used, as in the case of confidence intervals.

The two types of interval appear to be similar, but in fact they estimate different parameters. In order to clarify this point, let us consider the following simple example. Let us assume that we have a sample of n observations from the normal distribution $N(\mu, \sigma^2)$, and let $\bar{x}$ be the sample mean. Then, the confidence interval $\bar{x} \pm 1.96(\sigma / \sqrt{n})$ is an estimate of the interval within which the true population mean μ lies 95% of the time. That is, if we knew the true value of μ, then this value will lie within this interval with probability 0.95. The prediction interval, on the other hand, is an estimate of an interval within which a future observation will lie 95% of the time. That is, if a new observation is acquired, then there is a 0.95 chance that it will lie in this interval. Since the observations come from the normal distribution $N(\mu, \sigma^2)$, then 95% of them fall within the interval $\bar{x} \pm 1.96\sigma$, which is the prediction interval at 95% confidence. As we can see, the confidence interval is an interval estimate of the true population mean, whereas the prediction interval is an interval estimate of a single observation. An example of these two intervals is shown in Figure 5.20, where the red dots represent the sample data. We note that the confidence interval (C.I.) is not as wide as the prediction interval (P.I.), since $1.96\left(\sigma / \sqrt{n}\right) < 1.96\sigma$. This is also intuitively correct because the true population mean tends to be close to the sample mean, whereas an observation could be anywhere within the range of values of the population.

In the case of regression, a confidence and a prediction estimate can be constructed around the point estimate $\hat{y}$ given by the regression line for a given input value x. The confidence interval is given by the expression:

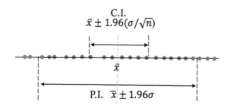

FIGURE 5.20 Confidence and prediction intervals.

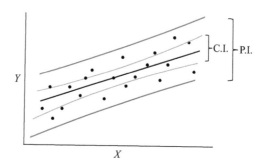

FIGURE 5.21 Confidence and prediction intervals for a regression model.

$$\hat{y} \pm t_{n-2} \times s \sqrt{\frac{1}{n} + \frac{(x-\bar{x})^2}{(n-1)s_x}},$$

where s is the standard deviation of the errors, s_x is the standard deviation of all the input values, n is the total number of observations used in the regression, and t_{n-2} is the t-student statistic with $n-2$ degrees of freedom. For $x = \bar{x}$, the term under the square root becomes equal to $\sqrt{1/n}$. For $x = \bar{x}+c$ or $x = \bar{x}-c$ the same term increases by the same amount since in both cases $(x-\bar{x})^2 = c^2$. In view of this, the confidence interval is symmetric around $\bar{x}$, as shown in Figure 5.21.

The prediction interval is given by the expression:

$$\hat{y} \pm t_{n-2} \times s \sqrt{1 + \frac{1}{n} + \frac{(x-\bar{x})^2}{(n-1)s_x}}$$

which is similar to the confidence interval with the exception that the term 1 is added to the term under the square root. The prediction interval is also symmetric around $\bar{x}$, as shown in Figure 5.21. It is also wider than the confidence interval for the reason explained above.

5.6 RIDGE, LASSO, AND ELASTIC NET REGRESSION

When we carry a multivariable regression, we can use the *stepwise selection* method to choose which of the independent variables to keep in the model. There are three ways that this can be done, namely, the *forward selection*, the *backward elimination*, and the *bidirectional elimination*. In the forward selection approach, we test sequentially whether adding an independent variable improves the regression model starting with no independent variables. In the backward elimination method, we start with a regression model with all the independent variables, and then remove one independent variable at a time testing whether this improves the model. Finally, the bidirectional elimination approach is a combination of the first two methods. At each step, the resulting regression model is tested using a sequence of *F*-tests or *t*-tests, the adjsusted R^2, *AIC*, *BIC*, and other techniques.

The stepwise selection method works well if there are a few independent variables which are strongly correlated with the dependent variable. However, it has been criticized for a number of issues, such as, incorrect results, an inherent bias in the process itself, and the necessity for significant computing power to develop complex regression models through iteration.

Alternative techniques to the stepwise regression method are the *ridge, lasso,* and *elastic net* regressions, described in this section. We note that in these techniques, the data has to be standardized. That is, let the data set consist of the data points $(y_i, x_{i1}, x_{i2}, ..., x_{iK})$, $i = 1, 2, ..., n$. Then each data point is transformed to

$$\left(\frac{y_i - \bar{y}}{s_y}, \frac{x_{i1} - \bar{x}_1}{s_{x_1}}, ..., \frac{x_{i1} - \bar{x}_K}{s_{x_K}} \right),$$

where $\bar{y}$ and s_y is the mean and standard deviation of all the y_i, $i = 1, 2, ..., n$, values, and $\bar{x}_k$ and s_{x_k}, $k - 1, 2, ..., K$, is the mean and standard deviation of all the x_{ik}, $i = 1, 2, ..., n$, values.

5.6.1 Ridge Regression

We recall that the coefficients $a_0, a_1, ..., a_k$ of the multivariable regression $Y = a_0 + a_1 X_1 + a_2 X_2 + \cdots + a_k X_k + \varepsilon$, model are estimated by minimizing the *SSE*:

$$SSE = \sum_{i=1}^{n} \left(y_i - \hat{y}_i \right)^2 = \sum_{i=1}^{n} \left(y_i - a_0 - \sum_{j=1}^{k} a_j x_{ij} \right)^2,$$

where *n* is the total number of observations, and $\hat{y}_i$ is the estimated value of the dependent variable *Y* using the *i*th observation, i.e., $\hat{y}_i = a_0 + a_1 x_{i1} + a_2 x_{i2} + \cdots + a_k x_{ik}$. The *SSE* cost function is minimized using the least squared estimate technique.

In ridge regression, the above cost function is modified by adding a penalty which is the squared sum of the coefficients, i.e.,

$$C_{ridge} = \sum_{i=1}^{n} \left(y_i - a_0 - \sum_{j=1}^{k} a_j x_{ij} \right)^2 + \lambda \sum_{i=1}^{n} a_i^2.$$

Setting λ to zero is the same as minimizing the *SSE* cost function, while the larger the value of λ the more the effect of the penalty. Parameter λ is known as the *regularization* parameter and the penalty is known as the *regularization penalty*. We note that regularization penalties are used in many techniques, such as, decision trees, Chapter 9, neural networks, Chapter 10, and support vector machines, Chapter 11.

The effect of the penalty is that it shrinks the magnitude of the regression coefficients, which in turn reduces the complexity of the model and consequently increases its accuracy. The complexity of a multivariable regression model is expressed in terms of the number of its independent variables. In general, the accuracy of a model is reduced as it becomes more complex, that is, it consists of more independent variables. Forcing the coefficients of some independent variables to become zero, removes them in effect from the model, and consequently the model's complexity is reduced. In ridge regression, the coefficients of the independent variables are not forced to become zero, but their magnitude is reduced. This reduces the model's complexity while keeping all the independent variables.

The regularization parameter λ is selected using two different methods. The first one uses Akaike's information criterion (*AIC*), and the Bayesian information criterion (*BIC*), and the second method uses *cross-validation*. The best value for λ is determined using both methods.

The *AIC* and *BIC* are given by the expressions:

$$AIC = nlog\left(\epsilon^T \epsilon\right) + 2df$$

$$BIC = nlog\left(\epsilon^T \epsilon\right) + 2df log\left(n\right),$$

where n is the number of observations, df is the degrees of freedom, and ϵ is the vector of the n residuals. We run the ridge regression for different values of λ and calculate the *AIC* and *BIC*. We then plot *AIC* as a function of λ and select the value of λ that corresponds to the lowest value of the curve. We do the same for *BIC*.

The cross-validation technique is a technique commonly used in many Machine Learning techniques. The data set used for the regression is randomly split into v equal non-overlapping subsets called *folds*, where v is typically set to 10. For a given λ, we use the combined folds 1, 2, …, 9 as the training data set and the 10th fold as the validation data set. We use the training data set to estimate the regression coefficients by minimizing C_{ridge} and then apply the resulting model to the validation data set and calculate the *SSE*. Subsequently, we use folds 1, 2, …, 8, 10 as the training data set and the 9th fold as the validation data set and calculate the *SSE*. We repeat this process until all folds have been used as a validation data set and average all the *SSE* values. We run this procedure for different values of λ and plot the average *SSE* versus λ. We select the value of λ that corresponds to the lowest value of the curve.

The λ values obtained from the *AIC*, *BIC*, and cross-validation curves are typically not the same, and one has to use one's judgment as to which value to select.

5.6.2 Lasso Regression

Lasso (least absolute shrinkage and selection operator) is similar to ridge regression, but instead of reducing the magnitude of the coefficients, it forces some of them to become zero, effectively creating a less complex model with fewer independent variables. This is achieved by adding a penalty which is the sum of the absolute values of the coefficients, that is,

$$C_{lasso} = \sum_{i=1}^{n} \left(y_i - a_0 - \sum_{j=1}^{k} a_j x_{ij} \right)^2 + \lambda \sum_{i=1}^{n} |a_i|.$$

The selection of the best value for λ is done using the methods described above for ridge regression. Figure 5.22 gives the results of applying cross-validation to a multivariable regression with five independent variables. The data set was generated as described in the regression project at the end of this chapter. The numbers given above the graph indicate the number of independent variables retained in the model as λ increases. A value of λ in-between the two vertical dotted lines is a good choice. Figure 5.22 gives the value of the coefficients of the five independent variables as a function of $\log(\lambda)$. As expected, they all converge to zero as λ gets very large (Figure 5.23).

5.6.3 Elastic Net Regression

Lasso regression works well in the case where only a few independent variables influence the dependent variable, and ridge regression works well when most of the independent variables influence the dependent variable. In view of this, the elastic network regression was proposed that combines both the ridge and lasso regularization penalties, i.e.,

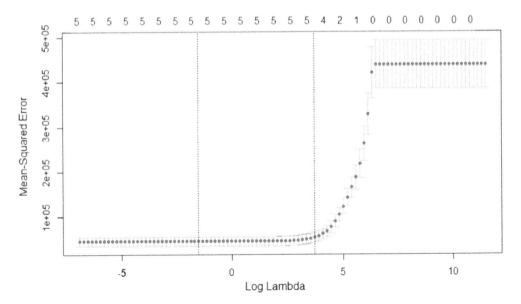

FIGURE 5.22 Mean squared error as a function of λ.

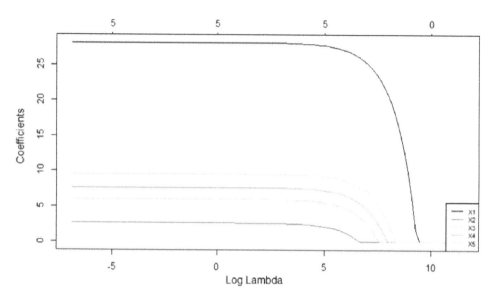

FIGURE 5.23 Coefficients as a function of λ.

$$C_{enet} = \sum_{i=1}^{n}\left(y_i - a_0 - \sum_{j=1}^{k} a_j x_{ij}\right)^2 + \lambda\left((1-\beta)\sum_{i=1}^{n}\alpha_i^2 + \beta\sum_{i=1}^{n}|a_i|\right),$$

where $0 \le \beta \le 1$. The two penalties are combined using a percentage parameter β, and the total penalty is multiplied by the regularization parameter λ. The elastic net regression becomes a lasso regression if we set $\beta = 1$, and a ridge regression if we set $\beta = 0$. The selection of the values of λ and β is determined using the cross-validation method described above for a grid of different combinations of (λ, β).

EXERCISES

1. Find the least square regression line for the points: $\{(-1,0), (0,2), (1,4), (2,5)\}$, and plot the points and the regression line.

2. What is the null hypothesis for the F-test when used in linear regression?

3. Draw a scatter plot where the residuals have unequal variance.

4. You are comparing a histogram drawn from a sample of observations against a theoretical distribution. The histogram and theoretical distribution are similar except that the histogram's right tail is shorter than that of the theoretical distribution. Draw the Q-Q plot.

5. We want to fit the non-linear function $Y = aX^b$ to a data set (x_i, y_i), $i = 1, 2, ..., n$. Show how this function can be linearized so that to estimate a and b using simple linear regression.

6. The logistic function is an S-shaped curve (sigmoid curve) given by the expression $Y = y_{max}/(1 + e^{-(a+bX)})$, where y_{max}, a, and b are constants. This function has an upper bound limit of y_{max} because when X tends to infinity the denominator tends to 1. Show how this function can be linearized so that to estimate a and b using simple linear regression.

REGRESSION PROJECT

The objective of this project is to develop a linear multivariable regression between a dependent variable Y and a 5-tuple of independent variables X_1, X_2, X_3, X_4 and X_5. For this analysis, you can use any software, such as, Python, R, MatLab, and SAS, or a combination of functions from different software.

Data Set Generation

First, we generate a data set of 300 tuples $(y_i, x_{i1}, x_{i2}, x_{i3}, x_{i4}, x_{i5})$, $i = 1, 2, ..., 300$, as follows:

We assume that each independent variable X_i, $i = 1, 2, ..., 5$, is normally distributed with a mean μ_i selected randomly from the uniform distribution $U((i-1)10, i10)$ and a variance σ_i^2 selected from $U(10,100)$. Accordingly, we first sample from the above two uniform distributions in order to fix the mean and variance of each independent variable. Then, we generate a tuple $(x_{i1}, x_{i2}, x_{i3}, x_{i4}, x_{i5})$ by randomly sampling from the normal distribution of each independent variable (see Section 3.3.2, Chapter 3). We repeat this 300 times.

Subsequently, we calculate y_i for each tuple $(x_{i1}, x_{i2}, x_{i3}, x_{i4}, x_{i5})$ using the multivariable expression:

$$y_i = a_0 + a_1 x_{i1} + a_2 x_{i2} + a_3 x_{i3} + a_4 x_{i4} + a_5 x_{i5} + \varepsilon,$$

where each coefficient α_i, $i = 0, 1, ..., 5$, is randomly selected from the uniform distribution $U((i-1)2, i2)$, and ε is a stochastic variate from the standard normal distribution with a variance of 100.

The above method for generating the data set can be modified in many different ways. For instance, you may assume that some of the independent variables follow a distribution other than normal. Also, you can generate one independent random variable from another, thus introducing positive or negative correlation between the independent variables. Feel free to experiment with the data generation! If the data set you generated is not good, that is, you cannot fit a good multivariable regression, generate a new one but remember to change the seed of the pseudo-random generator.

Now, forget (!) how the data set was generated and carry out the following tasks.

Task 1   Basic Statistics Analysis

1.1. For each variable X_i, i.e., column in the data set corresponding to X_i, calculate the following: histogram, mean, variance.

1.2. Use a box plot or any other function to remove outliers (do not over do it!). This can also be done during the model building phase (see tasks 2 and 3).

1.3. Calculate the correlation matrix for all variables, i.e., Y, X_1, X_2, X_3, X_4, and X_5. Draw conclusions related to possible dependencies among these variables.

1.4. Comment on your results.

Task 2   Simple Linear Regression

Before proceeding with the multivariable regression, carry out a simple linear regression to estimate the parameters of the model: $Y = a_0 + a_1 X_1 + \varepsilon$.

2.1. Determine the estimates for a_0, a_1, and σ^2.

2.2. Check the p-values, R-squared, and F value to determine if the regression coefficients are significant.

2.3. Plot the regression line against the data.

2.4. Do a residual analysis as per Section 5.2.2.

    a. Do a Q-Q plot of the pdf of the residuals against $N(0, s^2)$

    b. Draw the residuals histogram and carry out a χ^2 test to verify that it follows the normal distribution $N(0, s^2)$.

    c. Do a scatter plot of the residuals versus the predicted values and check for trends and constant variance. In addition, plot the correlogram of the residuals.

2.5. Use a higher-order polynomial regression, i.e., $Y = a_0 + a_1 X_1 + a_2 X_1^2 + \varepsilon$, to see if it gives better results.

2.6. Comment on your results.

Task 3   Linear Multivariable Regression

3.1. Carry out a multivariable regression on all the independent variables and determine the values for all the coefficients and σ^2.

3.2. Based on the p-values, R-squared, F value, and correlation matrix, identify which independent variables need to be removed (if any) and go back to step 3.1.

3.3. Do a residual analysis:

    (a) Do a Q-Q plot of the pdf of the residuals against $N(0, s^2)$. In addition, draw the residuals histogram and carry out a χ^2 test that it follows the normal distribution $N(0, s^2)$.

    (b) Do a scatter plot of the residuals to see if there are any trends.

3.4. Plot the confidence and prediction intervals.

3.5. Comment on your results.

Task 4 Lasso Regression

4.1. Apply the lasso regression technique. Plot the *AIC* and *BIC* curves and also the curve from the cross-validation, and discuss how you selected λ.

4.2. Carry out a residual analysis as described in task 3.3 and plot the confidence and prediction intervals.

4.3. Compare the model with the one obtained in task 3 and determine which one is better. Comment on your decision.

REFERENCES

1. H. Perros. *Computer Simulation Techniques: The Definitive Introduction!* https://people.engr. ncsu.edu/hp/files/simulation.pdf.

In addition, there are numerous references in the Internet on the topics discussed in this chapter, and the reader is strongly advised to look up some of this material.

CHAPTER 6

Time Series Forecasting

A TIME SERIES IS A SERIES OF DATA POINTS INDEXED ACCORDING to time. Typically, it is a sequence of data measured at successive equally spaced points in time, such as temperature reported every minute. Given a time series, forecasting techniques are used to determine future values. These techniques are useful in IoT since a lot of the data reported by sensors are time series.

In this chapter, we first discuss the stationarity of a time series, an important concept in time series, and then we describe the following forecasting models: the moving average or smoothing model, the moving average model MA(q), the autoregressive model AR(p), ARIMA (p, d, q), decomposition models, and the vector autoregressive model VAR(p). A set of exercises and a forecasting project are given at the end of the chapter.

6.1 A STATIONARY TIME SERIES

A stationary time series is one whose statistical properties do not change over time. Let X_t be a random variable that gives the value of a time series at time t. A *strict stationary process* is the one where the joint statistical distribution of $X_t, X_{t+1}, \ldots, X_{t+s}$ is the same as the joint statistical distribution of $X_{t+\tau}, X_{t+\tau+1}, \ldots, X_{t+\tau+s}$ for all s and τ. This is a very strong definition, and it means that all moments, i.e., expectations and variances of any order, are the same anywhere in time. It also means that the joint distribution of (X_t, X_{t+s}) is the same as that of $(X_{t+\tau}, X_{t+\tau+s})$ and it does not depend on s and τ.

A weaker definition is the *second-order stationarity* or *weak stationarity*. In this case, only the mean and variance are constant over time. In addition, the autocorrelation between X_t and X_{t+s} depends only on s and not on t; i.e., it only depends on the lag between the two values and not on the specific point of time.

An example of a stationary process is the random walk:

$$x_t = x_{t-1} + \varepsilon_t,$$

where x_{t-1} is the previous observation and ε_t is noise that is normally distributed with zero mean and variance σ^2. It can be shown that this process is strict stationary. Adding a constant gives us a closely related model of a random walk with drift, i.e.

$$x_t = c + x_{t-1} + \varepsilon_t.$$

A stationary time series has no predictable patterns in the long term such as the one shown in Figure 6.1. On the other hand, a non-stationary time series may display one or more of the following three features:

1. Trend (drift of the mean)

2. Variable (increasing/decreasing) variance

3. Seasonality

We say that a time series has trend, when the mean of the time series changes as a function of time. In Figure 6.2, we show a time series with no seasonality or variable variance, but whose mean decreases over time. This is in contrast to the one in Figure 6.1, where the mean is constant over time (shown by the red line).

A time series has variable (or unequal) variance when the spread of the observations about the mean changes. For instance, Figure 6.3 shows a time series with no trend or seasonality, but whose variance increases over time. Again, contrast this time series to the stationary one in Figure 6.1. We note that this increasing variance effect is the same as the funneling effect we discussed in the previous chapter in relation to the residuals.

Seasonality patterns may be confused with cyclic patterns which do not affect the stationarity of a time series. Both patterns have peaks and troughs, but the time between peaks (or troughs) is constant in seasonal patterns whereas it is irregular in cyclic ones.

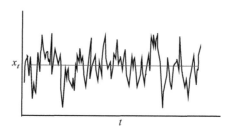

FIGURE 6.1 An example of a stationary time series.

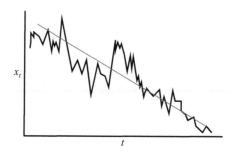

FIGURE 6.2 A time series with trend.

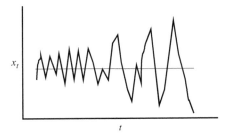

FIGURE 6.3 A time series with variable variance.

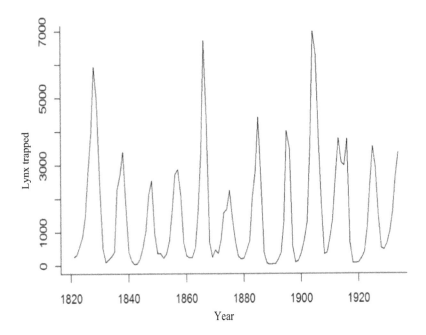

FIGURE 6.4 A time series with cyclic patterns (lynx data).

In addition, the time between successive peaks of a cyclic pattern is typically longer than that of the seasonal pattern.

A well-known example of a time series with cycles is the number of lynx trapped per year in the McKenzie river district of North-West Canada (1821–1934), shown in Figure 6.4. We observe aperiodic cycles of variable length between successive peaks. Some last 8 or 9 years and others more than 10 years.

An example of a non-stationary time series is shown in Figure 6.5. The time series gives the number of airline passengers flown from 1949 to 1960 (see [1]), and as we can see it displays all three features of non-stationarity, i.e., trend, variable variance, and seasonality. We note that the mean and variance increase over time, and there are peaks at fixed intervals. Another example of a non-stationary time series is given in Figure 6.6. The data shows the temperatures per minute in Raleigh, NC, during July 2014. The mean fluctuates over time, but the variance varies a little. The seasonality peaks are a day apart.

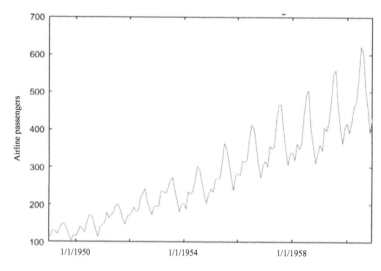

FIGURE 6.5 A non-stationary time series (airline passengers flown).

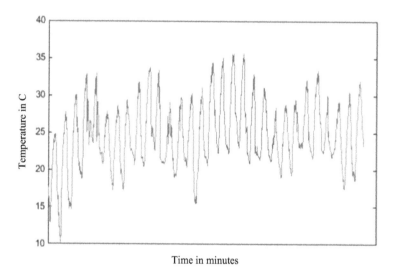

FIGURE 6.6 A non-stationary time series (temperatures during July).

6.1.1 How to Recognize Seasonality

There are several visual aids that can help us identify the presence of seasonality in a time series. Specifically, the original data set of a time series can be displayed using *seasonal plots* and *seasonal subseries plots* which show up seasonal patterns. In addition, seasonality patterns can be seen in the correlogram (see Section 3.5.5, Chapter 3) of the time series data.

We demonstrate these visual aids using some of the examples presented in [2]. Figure 6.7 gives a time series of antidiabetic drug sales. As we can see, it displays all three features of non-stationarity. The seasonal plot of this data is shown in Figure 6.8. This is obtained by

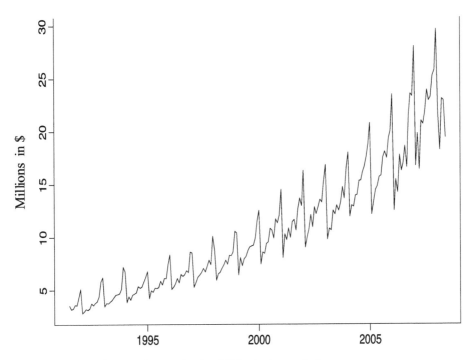

FIGURE 6.7 A non-stationary time series (antidiabetic drug sales).

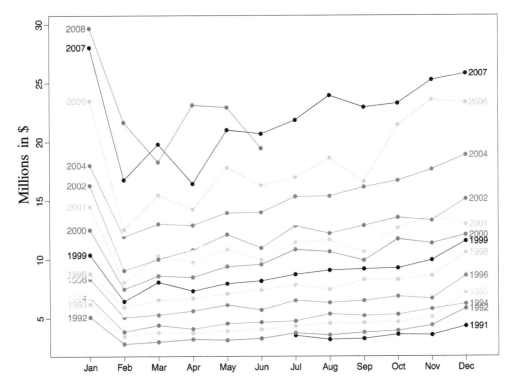

FIGURE 6.8 Seasonal plots.

plotting the data separately for each year. Consequently, we have as many plots as the number of the years in the original data set. We note that these plots have a similar pattern, with a peak in January followed by a trough in February, and then followed by a continuous increase in the sales. The January peak repeats every year. Figure 6.9 gives the seasonal subseries plot. This is a different way of plotting the data, whereby we first plot all the observations from all the years for January, then all the observations from all the years for February, and so on. The horizontal bar for each month gives the mean of the observations from all the years for the same month. Again, we observe that January has a higher mean than all the other months.

Seasonality is also reflected in the pattern of the correlogram of the time series data. Figure 6.10 gives a non-stationary time series with a time unit of one month, and its correlogram is given in Figure 6.11. The slow declining correlogram indicates the presence of trend, and the peaks are 12 months apart, indicating a seasonality effect with a period of 12.

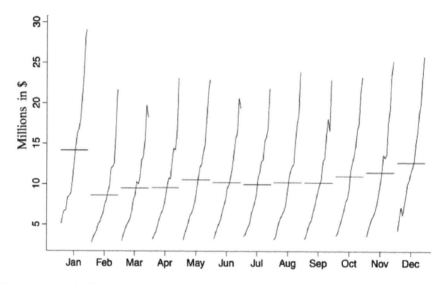

FIGURE 6.9 Seasonal subseries plots.

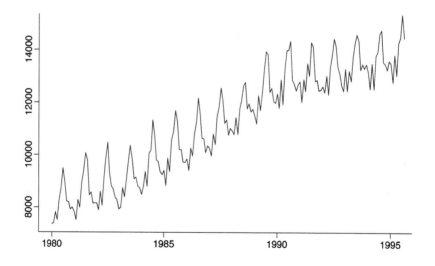

FIGURE 6.10 A non-stationary time series.

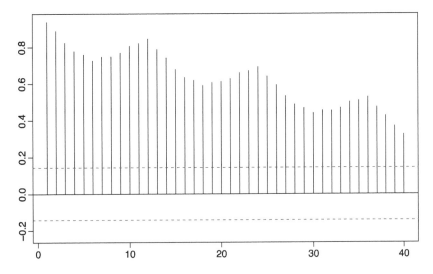

FIGURE 6.11 Correlogram of the time series data in Figure 6.10.

6.1.2 Techniques for Removing Non-Stationary Features

As will be seen, the forecasting models presented in this chapter require that the time series is stationary. In view of this, we have to first remove the non-stationary effects before we apply these models. There are several techniques that can be used in order to remove the non-stationary features of a time series, such as differencing, seasonal differencing, and data transformations, as discussed below.

a. Differencing

We compute the difference of the successive observations $x_t - x_{t-1}$ of a time series, and then use the resulting time series of the differences to forecast the next value. The forecasted value has to be transformed back to the data of the original time series. This method, often referred to as *first-order differencing*, typically removes the trend. *Second-order differencing* can also be applied by differencing an already first-order differenced time series.

b. Transformations

There is a number of transformations that can be applied to a time series. For instance, taking the log of the data, $log_e(x_t)$ may remove variability. Other transformations that can be used are $\sqrt{x_t}$ or $\sqrt[3]{x_t}$.

c. Seasonal differencing

Seasonal differencing is similar to first-order differencing only we take the difference between a value x_t and x_{t-m}, where m is the number of seasons, or more precisely the number of time units, in a year. For instance, if the time unit is a month, then $m = 12$, which means that we take the difference between the value for each month and the value for the same month of the previous year. Seasonal differencing can be combined with first-order differencing. It does not matter which one is done first, but it is preferably better to do the seasonal differencing first in case we do obtain a stationary time series, thus avoiding the extra step of first-order differencing.

In Figures 6.12 and 6.13 we give two examples reported in [2]. In Figure 6.12, we show how the non-stationary time series of the antidiabetic drug sales shown in Figure 6.7 can be transformed to a stationary time series. The top graph is the original time series x_t. Taking the logs of the time series data, $x_t' = \log_e(x_t)$, removes the variability as shown in the middle graph. Finally, taking seasonal differences, $x_t'' = x_t' - x_{t-m}'$, removes the seasonality effect as shown in the bottom graph.

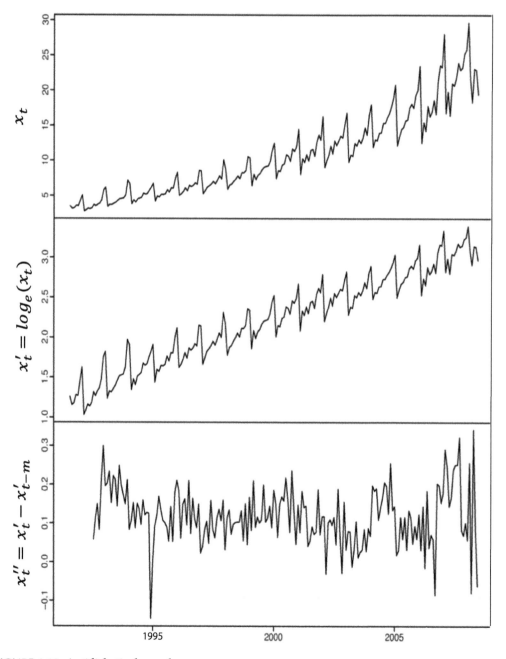

FIGURE 6.12 Antidiabetic drug sales.

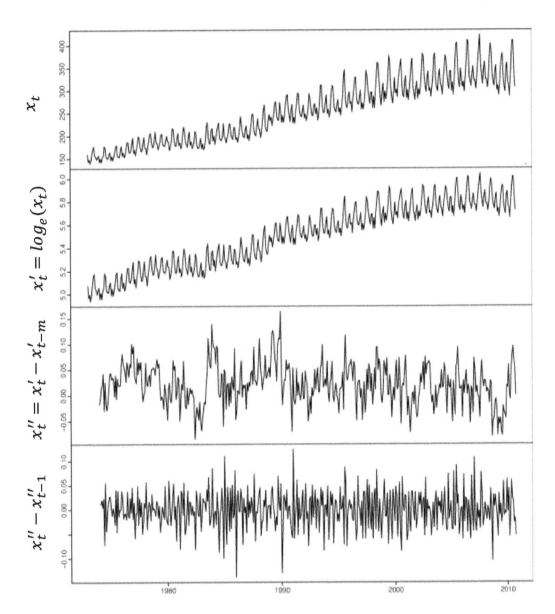

FIGURE 6.13 Monthly US net electricity generation.

The top graph in Figure 6.13 gives a time series of the monthly US net electricity generation x_t. The second graph shows the time series after we take the logs of the data, $x'_t = \log_e(x_t)$. The third graph shows the seasonal differenced time series, $x''_t = x'_t - x'_{t-m}$, and finally the last graph shows the first-order differenced time series $x''_t - x''_{t-1}$.

Finally, we note that we can determine more objectively if differencing is required by carrying out a statistical test, such as the augmented Dickey–Fuller test. If the time series is found to be non-stationary, then we apply differencing and repeat the test. This process is repeated until the time series becomes stationary.

6.2 MOVING AVERAGE OR SMOOTHING MODELS

These models should not be confused with the *moving average* (MA(q)) model presented in Section 6.3. They are applicable to time series which are locally stationary with slowly varying trend. These models are also known as *smoothing* models because they use a short-term average to forecast the value for the next period, thus smoothing out the variations in the previously observed values. Below, we present the simple average model and its variations, and the exponential moving average (or exponential smoothing) model.

6.2.1 The Simple Average Model

In this model, the forecasted value for the next time period t is the average of the previous k observations; i.e.,

$$\hat{x}_t = \frac{1}{k}\sum_{i=t-k}^{t-1} x_i,$$

where $\hat{x}_t$ is the forecasted estimate for the next period t, and x_t are the previous actual observations. For $k = 1$, we have $\hat{x}_t = x_{t-1}$, i.e., the estimate for the next period is the previously observed value. (Today's forecast is yesterdays' observed value!) This model is known as the *naïve* forecast, and surprisingly, it works well in many cases. When $k = \infty$, $\hat{x}_t$ becomes the long-term average of the time series. The following are some well-known variations of the simple average model.

a. The Weighted Average Model
   This is similar to the above model, only we assign weights to each previous observation; i.e.,

$$\hat{x}_t = \sum_{i=t-k}^{t-1} w_i x_i,$$

   where $0 \leq w_i \leq 1$, $i = 1, 2, \ldots, k$, and the sum of all the weights has to add up to 1. For instance, for a window size of $k = 3$, we can assign the weights 0.2, 0.3, 0.5, thus putting more emphasis towards the more recent observations within the window.

b. The Cumulative Average Model
   This is the average of all observations from the beginning to the current time. It is often used to demonstrate the performance of stocks.

c. The Median Model
   This is similar to the simple average model, only instead of calculating the mean over the k observations, we calculate the median. This can be easily computed by sorting out the k observations in an ascending order and then select the observation that corresponds to the position $\lceil k/2 \rceil$ (see also Section 3.5.4).

6.2.2 The Exponential Moving Average Model

In this model, also known as the *exponential smoothing model*, the forecasted value $\hat{x}_t$ for the next period t is obtained using the expression:

$$\hat{x}_t = ax_{t-1} + (1-a)\hat{x}_{t-1} \text{ (1st version)},$$

where a, $0 \leq a \leq 1$ is the smoothing factor, x_{t-1} is the previous observation, and $\hat{x}_{t-1}$ is the previous estimated value. There are two additional versions of the above expression:

$$\hat{x}_t = \hat{x}_{t-1} + ae_{t-1} \text{ (2nd version)},$$

where $e_{t-1} = x_{t-1} - s_{t-1}$ is the error made at time t, and

$$\hat{x}_t = a\left(x_{t-1} + (1-a)x_{t-2} + (1-a)^2 x_{t-3} + \cdots\right) \text{ (3rd version)}.$$

The three versions are computed differently but they are in fact the same expression. For instance, consider the 2nd version:

$$\hat{x}_t = \hat{x}_{t-1} + ae_{t-1}$$

$$= \hat{x}_{t-1} + a(x_{t-1} - \hat{x}_{t-1})$$

$$= ax_{t-1} + (1-a)\hat{x}_{t-1}.$$

Now consider the 1st version:

$$\hat{x}_1 = 0$$

$$\hat{x}_2 = ax_1 + (1-a)\hat{x}_1 = ax_1$$

$$\hat{x}_3 = ax_2 + (1-a)\hat{x}_2 = ax_2 + a(1-a)x_1$$

$$\hat{x}_4 = ax_3 + (1-a)\hat{x}_3 = ax_3 + a(1-a)x_2 + a(1-a)^2 x_1$$

$$\vdots$$

$$\hat{x}_t = a\left(x_{t-1} + (1-a)x_{t-2} + (1-a)^2 x_{t-3} + \cdots\right).$$

There are several extensions of this basic model which take into account trend and seasonal variations, see [2].

6.2.3 The Average Age of a Model

This is the period by which a forecasting model lags behind a turning point in the actual time series. For instance, let us consider the simple average model with $k = 3$. Figure 6.14 gives an example of how quickly this model adjusts to a down turn in the original data. Dots indicate the original data and squares the estimated values. We see that at time 4, the original time series begins to turn downwards, yet it is only at time 7 that the model begins to follow the downwards turn. The average age of this model is determined by the window length k, and it is equal to $(k + 1)/2$. If it is not an integer, it can be rounded up to the next nearest integer.

The average age of the exponential moving average model is $1/a$. This can be obtained by interpreting the 3rd version of the model in a manner similar to that of the hyper-exponential distribution (see Section 2.3.3); i.e., we can say that $\hat{x}_t = x_{t-1}$ with probability a, or that $\hat{x}_t = x_{t-2}$ with probability $a(1 - a)$, or that $\hat{x}_t = x_{t-3}$ with probability $a(1 - a)^2$ and so on; i.e., the model produces an estimate which is equal to the previous observation x_{t-1} with probability a, or otherwise stated, the model lags one period with probability a. Likewise, the model lags two periods with probability $a(1 - a)$, three periods with probability $a(1 - a)^2$ and so on. Therefore, the average age is the expectation:

$$\sum_{i=0}^{\infty} i\text{prob}\left(\text{model lags } i \text{ periods}\right)$$

$$= \sum_{i=0}^{\infty} ia\left(1-a\right)^{i-1} = a\sum_{i=0}^{\infty} i\left(1-a\right)^{i-1}$$

$$= a\sum_{i=0}^{\infty} \left(\left(1-a\right)^{i}\right)' = a\left(\sum_{i=0}^{\infty}\left(1-a\right)^{i}\right)'$$

$$= a\left(\frac{1}{1-\left(1-a\right)}\right)' = a\left(\frac{1}{a}\right)' = a\frac{1}{a^2} = \frac{1}{a}.$$

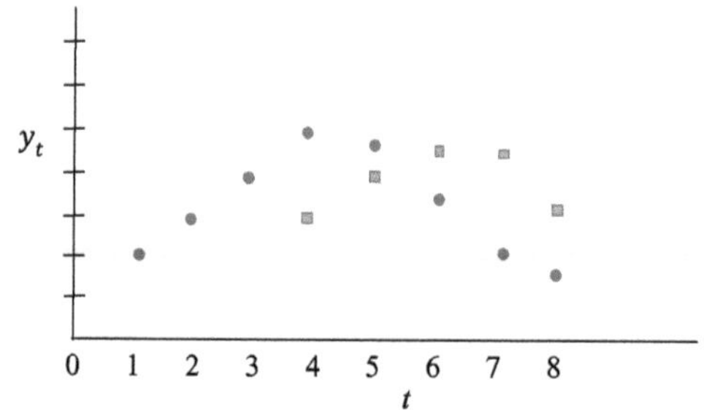

FIGURE 6.14 The simple average model.

6.2.4 Selecting the Best Value for k and a

The best value for k and a is obtained by trying different values on the same time series and then selecting the one that yields the smallest error. The following are some well-known metrics for measuring the error.

1. RMSE (Root Mean Squared Error)

$$\sqrt{\frac{1}{n}\sum_{t=1}^{n}(x_t - \hat{x}_t)^2},$$

where n is the total number of observations in the time series, x_t is the actual observation at time t, and $\hat{x}_t$ is the model's estimated value for time t. We note that the differences $x_t - \hat{x}_t$ are the same as the residuals in a regression, and the sum of the squared differences is the SSE in a regression; i.e., the above term is the squared root of the mean SSE. Taking the squared root of this term changes the unit of the $RMSE$ into the same as in the time series; i.e., if the time series gives temperatures in Celsius, then the terms $(\hat{x}_t - \hat{x}_t)^2$ are expressed in Celsius squared, and taking the square root changes the unit back to Celsius.

2. MAE (Mean Absolute Error)

$$\frac{1}{n}\sum_{t=1}^{n}|x_t - \hat{x}_t|.$$

3. Mean Error

$$\frac{1}{n}\sum_{t=1}^{n}(x_t - \hat{x}_t).$$

4. Mean Absolute Percentage Error

$$\frac{100}{n}\sum_{t=1}^{n}\left|\frac{x_t - \hat{x}_t}{x_t}\right|.$$

$RMSE$ is the most commonly used metric.

6.3 THE MOVING AVERAGE MA(q) MODEL

The moving average model is not to be confused with the simple moving average model described in Section 6.2.1. Let X_t be a random variable indicating the model's predicted value for the next period t. X_t is a moving average process of order q, referred to as MA(q), if

$$X_t = \mu + \varepsilon_t + \theta_1\varepsilon_{t-1} + \theta_2\varepsilon_{t-2} + \cdots + \theta_q\varepsilon_{t-q},$$

where μ is the mean of the time series, θ_i, $i = 1, 2, ..., q$, are weights, and ε_t, ε_{t-1}, ..., ε_{t-q}, are errors, i.e., white noise normally distributed with mean 0 and variance σ^2. For instance, for $q = 1$, we have the first-order MA(1) given by the expression:

$$X_t = \mu + \varepsilon_t + \theta_1 \varepsilon_{t-1}$$

and for $q = 2$, we have the second-order MA(2) given by the expression:

$$X_t = \mu + \varepsilon_t + \theta_1 \varepsilon_{t-1} + \theta_2 \varepsilon_{t-2}.$$

MA(q) is a stationary process, which means, that the time series we want to model by an MA(q) has to be stationary. Its parameters are estimated as follows.

The mean μ can be easily estimated by averaging out the time series data. The number of terms q is fixed by calculating the correlogram of the time series data. It can be shown that the first q terms of the correlogram of the MA(q) model are non-zero while the remaining terms are all zero. For instance, for the MA(1) model, it can be shown that

$$r_1 = \frac{\theta_1}{1 + \theta_1^2}, r_2 = r_3 = \cdots = 0;$$

i.e., the lag 1 autocorrelation r_1 is non-zero, while the autocorrelation of all the other lags r_i, $i \geq 2$, are zero. Similarly, for the MA(2) model it can be shown that

$$r_1 = \frac{\theta_1 + \theta_1 \theta_2}{1 + \theta_1^2 + \theta_2^2}, r_2 = \frac{\theta_2}{1 + \theta_1^2 + \theta_2^2}, r_3 = r_4 \cdots = 0;$$

i.e., the first two lags are non-zero, while the autocorrelation of all the higher lags are all zero. In view of this result, for a given time series we can obtain the value for q by simply computing the correlogram, and then setting q equal to the lag k after which the correlogram becomes zero. (Recall that any autocorrelation whose value falls within the confidence interval bounds, or in the absence of these bounds, its absolute value is less than 0.15 is considered as being zero.)

Having fixed μ and q, the remaining parameters are obtained using the maximum likelihood estimation (MLE) procedure, see Appendix B.

Once the model parameters are estimated, the model can be used by generating stochastic variates from $N(0, \sigma^2)$. For instance, for the MA(1) model, we start by generating a stochastic variate ε_1 from $N(0, \sigma^2)$. Then, the estimate for $t = 1$ is given by the expression $\hat{x}_1 = \mu + \varepsilon_1$. For the second period, we generate a stochastic variate ε_2 from $N(0, \sigma^2)$, and set $\hat{x}_2 = \mu + \varepsilon_2 + \theta_1 \varepsilon_1$. For the third period, we generate a stochastic variate ε_3 from $N(0, \sigma^2)$, and set $\hat{x}_3 = \mu + \varepsilon_3 + \theta_1 \varepsilon_2$, and so on.

Below, we derive the mean and variance of X_t and also we obtain the autocorrelation function of MA(1).

6.3.1 Derivation of the Mean and Variance of X_t

Let us consider the MA(q) model:

$$X_t = \mu + \varepsilon_t + \theta_1\varepsilon_{t-1} + \theta_2\varepsilon_{t-2} + \cdots + \theta_q\varepsilon_{t-q}$$

Then,

$$E(X_t) = E(\mu + \varepsilon_t + \theta_1\varepsilon_{t-1} + \theta_2\varepsilon_{t-2} + \cdots + \theta_q\varepsilon_{t-q}) = \mu,$$

since the mean of the error ε_t is zero. The variance is calculated as follows:

$$\mathrm{Var}(X_t) = E(X_t - \mu)^2$$

$$= E(\varepsilon_t + \theta_1\varepsilon_{t-1} + \theta_2\varepsilon_{t-2} + \cdots + \theta_q\varepsilon_{t-q})^2$$

$$= E(\varepsilon_t^2 + \theta_1^2\varepsilon_{t-1}^2 + \theta_2^2\varepsilon_{t-2}^2 + \cdots + \theta_q^2\varepsilon_{t-q}^2$$

$$+ 2\varepsilon_t\theta_1\varepsilon_{t-1} + 2\varepsilon_t\theta_2\varepsilon_{t-2} + \cdots + 2\varepsilon_t\theta_q\varepsilon_{t-q} + \cdots).$$

We observe that $E(\varepsilon_t\varepsilon_\tau) = 0$, $t \neq \tau$, since ε_t and ε_τ are independent of each other, and therefore $E(\varepsilon_t\varepsilon_\tau) = E(\varepsilon_t)E(\varepsilon_\tau) = 0$. In addition, we note that $\mathrm{Var}(\varepsilon_t) = E(\varepsilon_t - 0)^2 = E(\varepsilon_t)^2$; i.e., $\sigma^2 = E(\varepsilon_t)^2$. Hence, we have

$$\mathrm{Var}(X_t) = E(\varepsilon_t^2) + E(\theta_1^2\varepsilon_{t-1}^2) + E(\theta_2^2\varepsilon_{t-2}^2) + \cdots + E(\theta_q^2\varepsilon_{t-q}^2)$$

$$= \sigma^2 + \theta_1^2\sigma^2 + \theta_2^2\sigma^2 + \cdots + \theta_q^2\sigma^2$$

$$= \sigma^2(1 + \theta_1^2 + \theta_2^2 + \cdots + \theta_q^2).$$

The variance of X_t can be derived in an easier way as follows:

$$\mathrm{Var}(X_t) = \mathrm{Var}(\mu + \varepsilon_t + \theta_1\varepsilon_{t-1} + \theta_2\varepsilon_{t-2} + \cdots + \theta_q\varepsilon_{t-q})$$

$$= \mathrm{Var}(\mu) + \mathrm{Var}(\varepsilon_t) + \mathrm{Var}(\theta_1\varepsilon_{t-1}) + \cdots + \mathrm{Var}(\theta_q\varepsilon_{t-q})$$

$$= \sigma^2(1 + \theta_1^2 + \theta_2^2 + \cdots + \theta_q^2),$$

since $\mathrm{Var}(\mu) = 0$, $\mathrm{Var}(\varepsilon_t) = \sigma^2$, and $\mathrm{Var}(\theta_i\varepsilon_{t-i}) = \theta_i^2\,\mathrm{Var}(\varepsilon_{t-i}) = \theta_i^2\sigma^2$.

For MA(1), we have $\mathrm{Var}(X_t) = \sigma^2(1 + \theta_1^2)$, and for MA(2), $\mathrm{Var}(X_t) = \sigma^2(1 + \theta_1^2 + \theta_2^2)$.

6.3.2 Derivation of the Autocorrelation Function of the MA(1)

We recall that the MA(1) model is given by the expression:

$$X_t = \mu + \varepsilon_t + \theta_1 \varepsilon_{t-1}.$$

This expression can be re-written as follows:

$$X_t - \mu = \varepsilon_t + \theta_1 \varepsilon_{t-1}$$

or

$$\tilde{X}_t = \varepsilon_t + \theta_1 \varepsilon_{t-1},$$

where $\tilde{X}_t$ is the mean-adjusted random variable X_t, i.e., $\tilde{X}_t = X_t - \mu$. We can use either expressions to forecast a time series. The only difference being that when using the mean-adjusted expression, we need to add μ to the estimated value for the next period.

We first obtain the lag 1 $\mathrm{Cov}(X_t, X_{t-1})$ from which we can calculate the lag 1 autocorrelation. We have

$$\mathrm{Cov}\left(X_t, X_{t-1}\right) = E\left(X_t - \mu\right)\left(X_{t-1} - \mu\right) = E\left(\tilde{X}_t \tilde{X}_{t-1}\right).$$

Also,

$$\mathrm{Cov}\left(\tilde{X}_t, \tilde{X}_{t-1}\right) = E\left(\tilde{X}_t - \mu_{\tilde{X}_t}\right)\left(\tilde{X}_{t-1} - \mu_{\tilde{X}_t}\right) = E\left(\tilde{X}_t \tilde{X}_{t-1}\right)$$

since the mean $\mu_{\tilde{X}_t}$ of $\tilde{X}_t$ is zero. Therefore, both X_t and $\tilde{X}_t$ have the same covariance. We continue the derivation using the mean adjusted expression since it simplifies the mathematical derivation. We have

$$E\left(\tilde{X}_t \tilde{X}_{t-1}\right) = E\left(\left(\varepsilon_t + \theta_1 \varepsilon_{t-1}\right)\tilde{X}_{t-1}\right)$$

$$= E\left(\varepsilon_t \tilde{X}_{t-1} + \theta_1 \varepsilon_{t-1} \tilde{X}_{t-1}\right)$$

$$= E\left(\varepsilon_t \tilde{X}_{t-1}\right) + E\left(\theta_1 \varepsilon_{t-1} \tilde{X}_{t-1}\right).$$

Since ε_t and $\tilde{X}_{t-1}$ are independent of each other, we have that

$$E\left(\varepsilon_t \tilde{X}_{t-1}\right) = E\left(\varepsilon_t\right)E\left(\tilde{X}_{t-1}\right) = 0 \times E\left(\tilde{X}_{t-1}\right) = 0.$$

Now let us consider the second term $E\left(\theta_1 \varepsilon_{t-1} \tilde{X}_{t-1}\right)$. We note that in this case ε_{t-1} and $\tilde{X}_{t-1}$ are not independent. We have

$$E\left(\theta_1 \varepsilon_{t-1} \tilde{X}_{t-1}\right) = \theta_1 E\left(\varepsilon_{t-1} \tilde{X}_{t-1}\right)$$

$$= \theta_1 E\left(\varepsilon_{t-1}\left(\varepsilon_{t-1} + \theta_1 \varepsilon_{t-2}\right)\right)$$

$$= \theta_1 E\left(\varepsilon_{t-1}^2 + \theta_1 \varepsilon_{t-1} \varepsilon_{t-2}\right)$$

$$= \theta_1 E\left(\varepsilon_{t-1}^2\right) + \theta_1^2 E\left(\varepsilon_{t-1} \varepsilon_{t-2}\right).$$

ε_{t-1} and ε_{t-2} are independent with a zero mean, and therefore $E(\varepsilon_{t-1}\varepsilon_{t-2}) = 0$. Also, we note that $E\left(\varepsilon_{t-1}^2\right) = \text{Var}\left(\varepsilon_{t-1}\right) + E\left(\varepsilon_{t-1}\right)^2 = \sigma^2$.

Therefore,

$$\text{Cov}\left(\tilde{X}_t, \tilde{X}_{t-1}\right) = \theta_1 \sigma^2.$$

The autocorrelation of $\tilde{X}_t$ is

$$r_1 = \frac{\text{Cov}\left(\tilde{X}_t, \tilde{X}_{t-1}\right)}{\text{Var}\left(\tilde{X}_t\right)} = \frac{\theta_1 \sigma^2}{\text{Var}\left(X_t - \mu\right)} = \frac{\theta_1 \sigma^2}{\text{Var}\left(X_t\right)}$$

$$= \frac{\theta_1 \sigma^2}{\sigma^2 \left(1 + \theta_1^2\right)} = \frac{\theta_1}{1 + \theta_1^2}.$$

Now, let us consider the lag 2 autocovariance $\text{Cov}\left(\tilde{X}_t, \tilde{X}_{t-2}\right)$. Following similar arguments as above, we have

$$\text{Cov} = \left(\tilde{X}_t, \tilde{X}_{t-2}\right) = E\left(\tilde{X}_t - \mu_{\tilde{X}_t}\right)\left(\tilde{X}_{t-2} - \mu_{\tilde{X}_t}\right) = E\left(\tilde{X}_t \tilde{X}_{t-2}\right),$$

where

$$E\left(\tilde{X}_t \tilde{X}_{t-2}\right) = E\left(\left(\varepsilon_t + \theta_1 \varepsilon_{t-1}\right)\tilde{X}_{t-2}\right)$$

$$= E\left(\varepsilon_t \tilde{X}_{t-2} + \theta_1 \varepsilon_{t-1} \tilde{X}_{t-2}\right)$$

$$= E\left(\varepsilon_t \tilde{X}_{t-2}\right) + E\left(\theta_1 \varepsilon_{t-1} \tilde{X}_{t-2}\right).$$

Both terms in the above expression are zero since ε_t and $\tilde{X}_{t-2}$ are independent of each other and ε_{t-1} and $\tilde{X}_{t-2}$ are also independent of each other. Consequently, the lag 2 autocovariance is zero and so is the lag 2 autocorrleation. The same applies to all higher lags.

6.3.3 Invertibility of MA(q)

The mean-adjusted expression for MA(1) can be re-written as

$$\varepsilon_t = \tilde{X}_t - \theta_1 \varepsilon_{t-1}$$

$$= \tilde{X}_t - \theta_1 (\tilde{X}_{t-1} - \theta_1 \varepsilon_{t-2})$$

$$= \tilde{X}_t - \theta_1 \tilde{X}_{t-1} + \theta_1^2 \varepsilon_{t-2}$$

Working backwards, we can obtain the following after n steps:

$$\varepsilon_t = \sum_{j=0}^{n-1} \left(-\theta_1\right)^j \tilde{X}_{t-j} + \left(-\theta_1\right)^n \varepsilon_{t-n}.$$

Letting $n \to \infty$ and assuming $|\theta_1| < 1$, we have

$$\varepsilon_t = \sum_{j=0}^{\infty} \left(-\theta_1\right)^j \tilde{X}_{t-j}$$

which can be re-written as follows:

$$\tilde{X}_t = -\sum_{j=1}^{\infty} \left(-\theta_1\right)^j \tilde{X}_{t-j} + \varepsilon_t.$$

As we can see, if $|\theta_1| < 1$, then the time series converges to a finite value. Such process is called *invertible*. More complex constraints apply for higher orders of MA(q).

Invertibility is a restriction of the MA(q) model and not of the data. It is automatically checked by the software when it estimates the parameters of the MA(q) model.

6.4 THE AUTOREGRESSIVE MODEL

Let X_t be a random variable indicating the value of a time series at time t. The autoregressive model predicts the next value X_t for time t using the expression:

$$X_t = \delta + a_1 X_{t-1} + a_2 X_{t-2} + \cdots + a_p X_{t-p} + \varepsilon_t$$

where X_{t-1}, X_{t-2}, ..., X_{t-p} are the random variables indicating the previous p observed values of the time series, δ is a constant, a_i, $i = 1, 2, ..., p$, are weights, and ε_t is normally distributed white noise with 0 mean and variance σ^2. The autoregressive model is referred to as AR(p), where p is the order of the model. We note that the autoregressive model is a linear regression of the next value of the time series against one or more prior values of the time series.

6.4.1 The AR(1) Model

The AR(1) model is given by the expression:

$$X_t = \delta + a_1 X_{t-1} + \varepsilon_t.$$

Various simple models can be obtained for special values of δ and a_1. For instance, for $a_1 = 0$, we obtain the white noise model $X_t = \delta + \varepsilon_t$. For $a_1 = 1$ and $\delta = 0$, we obtain the random walk model $X_t = X_{t-1} + \varepsilon_t$, and for $a_1 = 1$ and $\delta \neq 0$, we obtain the random walk model with a drift $X_t = \delta + X_{t-1} + \varepsilon_t$.

Let μ be the expected value of X_t. Then, we have

$$\mu = E\left(X_t\right) = E\left(\delta + a_1 X_{t-1} + \varepsilon_t\right)$$

$$= \delta + a_1 E\left(X_t\right) = \delta + a_1 \mu.$$

Therefore,

$$\mu = \frac{\delta}{1 - a_1}.$$

We can convert the AR(1) expression to an equivalent mean-adjusted expression. Subtracting μ from both sides of the AR(1) expression gives

$$X_t - \mu = \delta - \mu + a_1 X_{t-1} + \varepsilon_t$$

$$X_t - \mu = \delta - \left(1 - a_1\right)\mu + a_1\left(X_{t-1} - \mu\right) + \varepsilon_t$$

$$X_t - \mu = a_1\left(X_{t-1} - \mu\right) + \varepsilon_t.$$

Let $\tilde{X}_t = X_t - \mu$ be the mean-adjusted X_t, then we have

$$\tilde{X}_t = a_1 \tilde{X}_{t-1} + \varepsilon_t.$$

As in the case of the MA(q) model, we can use either the expression for X_t or for $\tilde{X}_t$ to forecast the next value. Forecasted mean-adjusted values have to be converted back to the regular values by adding μ.

We now proceed to obtain the autocorrelation of X_t using the simpler mean adjusted expression, since both X_t and $\tilde{X}_t$ have the same variance, autocovarinace, and autocorrelation (see also Section 6.3.2).

Let $\sigma_{\tilde{X}_t}^2$ be the variance of $\tilde{X}_t$. We have

$$\text{Var}\left(\tilde{X}_t\right) = \text{Var}\left(a_1 \tilde{X}_{t-1}\right) + \sigma^2$$

$$\sigma_{\tilde{X}_t}^2 = a_1^2 \sigma_{\tilde{X}_t}^2 + \sigma^2.$$

Therefore,

$$\sigma_{\tilde{X}_t}^2 = \frac{\sigma^2}{1 - a_1^2}.$$

As in the case of MA(1), we have that

$$\text{Cov}\left(\tilde{X}_t, \tilde{X}_{t-1}\right) = E\left(\tilde{X}_t \tilde{X}_{t-1}\right)$$

$$= E\left(\left(a_1 \tilde{X}_{t-1} + \varepsilon_t\right) \tilde{X}_{t-1}\right)$$

$$= E\left(a_1 \tilde{X}_{t-1} \tilde{X}_{t-1}\right) + E\left(\varepsilon_t \tilde{X}_{t-1}\right)$$

$$= a_1 E\left(\tilde{X}_{t-1}^2\right).$$

The second term $E\left(\varepsilon_t \tilde{X}_{t-1}\right)$ is zero, because ε_t and $\tilde{X}_{t-1}$ are independent of each other, and therefore $E\left(\varepsilon_t \tilde{X}_{t-1}\right) = E\left(\varepsilon_t\right) E\left(\tilde{X}_{t-1}\right) = 0$, since $E(\varepsilon_t) = 0$.

We recall that $\text{Var}(X) = E(X^2) - E(X)^2$. For $X = \tilde{X}_{t-1}$ we have $\sigma_{\tilde{X}_t}^2 = E\left(\tilde{X}_{t-1}^2\right) - E\left(\tilde{X}_{t-1}\right)^2 = E\left(\tilde{X}_{t-1}^2\right)$ since $E\left(\tilde{X}_{t-1}\right) = 0$. Hence,

$$\text{Cov}\left(\tilde{X}_t, \tilde{X}_{t-1}\right) = a_1 \sigma_{\tilde{X}_t}^2$$

and the lag 1 autocorrelation of $\tilde{X}_t$ is

$$r_1 = \frac{\text{Cov}\left(\tilde{X}_t, \tilde{X}_{t-1}\right)}{\text{Var}\left(\tilde{X}_t\right)} = \frac{a_1 \sigma_{\tilde{X}_t}^2}{\sigma_{\tilde{X}_t}^2} = a_1.$$

The lag 2 autocorrelation of $\tilde{X}_t$ can be obtained similarly. We have

$$\text{Cov}\left(\tilde{X}_t, \tilde{X}_{t-2}\right) = E\left(\tilde{X}_t \tilde{X}_{t-2}\right)$$

$$= E\left(\left(a_1 \tilde{X}_{t-1} + \varepsilon_t\right) \tilde{X}_{t-2}\right)$$

$$= E\left(a_1 \tilde{X}_{t-1} \tilde{X}_{t-2}\right) + E\left(\varepsilon_t \tilde{X}_{t-2}\right)$$

$$= a_1 E\left(\tilde{X}_{t-1} \tilde{X}_{t-2}\right)$$

$$= a_1 \text{Cov}\left(\tilde{X}_t \tilde{X}_{t-1}\right)$$

$$= a_1^2 \sigma_{\tilde{X}_t}^2.$$

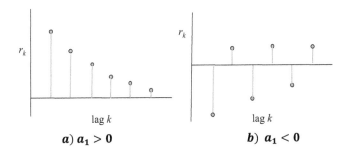

FIGURE 6.15 The autocorrelation function of MA(1), $k = 0, 1, 2, \ldots(a)a_1 > 0$; (b) $a_1 < 0$.

Therefore

$$r_2 = \frac{\mathrm{Cov}\left(\tilde{X}_t, \tilde{X}_{t-2}\right)}{\mathrm{Var}\left(\tilde{X}_t\right)} = \frac{\alpha_1^2 \sigma_{\tilde{X}_t}^2}{\sigma_{\tilde{X}_t}^2} = \alpha_1^2.$$

In general, it can be shown that the lag k autocorrelation is

$$r_k = \alpha_1^k.$$

If $a_1 > 0$, then all the autocorrelations are positive. If $a_1 < 0$ then the autocorrelation alternates in sign (see Figure 6.15).

6.4.2 Stationarity Condition of AR(p)

The mean-adjusted expression of AR(p) is

$$\tilde{X}_t = a_1 \tilde{X}_{t-1} + a_2 \tilde{X}_{t-2} + \cdots + a_p \tilde{X}_{t-p} + \varepsilon_t.$$

This expression can be re-written using the backshift (or lag) operator B, defined as follows: $B\tilde{X}_t = \tilde{X}_{t-1}$; i.e., if we apply B to a time series value at time t, we obtain the previous time series value at time $t - 1$. If we apply it twice, then we get the time series value at time $t - 2$, i.e., $B^2 \tilde{X}_t = \tilde{X}_{t-1}$, and so on. We have

$$\tilde{X}_t = a_1 B\tilde{X}_t + a_2 B^2 \tilde{X}_t + \cdots + a_p B^p \tilde{X}_t + \varepsilon_t$$

or

$$\tilde{X}_t \left(1 - a_1 B - a_2 B^2 - \cdots - a_p B^p \right) = \varepsilon_t.$$

The polynomial in the parenthesis is known as the *characteristic polynomial*. Let

$$\varphi_p \left(B\right) = 1 - a_1 B - a_2 B^2 - \cdots - a_p B^p.$$

Then, the AR(p) expression can be written as

$$\varphi_p(B)\tilde{X}_t = \varepsilon_t.$$

Now, let z_i be the roots of $\varphi_p(z) = 0$. Then, it can be shown that an AR(p) process is stationary if $|z_i| > 1$, for all i. For instance, for the case of AR(1), we have that $\varphi_1(z) = 1 - a_1 z$, and the root of $\varphi_1(z)$ is $1/a_1$. In order for AR(1) to be stationary $|1/a_1| > 1$ or $|a_1| < 1$. The following two conditions have to hold in order for AR(2) to be stationary: $|a_2| < 1$ and $a_2 \pm a_1 < 1$.

6.4.3 Derivation of the Coefficients $a_i, i = 1, 2, ..., p$

Following similar arguments as in the case of AR(1), we have

$$\mathrm{Cov}\left(\tilde{X}_t, \tilde{X}_{t-k}\right) = E\left(\tilde{X}_t \tilde{X}_{t-k}\right)$$

$$= E\left(\left(a_1 \tilde{X}_{t-1} + \cdots + a_p \tilde{X}_{t-p} + \varepsilon_t\right)\tilde{X}_{t-k}\right)$$

$$= E\left(a_1 \tilde{X}_{t-1} \tilde{X}_{t-k}\right) + \cdots + E\left(a_p \tilde{X}_{t-p} \tilde{X}_{t-k}\right) + E\left(\varepsilon_t \tilde{X}_{t-k}\right)$$

The last term $E\left(\varepsilon_t \tilde{X}_{t-k}\right)$ is zero, because ε_t and $\tilde{X}_{t-k}$ are independent of each other, and therefore $E\left(\varepsilon_t \tilde{X}_{t-k}\right) = E\left(\varepsilon_t\right)E\left(\tilde{X}_{t-k}\right) = 0$, since $E(\varepsilon_t) = 0$. Therefore, we have

$$\mathrm{Cov}\left(\tilde{X}_t, \tilde{X}_{t-k}\right) = a_1 E\left(\tilde{X}_{t-1} \tilde{X}_{t-k}\right) + \cdots + a_p E\left(\tilde{X}_{t-p} \tilde{X}_{t-k}\right)$$

$$= a_1 \mathrm{Cov}\left(\tilde{X}_{t-1}, \tilde{X}_{t-k}\right) + \cdots + a_p \mathrm{Cov}\left(\tilde{X}_{t-p}, \tilde{X}_{t-k}\right).$$

Dividing both sides of the above expression by $\mathrm{Var}\left(\tilde{X}_t\right)$, we get

$$r_k = a_1 r_{k-1} + a_2 r_{k-2} + \cdots + a_p r_{k-p},$$

where

$$r_k = \frac{\mathrm{Cov}\left(\tilde{X}_t, \tilde{X}_{t-k}\right)}{\mathrm{Var}\left(\tilde{X}_t\right)}, \quad k = 0, 1, \ldots$$

is the lag k autocorrelation. Recall that $r_0 = 1$, since $\mathrm{Cov}\left(\tilde{X}_t, \tilde{X}_t\right) = E\left(\tilde{X}_t \tilde{X}_t\right) = \mathrm{Var}\left(\tilde{X}_t\right)$. Because of the different values that k and p may take, it is possible that a lag may be negative. However, this does not pause a problem since $r_{-l} = r_l$.

In summary, for the AR(p) model, the lag k autocorrelation is expressed as a linear weighted combination of the previous p autocorrelations, where the weights are the coefficients $a_i, i = 1, 2, ..., p$.

Let us consider the case of AR(2), i.e., $p = 2$. We have

$$r_k = a_1 r_{k-1} + a_2 r_{k-2}, k \geq 1.$$

For $k = 1, 2$, we have

$$r_1 = a_1 r_0 + a_2 r_{-1}$$

$$r_2 = a_1 r_1 + a_2 r_0.$$

The autocorrelations r_1 and r_2 can be calculated from the time series, which allows us to solve the above system of equations for the AR(2) coefficients a_1 and a_2. Since $r_0 = 1$ and $r_{-1} = r_1$, we have

$$r_1 = a_1 + a_2 r_1$$

$$r_2 = a_1 r_1 + a_2.$$

From where we obtain

$$a_1 = \frac{r_1 - r_1 r_2}{1 - r_1^2}, \quad a_2 = \frac{r_2 - r_1^2}{1 - r_1^2}.$$

In the general case, the first p equations are as follows:

$$r_1 = a_1 r_0 + a_2 r_{-1} + \cdots + a_p r_{-p}$$

$$\vdots$$

$$r_p = a_1 r_{p-1} + a_2 r_{p-2} + \cdots + a_p r_0$$

or

$$\begin{pmatrix} 1 & r_1 & \cdots & r_{-p} \\ r_1 & 1 & \cdots & r_{1-p} \\ \vdots & \vdots & \cdots & \vdots \\ r_{p-1} & r_{p-2} & \cdots & 1 \end{pmatrix} \begin{pmatrix} a_1 \\ a_2 \\ \vdots \\ a_p \end{pmatrix} = \begin{pmatrix} r_1 \\ r_2 \\ \vdots \\ r_p \end{pmatrix}.$$

The above system of equations is known as the Yule-Walker equations. The coefficients a_i, $i = 1, 2, \ldots, p$, can be calculated from these equations once the first p autocorrelations $r_1, r_2, \ldots, r_p$ are computed from the time series. The solution can be obtained by inversion, i.e.,

$$\begin{pmatrix} a_1 \\ a_2 \\ \vdots \\ a_p \end{pmatrix} = \begin{pmatrix} 1 & r_1 & \cdots & r_{p-1} \\ r_1 & 1 & \cdots & r_{p-2} \\ \vdots & \vdots & \cdots & \vdots \\ r_{p-1} & r_{p-2} & \cdots & 1 \end{pmatrix}^{-1} \begin{pmatrix} r_1 \\ r_2 \\ \vdots \\ r_p \end{pmatrix}.$$

The coefficients of an AR(p) can be also estimated using MLE.

Finally, we note that in addition to estimating the AR(p) coefficients, we also need to estimate the variance σ^2 of the error ε_t. This can be done in the same way as in regression (see Section 5.1). Specifically, we run the AR(p) model against the same time series we used to estimate the coefficients, and for each time t we calculate the difference between the forecasted value and the actual value. The sample variance of these differences is the estimate of σ^2.

6.4.4 Determination of the Order of AR(p)

We recall that the order of an MA(q) model is determined by calculating the autocorrelation function (ACF) and then set q to the lag k past which all lags are zero. There is a similar method for the AR(p) but based on the partial autocorrelation function (PACF); i.e., the value of p is set to the lag k past which all lags of the PACF are zero.

When we calculate the autocorrelation between X_t and X_{t-k} we do not account for the intermediate values $X_{t-1}, X_{t-2}, \ldots, X_{t-k+1}$. In the partial autocorrelation, these intermediate values are taken into account. The lag k partial autocorrelation $\alpha(k)$ is given by the expression:

$$\alpha(k) = \frac{\mathrm{Cov}\left(X_t,\, X_{t-k} \mid X_{t-1},\, X_{t-2},\, \ldots,\, X_{t-k+1}\right)}{\sqrt{\mathrm{Var}\left(X_t \mid X_{t-1},\, \ldots,\, X_{t-k+1}\right)\mathrm{Var}\left(X_{t-k} \mid X_{t-1},\, \ldots,\, X_{t-k+1}\right)}}$$

For instance, for $k = 2$, we have

$$\alpha(2) = \frac{\mathrm{Cov}\left(X_t,\, X_{t-2} \mid X_{t-1}\right)}{\sqrt{\mathrm{Var}\left(X_t \mid X_{t-1}\right)\mathrm{Var}(X_{t-2} \mid X_{t-1})}},$$

and for $k = 3$

$$\alpha(3) = \frac{\mathrm{Cov}\left(X_t, X_{t-3} \mid X_{t-1}, X_{t-2}\right)}{\sqrt{\mathrm{Var}\left(X_t \mid X_{t-1}, X_{t-2}\right)\mathrm{Var}\left(X_{t-3} \mid X_{t-1}, X_{t-2}\right)}}.$$

We note that for $k = 1$, the lag 1 partial autocorrelation is the same as the lag 1 autocorrelation; i.e.,

$$\alpha(1) = \frac{\mathrm{Cov}\left(X_t, X_{t-1}\right)}{\sqrt{\mathrm{Var}\left(X_t\right)\mathrm{Var}\left(X_{t-1}\right)}} = r_1$$

It can be shown that the partial autocorrelations $\alpha(k)$ can be calculated using the Yule-Walker equations:

$$
\begin{pmatrix}
1 & r_1 & \cdots & r_{k-1} \\
r_1 & 1 & \cdots & r_{k-2} \\
\vdots & \vdots & \cdots & \vdots \\
r_{k-1} & r_{k-2} & \cdots & 1
\end{pmatrix}
\begin{pmatrix}
\alpha(1) \\
\alpha(2) \\
\vdots \\
\alpha(k)
\end{pmatrix}
=
\begin{pmatrix}
r_1 \\
r_2 \\
\vdots \\
r_k
\end{pmatrix}
$$

Specifically, in order to obtain $\alpha(k)$ we set up the above system of equations from which we only solve for $\alpha(k)$. The remaining solutions for $\alpha(1)$ to $\alpha(k-1)$ are discarded. In view of this, we do not solve the entire system of equations. Rather, we obtain the solution for $\alpha(k)$ using Cramer's rule; i.e.,

$$
\alpha(k) = \frac{\left|P_k^*\right|}{\left|P_k\right|},
$$

where $|P_k|$ is the determinant of the $k \times k$ matrix on the left-hand side of the Yule-Walker equations obtained by setting $p = k$, and $\left|P_k^*\right|$ is the determinant of matrix P_k where the kth column, i.e., the last column, has been replaced by $(r_1, r_1, \ldots, r_k)^T$.

For $k = 1$, we have $1 \text{x } \alpha(1) = r_1$, i.e., $\alpha(1) = r_1$ which is what we obtained above. For $k = 2$, we have

$$
P_2 = \begin{pmatrix} 1 & r_1 \\ r_1 & 1 \end{pmatrix}, \; P_2^* = \begin{pmatrix} 1 & r_1 \\ r_1 & r_2 \end{pmatrix}
$$

Applying Cramer's rule, we have

$$
\alpha(2) = \frac{r_2 - r_1^2}{1 - r_1^2}.
$$

We continue to calculate the PACF coefficients until we get the first zero coefficient. We then fix the order p of the AR(p) model to the lag prior to getting the zero coefficient. As in the case of the autocorrelation, a partial autocorrelation value is zero if it falls within the confidence interval bounds calculated by the software, or in the absence of these bounds, its absolute value is less than 0.15. An example of a PACF is given in Figure 6.16. We see that at $k = 4$, $\alpha(k) < 0.15$, and therefore we select an AR(3).

6.5 THE NON-SEASONAL ARIMA (p,d,q) MODEL

This is the most general class of forecasting models for non-seasonal time series. The ARIMA model is an extension of the ARMA model, which combines an AR and an MA model; i.e.,

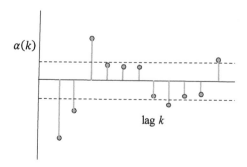

FIGURE 6.16 An example of a PACF plot

$$X_t = \delta + a_1 X_{t-1} + a_2 X_{t-2} + \cdots + a_p X_{t-p}$$

$$+ \varepsilon_t + \theta_1 \varepsilon_{t-1} + \theta_2 \varepsilon_{t-2} + \cdots + \theta_q \varepsilon_{t-q}.$$

The name ARMA is made up by combining the terms AR and MA, and the name ARIMA is a combination of AR, I, and MA, where I stands for integrated and indicates the differencing in the observations necessary to turn a time series into a stationary one. An ARIMA model is defined by p, d, and q, where p is the order of the AR model, d is the degree of differencing, and q is the order of the MA model. Thus, the ARIMA model is given by the expression:

$$\tilde{X}_t = \delta + a_1 \tilde{X}_{t-1} + a_2 \tilde{X}_{t-2} + \cdots + a_p \tilde{X}_{t-p}$$

$$+ \varepsilon_t + \theta_1 \varepsilon_{t-1} + \theta_2 \varepsilon_{t-2} + \cdots + \theta_q \varepsilon_{t-q},$$

where $\tilde{X}_t$ is the d-differenced series, i.e., $\tilde{X}_t = X_t - X_{t-d}$.

The following are some examples of ARIMA models (note that the presence of δ is indicated separately).

ARIMA $(0, 0, 0)$, $\delta \neq 0$: This is the white noise model

$$X_t = \delta + \varepsilon_t.$$

ARIMA $(0, 1, 0)$, $\delta \neq 0$: This is the random walk model with drift

$$\tilde{X}_t = \delta + \varepsilon_t,$$

where $\tilde{X}_t = X_t - X_{t-1}$, since the data is first differenced ($d = 1$). We have

$$X_t - X_{t-1} = \delta + \varepsilon_t$$

$$X_t = \delta + X_{t-1} + \varepsilon_t.$$

ARIMA $(0, 1, 0)$, $\delta = 0$: This is the random walk model

$$X_t = X_{t-1} + \varepsilon_t.$$

ARIMA $(p, 0, 0)$, $\delta \neq 0$: This is an AR(p) model.
ARIMA $(0, 0, q)$, $\delta \neq 0$: This is an MA(q) model.
ARIMA $(1, 1, 0)$, $\delta \neq 0$: This is an AR(1) model with first differencing

$$\tilde{X}_t = \delta + a_1 \tilde{X}_{t-1} + \varepsilon_t.$$

ARIMA $(0, 1, 1)$, $\delta = 0$: This is an MA(1) model with first differencing

$$\tilde{X}_t = \varepsilon_t + \theta_1 \varepsilon_{t-1}.$$

It can be shown that this ARIMA model is the same as the exponential moving average model (see Section 6.2.2).

ARIMA $(1, 0, 1)$, $\delta \neq 0$: This is a combined model of an AR(1) and an MA(1); i.e.,

$$X_t = \delta + a_1 X_{t-1} + \varepsilon_t + \theta_1 \varepsilon_{t-1}.$$

ARIMA $(1, 1, 1)$, $\delta \neq 0$: This is a combined model of an AR(1) and an MA(1) with first differencing

$$\tilde{X}_t = \delta + a_1 \tilde{X}_{t-1} + \varepsilon_t + \theta_1 \varepsilon_{t-1}.$$

6.5.1 Determination of the ARIMA Parameters

The coefficients of an ARIMA model are typically estimated using MLE.

The time series has to be stationary. If this is not the case, we can use a transformation to stabilize the variance and differencing to remove trend, see Section 6.1.2. If the trend is constant, then a first-order differencing can be used, and if the trend varies with time, then a second-order differencing is necessary.

The selection of p and q can be aided by examining the ACF and PACF plots together. There are several patterns of these two plots that reveal a specific model. Some of these are easily identifiable, but others are more difficult.

For instance, if the ACF plot cuts off sharply at lag k, i.e., it is significantly non-zero up to lag k and extremely low at lag $k + 1$ and afterwards, and the PACF tails off to zero, i.e., it decays to zero asymptotically, then we have an *MA signature*, and in this case we use an MA(q) model with $q = k$. An example of the MA signature is given in Figure 6.17. We see that the ACF plot r_k cuts off at $k = 1$ and the PACF $\alpha(k)$ plot decays gradually. Therefore, we will choose MA(1).

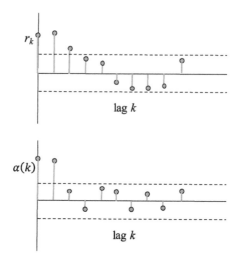

FIGURE 6.17 An example of an MA signature.

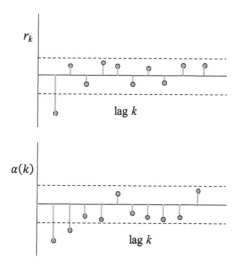

FIGURE 6.18 An example of an AR signature, $k = 0, 1, 2, \ldots$

If the PACF plot $\alpha(k)$ cuts off sharply at lag k and the ACF plot r_k tails off to zero, then we have an *AR signature*, and in this case, we use an AR(p) with $p = k$. An example of the AR and signature is given in Figure 6.18. We see that the PACF plot cuts off at $k = 2$, whereas the ACF plot decays exponentially. In this case, we will choose an AR(2).

If both the ACF and PACF tail off, then both AR and MA terms should be included in the model. However, it is not clear how many terms of each model to use.

In general, the selection of p and q can be done using the Akaike's information criterion (*AIC*), given by the expression:

$$AIC = -2\log(L) + 2(p + q + k),$$

where L is the maximized likelihood of the model, p and q are the order of AR and MA, respectively, and $k = 1$ if $\delta \neq 0$ or $k = 0$ if $\delta = 0$. The idea behind it is that we fit different ARIMA models to a time series and compute the AIC along with all the other parameters. We select the model that minimizes AIC.

The following expression corrected for ARIMA models can also be used:

$$AIC_c = AIC + \frac{2(p+q+k)(p+q+k+1)}{n-p-q-k-1},$$

where n is the sample size. Another criterion that can be used is the Bayesian information criterion (BIC) given by the expression:

$$BIC = AIC + (\log n - 2)(p+q+k).$$

We select the best ARIMA model that minimize one of the above three criteria, preferably AIC_c.

We note that some software packages automatically determine the best values of the parameters p, d, and q.

6.6 DECOMPOSITION MODELS

This method is typically used in forecasting sales and other business-related time series. It assumes that a time series is affected by four factors: trend, cyclical patterns, seasonality, and random occurrences. Consequently, each of these components is forecasted separately and then all the forecasts are combined together to obtain a forecast of the original time series.

Decomposition models are popular among forecasters because they are easy to use and understand. For seasonal (quarterly, monthly, weekly) data, decomposition methods can be very accurate and they also provide additional information about the trend and cyclical patterns.

Two decomposition models are used, namely, the *additive model* and the *multiplicative model*, defined as follows. Let T_t, C_t, S_t, and R_t be the value of the trend, cyclical, seasonal and random component at time t. Then,

- Additive model: $X_t = T_t + C_t + S_t + R_t$

- Multiplicative model: $X_t = T_t \times C_t \times S_t \times R_t$

The additive model is used when the seasonal variation is relatively constant (see Figure 6.19). The multiplicative model is used when the seasonal variation increases (see Figure 6.20). We note that the multiplicative model becomes an additive model by taking logs, i.e.,

$$\log(X_t) = \log(T_t) + \log(C_t) + \log(S_t) + \log(R_t).$$

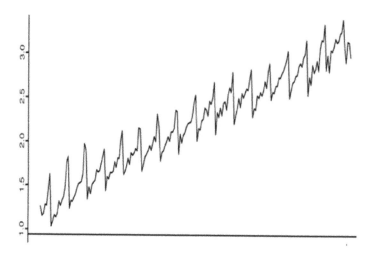

FIGURE 6.19 An example of constant seasonal variation.

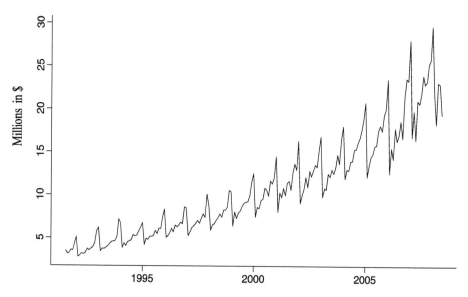

FIGURE 6.20 An example of increasing seasonal variation.

The trend and cyclical components are often grouped into a single trend-cycle component, typically referred to as the trend. In this case, the above two models consist of only three components T_t, S_t, and R_t. This is the case considered in this section.

6.6.1 Basic Steps for the Decomposition Model

The following steps are carried out in order to apply the decomposition model.

a. Estimate the Trend

Fit a smoothing model like the simple moving average model. Set $k = 12$ for monthly data, and $k = 4$ for quarterly data. Alternatively, fit a regression model to the time series.

b. De-trend the Time Series

For the additive model, the de-trended time series Y_t is obtained by subtracting the estimated trend value T_t from the time series X_t, ie., $Y_t = X_t - T_t$. For the multiplicative model, divide the series X_t by the trend values T_t, i.e., $Y_t = X_t/T_t$.

c. Estimate the Seasonal Factor Using the De-trended Time Series

Estimate the seasonal component S_t for each time unit, such as, month, quarter, etc. For instance, if we have 10 years of monthly data, then we average the de-trended values Y_t for all the January months over the 10 years. We do the same for all the February months, and so on. Eventually, we will construct 12 averages, one per month. This is similar to the seasonal subseries plot given in Figure 6.9.

d. Calculate the Random Component

For each time unit, such as, month, quarter, etc., calculate the random component as follows:

- Additive model: $R_t = Y_t - S_t$

- Multiplicative model: $R_t = Y_t/S_t$

In the additive model, we subtract the corresponding seasonal component from each value of the de-trended series. For the multiplicative model, we divide the corresponding seasonal component from each value of the de-trended series. The random component represents the noise or the residuals and it can be modeled using one of the models presented in the previous sections of this chapter.

Figure 6.21 gives an example of the application of the additive model. The top figure gives the original time series X_t. The second figure from the top gives the trend values T_t obtained by applying a simple moving average model. As can be seen, the model follows the trend in the series. For each time point t, the trend value T_t is deducted from its

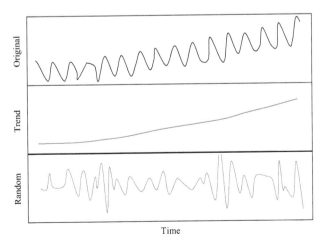

FIGURE 6.21 An example of applying the additive model.

corresponding original time series value X_t. The resulting series Y_t is the de-trended series, not shown in Figure 6.21. Subsequently, we calculate the seasonal component S_t for each time unit which is then deducted from the de-trended series in order to obtain the random component R_t, shown in the bottom graph of Figure 6.21. This component can be forecasted using one of the models presented in this chapter. Working backwards, we first forecast the random component R_t, then add up the seasonal component S_t and the trend estimate T_t, in order to obtain the final forecasted value.

Depending upon the problem, we may not be interested in forecasting the value for the next time period t. Instead, we may be interested in forecasting the trend for the next period. In this case, we first de-seasonalize the original time series and then estimate the trend; i.e., we estimate the trend after the seasonal variation has been removed.

6.7 FORECAST ACCURACY

Depending upon the forecast model and the software package, one may be able to obtain the p-values of the coefficients of the model. The p-value of a coefficient is crucial in determining whether the coefficient is zero or not. We recall that if the p-value of a coefficient is greater than 0.01 then the hypothesis that the coefficient is zero is rejected at 99% confidence.

The next step in checking the accuracy of a forecasting model is to carry out a residual analysis in order to verify that the model has adequately captured the information in the data series. We recall that a residual is the difference $x_t - \hat{x}_t$, where x_t is the actual value of the time series at time t and $\hat{x}_t$ its forecasted value. A minimum requirement is that the residuals should be uncorrelated and have a zero mean. The correlation of the residuals can be checked by plotting their ACF. All autocorrelation values should be zero. We recall that any autocorrelation whose value falls within the confidence interval bounds, or in the absence of these bounds, its absolute value is less than 0.15 is considered as being zero. In addition, the Ljung–Box test can be used. Calculating the mean of all the residuals will allow us to determine if it is close to zero or not. In addition, a hypothesis test can be carried out with the null hypothesis that the population mean of the residuals is zero versus the alternative hypothesis that it is not.

Additionally, it is useful but not necessary, that the residuals are normally distributed and their variance is constant. We can test that the residuals are normally distributed by carrying out a χ^2 test and a Q-Q plot using the histogram of the residuals. Finally, we can check for constant variance by plotting the fitted values versus the residuals. For details see Section 5.2.2.

If the residuals do not behave satisfactorily, then a different forecasting model should be developed. In general, one should develop a number of different forecasting models. For this, the time series $x_1, x_2, ..., x_T$ should be divided into two contiguous sets, a training set $x_1, x_2, ..., x_\tau$ and a test set $x_{\tau+1}, x_{\tau+2}, ..., x_T$, with the training set being around 80% of the entire time series. Each model should be developed using the training set, and then all the models should be compared using the test set so that the best one can be selected. The models are compared using a metric such as the RMSE of the errors $\varepsilon_t = x_t - \hat{x}_t, t = \tau + 1, \tau + 2, ..., T$, calculated using the test set. These errors are known as *forecast errors*, as opposed to the residuals that are computed using the training set. Other metrics of forecast errors can be used, such as, the MAE (mean absolute error), the mean error, and the mean absolute percentage error (see Section 6.2.4 for more details).

6.8 PREDICTION INTERVALS

A forecasting model is developed so that we can predict future values. For instance, let us assume that we have developed an AR(1) model using the time series $x_1, x_2, \ldots, x_t$. Then, the point estimate for the future period $t + 1$ is $\hat{x}_{t+1} = \delta + \alpha_1 x_t$. Likewise, if we have developed a simple average model with, say, $k = 2$, the point estimate for $t + 1$ is $\hat{x}_{t+1} = (x_t + x_{t-1})/2$.

A point estimate $\hat{x}_{t+1}$ is not very useful by itself since there is uncertainty about its predicted value. A better prediction can be obtained using a prediction interval $[a, b]$, which is an interval constructed around a point estimate $\hat{x}_{t+1}$ within which we expect the actual observation x_{t+1} to lie 95% of the time. The values a and b are calculated in the same way as the lower and upper bound of the 95% confidence interval, and they correspond to the 2.5th and 97.5th percentile, respectively. Other prediction intervals can be used such as the 80% interval, in which case, a and b correspond to the 10th and 90th percentile, respectively. As discussed in Section 5.5, prediction and confidence intervals appear to be similar, but in fact they estimate different parameters. A prediction interval is an estimate of an interval within which lies the value x_{t+1} that will be observed in the future. A confidence interval is an estimate of an interval within which a parameter that cannot be observed, such as a population mean, lies.

Assuming that the forecast errors are normally distributed, the 95% prediction interval for the time period $t + 1$ is $\hat{x}_{t+1} \pm 1.96 s_1$, where $\hat{x}_{t+1}$ is the estimated value at time $t + 1$, and s_1 is the standard deviation of the estimates at time $t + 1$. Other prediction intervals, such as the 80% can be used. In this case, we have $\hat{x}_{t+1} \pm 1.28 s_1$. The constant with which we multiply the standard deviation s_1 is calculated in the same way as in confidence intervals. In general, the h-step 95% prediction interval is $\hat{x}_{t+h} \pm 1.96 s_h$, where $\hat{x}_{t+h}$ is the estimated value at time $t + h$, and s_h is the standard deviation of the h-step predictions. Below, we obtain an expression for s_h for an ARIMA model.

We recall that the mean-adjusted MA(q) model is given by the expression:

$$\tilde{X}_t = \varepsilon_t + \theta_1 \varepsilon_{t-1} + \theta_2 \varepsilon_{t-2} + \cdots + \theta_q \varepsilon_{t-q},$$

where θ_i, $i = 1, 2, \ldots, q$, are the weights, and $\varepsilon_t, \varepsilon_{t-1}, \ldots, \varepsilon_{t-q}$ are errors normally distributed with mean 0 and variance σ^2. Any ARIMA model can be converted to an MA model with an infinite order given by the expression:

$$\tilde{X}_t = \varepsilon_t + \sum_{i=1}^{\infty} \theta_i \varepsilon_{t-i},$$

where $\sum_{i=1}^{\infty} |\theta_i| < \infty$ so that the series converges to a finite value. We assume that $\theta_0 = 0$. We demonstrate this by considering the AR(1) model. We have the following equations for t, $t - 1, t - 2$, etc.:

$$\tilde{X}_t = a_1 \tilde{X}_{t-1} + \varepsilon_t$$

$$\tilde{X}_{t-1} = a_1 \tilde{X}_{t-2} + \varepsilon_{t-1}$$

$$\tilde{X}_{t-2} = a_1 \tilde{X}_{t-3} + \varepsilon_{t-2}$$

$$\vdots$$

Substituting the expression for $\tilde{X}_{t-1}$ from the second equation into the first one, we obtain

$$\tilde{X}_t = a_1 \left(a_1 \tilde{X}_{t-2} + \varepsilon_{t-1} \right) + \varepsilon_t$$

$$= a_1^2 \tilde{X}_{t-2} + a_1 \varepsilon_{t-1} + \varepsilon_t.$$

Likewise, substituting the expression for $\tilde{X}_{t-2}$ from the third expression into the above equation gives

$$\tilde{X}_t = a_1^2 \left(a_1 \tilde{X}_{t-3} + \varepsilon_{t-2} \right) + a_1 \varepsilon_{t-1} + \varepsilon_t$$

$$= a_1^3 \tilde{X}_{t-3} + a_1^2 \varepsilon_{t-2} + a_1 \varepsilon_{t-1} + \varepsilon_t.$$

Continuing in this fashion, we obtain

$$\tilde{X}_t = \sum_{i=0}^{\infty} a_1^i \varepsilon_{t-i}.$$

It can be shown that the standard deviation s_h of the forecast error at time $t + h$ is given by the expression:

$$s_h = s \sqrt{\sum_{i=0}^{h-1} \theta_t^2},$$

where s is the standard deviation of the residuals. For $h = 1$, we have $s_1 = s\sqrt{1} = s$. For $h = 2$, we have $s_2 = s\sqrt{1 + \theta_1^2}$, and so on. Therefore, the 95% prediction interval is given by the expression:

$$\hat{x}_{t+h} \pm 1.96 s \sqrt{\sum_{i=0}^{h-1} \theta_t^2}.$$

For instance, given we have estimated the parameters of an AR(1) model using the sequence $x_1, x_2, \ldots, x_t$, the 95%prediction interval for the next period is

$$\hat{x}_{t+1} = \delta + a_1 x_t \pm 1.96s.$$

For the period $t + 1$, we have

$$\hat{x}_{t+2} = \delta + a_1 \hat{x}_{t+1} \pm 1.96s\sqrt{1 + \theta_1^2},$$

i.e., we use the estimate $\hat{x}_{t+1}$ in place of x_{t+1} since we do not have the real observation. Likewise,

$$\hat{x}_{t+3} = \delta + a_1 \hat{x}_{t+2} \pm 1.96s\sqrt{1 + \theta_1^2 + \theta_2^2},$$

and so on.

As we predict further into the future, the standard deviation s_h increases, and consequently the prediction interval increases as well. This makes sense intuitively since the uncertainty of the estimation increases as well. As h increases, the point estimates will converge to a fixed value.

The above construction of the prediction interval assumes that the forecast errors are normally distributed. If this is not a good assumption, we can construct a prediction interval using bootstrapping which only assumes that the forecast errors are uncorrelated. Let us assume that we have developed a forecasting model using the time series $x_1, x_2, \ldots, x_t$. Then, for the period t we have: $\varepsilon_t = x_t - \hat{x}_t$, or $x_t = \hat{x}_t + \varepsilon_t$. Consequently, for the future period $t + 1$, we have: $x_{t+1} = \hat{x}_{t+1} + \varepsilon_{t+1}$. Assuming that the unknown future errors are similar to past errors, we can replace ε_{t+1} by randomly selecting with replacement a residual from the set of past residuals $\{\varepsilon_1, \varepsilon_1, \ldots, \varepsilon_t\}$. Likewise, for the future period $t + 2$, we have $x_{t+2} = \hat{x}_{t+2} + \varepsilon_{t+2}$. Repeating this for h times, we generate the h-step future observation x_h. Now, if we repeat this process a large number of times, we will obtain a set of h-step future observations x_h. The 95th prediction interval can then be constructed by calculating the 2.5th and 97.5th percentiles of the generated set of x_h values, as described in Section 3.5.4.

6.9 VECTOR AUTOREGRESSION

So far, we have considered different models for forecasting a *univariate* variable, i.e., a variable that takes scalar values. However, often, we may have to deal with a cluster of sensors that produce different types of measurements. For instance, a weather station is typically equipped with a variety of different sensors each capturing a different univariate variable, such as, air temperature (A), humidity (H), barometric pressure (B), wind speed (S), wind direction (D), and hourly precipitation (P). Each of these variables is associated with a time series of past observations; i.e., we have the time series: A_t, H_t, B_t, S_t, D_t, and $P_t, t \geq 1$. Now, if these time series do not affect one another, then each variable can be forecasted separately. However, if the time series affect one another, as is the case with the weather sensors, then all the univariate variables should be forecasted as a single vector variable, i.e., a *multivariate* variable. For instance, in the case of the weather station, we treat the individual variables as a single multivariate variable W_t, where $W_t = (A_t, H_t, B_t, S_t, D_t, P_t)^T$, and then build a forecasting model for W_t, using vector forecasting techniques. To that effect, vector extensions of the MA(q), AR(p), and ARMA models have been developed. In this section, we describe the vector AR(p) model, notated as VAR(p).

The simplest model of a VAR(p) is VAR(1) for a two-dimensional vector. In this case, the vector time series is in the form $(x_{1t}, x_{2t})^T$, $t \geq 1$, and the order of the autoregression is 1. VAR(1) is given by the expression:

$$\begin{pmatrix} X_{1t} \\ X_{2t} \end{pmatrix} = \begin{pmatrix} \delta_1 \\ \delta_2 \end{pmatrix} + \begin{pmatrix} a_{11} & a_{12} \\ a_{21} & a_{22} \end{pmatrix} \begin{pmatrix} X_{1t-1} \\ X_{2t-1} \end{pmatrix} + \begin{pmatrix} \varepsilon_{1t} \\ \varepsilon_{2t} \end{pmatrix},$$

where $\varepsilon_{1t} \sim N\left(0, \sigma_1^2\right)$ and $\varepsilon_{2t} \sim N\left(0, \sigma_2^2\right)$. As in the univariate case, the successive values of ε_{1t}, i.e., $\varepsilon_{1t}, \varepsilon_{1t-1}, \varepsilon_{1t-3} \ldots$, are independent of each other. The same holds for the successive values of ε_{2t}, i.e, $\varepsilon_{2t}, \varepsilon_{2t-1}, \varepsilon_{2t-3} \ldots$. However, ε_{1t} and ε_{2t}, $t \geq 1$, may be correlated with a covariance matrix

$$\Sigma = \begin{pmatrix} \sigma_1^2 & \sigma_{12} \\ \sigma_{21} & \sigma_2^2 \end{pmatrix},$$

where $\sigma_{21} = \sigma_{12}$ is the covariance of ε_{1t} and ε_{2t}. The above vector representation of the AR(1) model can be re-written as a system of equations as follows:

$$\begin{cases} X_{1t} = \delta_1 + a_{11} X_{1t-1} + a_{12} X_{2t-1} + \varepsilon_{1t} \\ X_{2t} = \delta_2 + a_{21} X_{1t-1} + a_{22} X_{2t-1} + \varepsilon_{2t} \end{cases}$$

In the univariate AR(1), X_{1t} is predicted using the previous observation X_{1t-1}. However, in the VAR(1) case we see that we also use the previous value X_{2t-1} of the other time series as well. The same holds for the prediction of X_{2t}.

Let us now consider the VAR(2) model, for a two-dimesional vector. In this case, the vector time series is in the form $(x_{1t}, x_{2t})^T$, $t \geq 1$ and the order of the autoregression is 2. VAR(2) is given by the expression:

$$\begin{pmatrix} X_{1t} \\ X_{2t} \end{pmatrix} = \begin{pmatrix} \delta_1 \\ \delta_2 \end{pmatrix} + \begin{pmatrix} a_{11}^{(1)} & a_{12}^{(1)} \\ a_{21}^{(1)} & a_{22}^{(1)} \end{pmatrix} \begin{pmatrix} X_{1t-1} \\ X_{2t-1} \end{pmatrix} + \begin{pmatrix} a_{11}^{(2)} & a_{12}^{(2)} \\ a_{21}^{(2)} & a_{22}^{(2)} \end{pmatrix} \begin{pmatrix} X_{1t-2} \\ X_{2t-2} \end{pmatrix} + \begin{pmatrix} \varepsilon_{1t} \\ \varepsilon_{2t} \end{pmatrix},$$

where $\varepsilon_{1t} \sim N\left(0, \sigma_1^2\right)$ and $\varepsilon_{2t} \sim N\left(0, \sigma_2^2\right)$. The general expression of VAR(p) for a d-dimensional vector variable is as follows:

$$\begin{pmatrix} X_{1t} \\ X_{2t} \\ \vdots \\ X_{dt} \end{pmatrix} = \begin{pmatrix} \delta_1 \\ \delta_2 \\ \vdots \\ \delta_d \end{pmatrix} + \begin{pmatrix} a_{11}^{(1)} & a_{12}^{(1)} & \cdots & a_{1d}^{(1)} \\ a_{21}^{(1)} & a_{22}^{(1)} & \cdots & a_{2d}^{(1)} \\ \vdots & \vdots & & \vdots \\ a_{d1}^{(1)} & a_{2d}^{(1)} & \cdots & a_{dd}^{(1)} \end{pmatrix} \begin{pmatrix} X_{1t-1} \\ X_{2t-1} \\ \vdots \\ X_{dt-1} \end{pmatrix} + \cdots$$

$$+ \begin{pmatrix} a_{11}^{(p)} & a_{12}^{(p)} & \cdots & a_{1d}^{(p)} \\ a_{21}^{(p)} & a_{22}^{(p)} & \cdots & a_{2d}^{(p)} \\ \vdots & \vdots & & \vdots \\ a_{d1}^{(p)} & a_{2d}^{(p)} & \cdots & a_{dd}^{(p)} \end{pmatrix} \begin{pmatrix} X_{1t-1} \\ X_{2t-1} \\ \vdots \\ X_{dt-1} \end{pmatrix} + \begin{pmatrix} \varepsilon_{1t} \\ \varepsilon_{2t} \\ \vdots \\ \varepsilon_{dt} \end{pmatrix},$$

where $\varepsilon_{it} \sim N\left(0, \sigma_i^2\right)$, $i = 1, 2, \ldots d$, with a covariance matrix:

$$\Sigma = \begin{pmatrix} \sigma_1^2 & \sigma_{12} & \cdots & \sigma_{1d} \\ \sigma_{21} & \sigma_2^2 & \cdots & \sigma_{2d} \\ \vdots & \vdots & & \vdots \\ \sigma_{d1} & \sigma_{d2} & \cdots & \sigma_d^2 \end{pmatrix}.$$

We note that in the VAR(p) model, all the time series have the same order p.

Let, $X_t = (X_{t1}, \ldots, X_{dt})^T$, $\Delta = (\delta_1, \ldots, \delta_d)^T$, $E = (\varepsilon_1, \ldots, \varepsilon_d)^T$, and

$$A_i = \begin{pmatrix} a_{11}^{(i)} & a_{12}^{(i)} & \cdots & a_{1d}^{(i)} \\ a_{21}^{(i)} & a_{22}^{(i)} & \cdots & a_{2d}^{(i)} \\ \vdots & \vdots & & \vdots \\ a_{d1}^{(i)} & a_{2d}^{(i)} & \cdots & a_{dd}^{(i)} \end{pmatrix}, i = 1, 2, \ldots p.$$

Then, the above VAR(p) can be written as

$$X_t = \Delta + A_1 X_{t-1} + \cdots + A_p X_{t-p} + E.$$

Now, let B be the backshift operator, i.e., $BX_t = X_{t-1}$, $B^2 X_t = X_{t-2}$, ..., $B^p X_t = X_{t-p}$. Then, the above equation can be re-written as follows:

$$X_t = \Delta + A_1 B X_t + \cdots + A_p B^p X_t + E$$

or

$$\left(I - A_1 B - \cdots - A_p B^p\right) X_t = \Delta + E.$$

VAR(p) is stable if the roots of $det(I - A_1 z - \cdots - A_p z^p) = 0$ lie outside the unit circle.

6.9.1 Fitting a VAR(p)

We now proceed to describe how to fit a VAR(p) to a vector time series. We consider a d-dimensional time series X_t, $t \geq 1$, where $X_t = (X_{1t}, X_{2t}, \ldots, X_{dt})^T$, which can also be seen as consisting of d different univariate time series X_{it}, $i = 1, 2, \ldots, d$. The vector time series is in the form: $(x_{1t}, x_{2t}, \ldots, x_{dt})^T$, $t \geq 1$, and the individual univariate time series are x_{it}, $t > 1$, $i = 1$, 2, ..., d.

The first step is to make sure that the individual univariate time series are stationary. This can be done by visual inspection together with using a test such as the augmented Dickey–Fuller test. A non-stationary univariate time series can be transformed to a stationary one using transformations and differencing as discussed in Section 6.1.2. Note that the same order differencing has to be applied to all the univariate time series so that they retain their time synchronization with each other.

In order to construct a meaningful VAR(p) model, the univariate time series should influence one another. For this, Granger's causality test can be used. Let X and Y be two univariate stationary time series. X is said to *Granger-cause Y*, or simply to *cause Y*, if predictions of Y based on its own past values and on the past values of X are better than predictions of Y based only on its own past. This is known as *unidirectional causality*. If Y also causes X, then both time series influence one another, and we have a *bidirectional causality*. For instance, let us assume that we have two time series, one for the outside temperature X and another for the electricity consumption Y in a house which is heated and cooled with electricity. Obviously, the outside temperature affects the electrical consumption in the house, but not the other way around. In this case, we have a unidirectional causality whereby X causes Y.

The null hypothesis of the Granger's causality test is that X does not cause Y. The test can be implemented as follows. For a given lag p, we fit the following two autoregressive models:

$$Y_t = \delta + a_1 Y_{t-1} + \cdots + a_p Y_{t-p} + u_t$$

$$Y_t = \delta + a_1 Y_{t-1} + \cdots + a_p Y_{t-p} + \beta_1 X_{t-1} + \cdots + \beta_p X_{t-p} + \varepsilon_t$$

The first one is an AR(p) model of Y, and the second one is an AR(p) model of Y augmented with p lags of X. The null hypothesis that X does not cause Y is expressed mathematically as follows: $H_0: \beta_1 = \beta_2 \ldots = \beta_p = 0$, and it is tested using the F-test. Let $RSS1$ and $RSS2$ be the residual sum of squares obtained from the first and second autoregression, respectively. Then, the ratio

TABLE 6.1   p-values for Granger's causality test

	X_1	X_2	X_3	X_4
X_1	0.00	0.45	0.00	0.37
X_2	0.02	0.00	0.00	0.23
X_3	0.40	0.005	0.00	0.38
X_4	0.15	0.25	0.35	0.00

$$\frac{(RSS1 - RSS2)/p}{RSS1/(n-2p-1)},$$

where n is the number of data points, follows the F-distribution. This is one-tailed test and the hypothesis is rejected if the above ratio is greater than the critical value $F_{p,\ n-2p-1}$. The Granger causality test is implemented in many econometrics software packages, which typically report the p-value. The test should be run for different values of p, but the decision whether to accept or reject the null hypothesis should not be sensitive on p. The same procedure can be applied to the X series to test whether Y causes X.

For more than two time series, we test all possible combinations of any two of them and then we can decide accordingly which ones to include in the VAR(p) model. For instance, Table 6.2 gives an example of the p-value for Granger's causality test for all possible combinations of the time series X_1, X_2, X_3, X_4. Each entry is the p-value of the null hypothesis that a time series listed in the first column does not cause a time series listed in the top row. For instance, let us consider X_2 in the first column. Then, looking at the p-values on the same row we see that X_2 causes X_3, does not cause X_4, and causes X_1 at 5% confidence level but not at 1% confidence level. This table can also help us to identify univariate and bivariate causalities. For instance, we see that X_1 causes X_3, but X_3 does not cause X_1. In this case, we have a unidirectional causality between X_1 and X_3. A bidirectional causality exists between X_2 and X_3. Based on the p-values, we conclude that all four time series should be included in the VAR(p) model.

Given a value for p, the set of coefficients $a_{ij}^{(k)}$, $i, j = 1, 2, \ldots, d, k = 1, 2, \ldots, p$ can be determined using the ordinary least squares method, i.e., by minimizing the sum of squared errors:

$$\underset{\Theta}{\text{argmin}} \sum_{i=1}^{n} \left(\hat{x}_t - x_t \right)^2,$$

where x_t is the observed vector at time t, $\hat{x}_t$ is its estimated value, and Θ is the set of all the coefficients.

To select the best order p, we fit a VAR(p) for various values of p and select the value that minimizes a criterion, such as the AIC and the BIC, given by the expressions:

$$AIC = \log(|\Sigma|) + \frac{2d^2 p}{n}$$

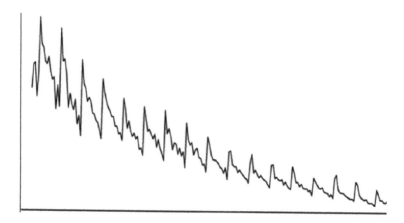

FIGURE 6.22 Time series for problem 1.

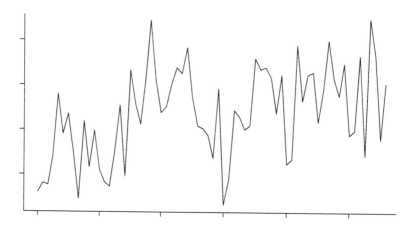

FIGURE 6.23 Time series for problem 2.

$$BIC = \log\left(\left|\Sigma\right|\right) + \frac{d^2 p \log(n)}{n},$$

where Σ is an estimate of the covariance matrix of the errors ε_{it}, $i = 1, 2, \ldots, d$, n is the number of data points, d the number of univariate variables, and p the order of the VAR. In practice, one should calculate both criteria for different values of p and then make an educated decision.

The accuracy of the fitted model is checked by first examining the p-values of the coefficients of the model. The p-value of a coefficient allows us to determine whether the coefficient is zero or not. We recall that if the p-value is greater than 0.01, then the hypothesis that the coefficient is zero is rejected at 99% confidence. The next step in checking the accuracy of the model is to carry out a residual analysis. This is carried out in the same way as described in Section 6.7 on each individual univariate time series. Finally, prediction intervals can be computed using different methods, including the bootstrapping method described in Section 6.8.

EXERCISES

1. Consider the time series in Figure 6.22. Is it stationary? Why?

2. Same as above for the time series in Figure 6.23.

TABLE 6.2   Data for problem 3

t	1	2	3	4	5	6	7	8	9	10
Actual	20	21	15	14	13	16	19	23	25	22
Forecasted	21									

3. Use exponential smoothing to forecast the observations generated by a sensor given in Table 6.2. Two values of a are considered $a = 0.8$ and $a = 0.5$. Use the RMSE (root mean squared error) to determine which value of the two values for α works better. The initial forecasted value for time $t = 1$ is 21.

4. Same as in problem 3, but now use the other metrics (one at a time) described in Section 6.2.4 to determine which value of the two values for α works better. Comment on your conclusions. Which metric works the best?

5. For the MA(2) model, show that

$$r_1 = \frac{\theta_1 + \theta_1 \theta_2}{1 + \theta_1^2 + \theta_2^2}, r_2 = \frac{\theta_2}{1 + \theta_1^2 + \theta_2^2}, r_3 = r_4 \ldots = 0$$

6. For the AR(1) model, show that the lag k autocorrelation r_k is given by the expression: $r_k = \alpha_1^k$.

FORECASTING PROJECT

The objective of this project is to develop various forecasting models for a time series in order to determine the best one. For this analysis, you can use any software, such as, Python, R, MatLab, and SAS, or a combination of functions from different software.

Data Set

In this project, you will develop different forecasting models of a time series of air temperature observations. There are several sites from where such a time series can be obtained, such as the North Carolina State Climate Data (https://climate.ncsu.edu). It contains data from several weather stations across the state of North Carolina, such as the Lake Wheeler Rd Field Lab (LAKE) station. The air temperatures are averaged hourly. Retrieve 2000 contiguous observations and check the time series for missing data points. Approximated a missing data point at time t by the average of the two adjacent data points at times $t - 1$ and $t + 1$.

Task 1   Check for Stationarity

Plot the entire time series of 2000 observations and check it visually for stationarity. Use also the augmented Dickey–Fuller test. Make the necessary transformations as discussed in Section 6.1.2, to transform the series into a non-seasonal stationary one. Subsequently, split the data into two contiguous sets; a training data set of the first 1500 observations and a test data set of the remaining 500 observations. Tasks 2, 3 and 4 use the training set and task 5 uses the test set.

Task 2   Fit a Simple Moving Average Model

2.1. Apply the simple moving average model $\hat{x}_t = (1/k)\sum_{i=t-k}^{t-1} x_i$ to the training data set, for a given k.

2.2. Calculate the residuals and compute the RMSE.

2.3. Repeat the above two steps by varying k and calculate the RMSE.

2.4. Plot RMSE vs k. Select k based on the lowest RMSE value.

2.5. For the best value of k plot the predicted values against the original values, and visually inspect the accuracy of the forecasting model

2.6. Carry out a residual analysis.

- Plot the correlogram of the residuals and check if the residuals are uncorrelated. In addition, do the Ljung–Box test.
- Check if the residuals sum up to zero.
- Do a Q-Q plot of the pdf of the residuals against $N(0, s^2)$.
- Carry out a χ^2 test that the residuals follow the normal distribution $N(0, s^2)$.
- Plot the scatter diagram of the residuals versus the predicted values to check for trends and constant variance.

2.7. Comment on your results.

Task 3   Fit an Exponential Smoothing Model

3.1. Apply the exponential smoothing model $\hat{x}_t = ax_{t-1} + (1-a)x_{t-1}$ to the training data set for $\alpha = 0$.

3.2. Calculate the residuals and compute the RMSE.

3.3. Repeat steps 3.1 and 3.2 by increasing a each time by 0.1, until $a = 1$.

3.4. Plot RMSE vs a. Select a based on the lowest RMSE value.

3.5. For the selected value of a plot the predicted values against the original values and visually inspect the accuracy of the forecasting model.

3.6. Carry out a residual analysis following the instructions in task 2.6.

3.7. Comment on your results.

Task 4 Fit an Arma(p, q) Model

Plot the ACF and PACF, and check for an AR or an MA signature using the training time series.

4.1. If an AR signature is identified, then:
   - Fit an AR model with order p set to the lag k at which the PACF cuts off.
   - Calculate the RMSE value and plot the predicted values against the original values.

4.2. If an MA signature is identified, then:
   - Fit an MA model with order q equal to the lag k at which the ACF cuts off.
   - Calculate the RMSE value and a plot the predicted values against the original values.

4.3. If both ACF and PACF tail off, then:
   - Fit an ARMA model using a software package that searches automatically for the best model.
   - Calculate the RMSE value and plot the predicted values against the original values.

4.4. Carry out a residual analysis following the instructions in Task 2.6.

4.5. Comment on your results.

Task 5 Comparison of All the Models

5.1. Run the three models obtained in Tasks 2, 3 and 4 on the test data and calculate the RMSE for each model. Choose the best model based on their RMSE value. Comment on your decision.

5.2. Calculation of prediction intervals.
   - Use the best model to construct the point estimate and its prediction interval for the future period $t = 2001$. If you have transformed the time series to make it stationary, transform the point estimate and its prediction interval back to the original values. Comment on your results.
   - Obtain the prediction intervals for many subsequent points to see where it converges to. Comment on your results.

REFERENCES

1. E.P. Box, G.M. Jenkins, G.C. Reinsel, and G.M. Ljung. *Time Series: Forecast and Control*. 2015. Wiley.
2. R.J. Hyndman and G. Athanasopoulos. *Forecasting: Principles and Practice*, OTexts.org/fpp/.

CHAPTER 7

Dimensionality Reduction

O NE OF THE PROBLEMS THAT ARISES WHEN DEALING WITH A large data set is that the number of variables involved may be very large. A data set is stored as a matrix where each column represents a variable and each row represents a data point. For instance, if we have d variables and n data points, then the data set is stored in a $n \times d$ matrix, and each data point can be seen as a point in a d-dimensional space. In data analysis, variables are also known as *features* or *attributes*, since each one represents an aspect of the system under study. Often, the dimensionality of a data set can be very large, which makes the analysis difficult. A common approach to address this issue is to first reduce the dimensionality to a smaller set of variables before applying the appropriate Machine Learning algorithm. Reducing the dimensionality helps to reduce the storage space and improves the time required to run a Machine Learning algorithm. In addition, it takes care of multi-collinearity, and if the data is reduced to two or three dimensions it can be visualized as well.

There are several techniques for reducing the dimensionality of a data set, and they can be classified into two groups, namely, *feature selection* and *feature projection* or *feature extraction*.

In feature selection, we determine a subset of the original set of variables which best describes the data set. The main idea is that the data set may contain features that are irrelevant, or features that are correlated. If a number of variables are correlated, then they can be all represented by one of the variables, which helps to reduce the dimensionality of the data.

In feature projection, the reduction is achieved by transforming the data into a space with fewer dimensions than originally. There are two classical approaches for feature projection, namely, *principal component analysis* (PCA), and *linear discriminant analysis* (LDA) and its generalization to *multiple discriminant analysis* (MDA). In this chapter, we describe these algorithms in detail. A set of exercises is given at the end of the chapter.

We note that both algorithms make use of eigenvalues and eigenvectors. In view of this, we briefly review the eigenvalues and eigenvectors of a square matrix in the following section, before we proceed to describe the algorithms.

7.1 A REVIEW OF EIGENVALUES AND EIGENVECTORS

Let A be a square matrix, e a non-zero vector with the same number of rows as A, and λ a scalar. Then, λ is called the *eigenvalue* and e is called the *eigenvector* of A, if $Ae = \lambda e$.

Multiplying the eigenvector e by a constant c gives a vector of different length which is also an eigenvector with the same eigenvalue since the relation $Ae = \lambda e$ still holds. To avoid ambiguity regarding the length of the eigenvector, we require that every eigenvector is a unit vector; i.e., the sum of the squares of the components of the vector is 1. However, this requirement still does not make the eigenvector unique, since we can multiply it by -1 without changing the sum of the squares of the components. In view of this, we require that the first non-zero component of an eigenvector be positive.

The above expression, $Ae = \lambda e$, can be re-written as $(A - \lambda I)e = 0$, where I is the identity matrix. This linear system has a non-zero solution if and only if $det(A - \lambda I) = 0$. Therefore, we can use the expression $det(A - \lambda I) = 0$ to obtain the eigenvalues. For example, let us consider the matrix

$$A = \begin{pmatrix} 2 & 1 \\ 1 & 2 \end{pmatrix}.$$

We have

$$det(A - \lambda I) = \begin{vmatrix} 2 - \lambda & 1 \\ 1 & 2 - \lambda \end{vmatrix} = (2 - \lambda)^2 - 1 = 0$$

or

$$\lambda^2 - 4\lambda + 3 = 0.$$

The roots of the above second-order polynomial are the two eigenvalues $\lambda_1 = 1$ and $\lambda_2 = 3$. The eigenvectors can then be obtained as follows. For $\lambda_1 = 1$ we have $(A - I)e_1 = 0$, or

$$\left(\begin{pmatrix} 2 & 1 \\ 1 & 2 \end{pmatrix} - \begin{pmatrix} 1 & 0 \\ 0 & 1 \end{pmatrix} \right) \begin{pmatrix} e_{11} \\ e_{12} \end{pmatrix} = \begin{pmatrix} 0 \\ 0 \end{pmatrix}$$

or

$$e_1 = \begin{pmatrix} e_{11} \\ e_{12} \end{pmatrix} = \begin{pmatrix} 1 \\ -1 \end{pmatrix}.$$

In order for e_1 to be a unit vector, we have to divide each element by $\sqrt{2}$, since $\left(1/\sqrt{2}\right)^2 + \left(-1/\sqrt{2}\right)^2 = 1$; i.e., the first eigenvector is

$$e_1 = \begin{pmatrix} 1/\sqrt{2} \\ -1/\sqrt{2} \end{pmatrix}.$$

For $\lambda_2 = 3$, we have $(A - 3I)e_2 = 0$, or

$$\left(\begin{pmatrix} 2 & 1 \\ 1 & 2 \end{pmatrix} - 3 \begin{pmatrix} 1 & 0 \\ 0 & 1 \end{pmatrix} \right) \begin{pmatrix} e_{21} \\ e_{22} \end{pmatrix} = \begin{pmatrix} 0 \\ 0 \end{pmatrix}$$

or

$$e_2 = \begin{pmatrix} e_{21} \\ e_{22} \end{pmatrix} = \begin{pmatrix} 1 \\ 1 \end{pmatrix}.$$

Dividing each element by $\sqrt{2}$ we obtain the second unit vector eigenvector

$$e_2 = \begin{pmatrix} 1/\sqrt{2} \\ 1/\sqrt{2} \end{pmatrix}.$$

Now let us assume that matrix A is symmetric; i.e. the element in the ith row and jth column is the same as the one in the jth row and ith column. Define E to be a matrix where each column is equal to an eigenvector; i.e., the ith column is equal to e_i. Then $EE^T = E^TE = I$; i.e., the eigenvectors are orthogonal to each other.

For example, let us consider matrix A given above, which is symmetric. Then,

$$E = \begin{pmatrix} 1/\sqrt{2} & 1/\sqrt{2} \\ -1/\sqrt{2} & 1/\sqrt{2} \end{pmatrix},$$

and

$$EE^T = \begin{pmatrix} 1/\sqrt{2} & 1/\sqrt{2} \\ -1/\sqrt{2} & 1/\sqrt{2} \end{pmatrix} \begin{pmatrix} 1/\sqrt{2} & -1/\sqrt{2} \\ 1/\sqrt{2} & 1/\sqrt{2} \end{pmatrix} = \begin{pmatrix} 1 & 0 \\ 0 & 1 \end{pmatrix}$$

We obtain the same when calculating E^TE. The ith diagonal element of EE^T is the dot product $e_i^T e_i$ which is 1 since it is the sum of the components of unit vector e_i. The (i, j) off diagonal element is the dot product $e_i^T e_j$. Since it is zero, e_i and e_j are orthogonal.

7.2 PRINCIPAL COMPONENT ANALYSIS (PCA)

PCA reduces the dimensionality of a data set by identifying a set of new dimensions on which the data has the largest variation. Specifically, it uses an orthogonal transformation to convert a set of observations of possibly correlated variables into a smaller set of uncorrelated variables called *principal components* (pc). The transformation is such that the data onto the first pc has the highest variance. For each succeeding *pc*, the data has the next highest variance under the constraint that the pc is orthogonal to the previous pc. PCA has many applications. It can be used for exploratory data analysis, cluster analysis, and constructing predictive models.

As an example, let us consider the data set $\{(2,2), (4,3), (5,5.5), (6,4), (7,6.5)\}$ shown in Figure 7.1. The projections of the data points onto the x and y axes are also given in Figure 7.1. PCA puts a line through the data so that the variance of the projections of the data points onto this line is the highest; i.e., the projected points are spread as much as possible.

After running the PCA algorithm, the projections onto the first principal component (pc1) are −3.547, −1.405, 1.036, 0.737, 3.178, and onto the second principal component pc2 are −0.316, 0.326, −0.81, 0.968, −0.168. The two principal components and the projections of the data points onto them are shown in Figure 7.2. As can be seen, we get a good spread of the data onto pc1 as opposed to the projections of the data on pc2.

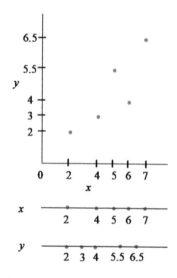

FIGURE 7.1 The data points and their projection on the X and Y axes.

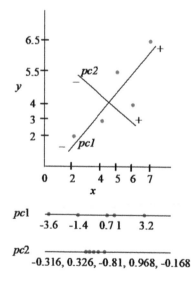

FIGURE 7.2 The two principal components and projections of the data.

7.2.1 The PCA Algorithm

Let us consider n d-dimensional data points $x_i = (x_{i1}, x_{i2}, ..., x_{id})$, $i = 1, 2, ..., n$. The n data points can be seen as an $n \times d$ matrix A where each row contains a single d-dimensional data point and each column is a feature. The data is mean adjusted; i.e., the sample mean of each column is subtracted from each data point in the column. The advantage of mean-adjusted data is that it simplifies the calculation of the covariance matrix. Let X and Y be two random variables, and let $\tilde{X}$ and $\tilde{Y}$ be their corresponding mean-adjusted random variables. Then, we have

$$\mathrm{Var}\left(\tilde{X}\right) = E\left(\tilde{X} - 0\right)^2 = E\left(\tilde{X}\right)^2$$

$$\mathrm{Var}\left(\tilde{Y}\right) = E\left(\tilde{Y} - 0\right)^2 = E\left(\tilde{Y}\right)^2.$$

and

$$\mathrm{Cov}\left(\tilde{X}, \tilde{Y}\right) = E\left(\tilde{X} - 0\right)\left(\tilde{Y} - 0\right) = E\left(\tilde{X}\tilde{Y}\right).$$

As a result, the covariance matrix of the two mean-adjusted random variables can be written as

$$\Sigma = \begin{pmatrix} E\left(\tilde{X}\right)^2 & E\left(\tilde{X}\tilde{Y}\right) \\ E\left(\tilde{X}\tilde{Y}\right) & E\left(\tilde{Y}\right)^2 \end{pmatrix}.$$

For example, let us consider the data set A with two mean-adjusted variables X and Y shown in Table 7.1. Columns 1 and 2 give the original data, and the remaining columns give results for the calculation of the two variances and the covariance.

TABLE 7.1   Calculation of the covariance

x_i	y_i	x_i^2	y_i^2	$x_i y_i$
−3	−15	9	225	45
−2	−10	4	100	20
−1	−5	1	25	5
0	0	0	0	0
1	5	1	25	5
2	10	4	100	20
3	15	9	225	45
		$\Sigma = 28$	$\Sigma = 700$	$\Sigma = 140$

We have $s_X^2 = (1/7) \sum_{i=1}^{7} \sum x_i^2 = 28/7 = 4$, $s_Y^2 = (1/7) \sum_{i=1}^{7} \sum y_i^2 = 700/7 = 100$, and $s_{XY} = (1/7) \sum_{i=1}^{7} x_i y_i = 140/7 = 20$. Hence, the covariance matrix of X and Y is

$$\Sigma = \begin{pmatrix} s_Y^2 & s_{XY} \\ s_{XY} & s_Y^2 \end{pmatrix} = \begin{pmatrix} 4 & 20 \\ 20 & 100 \end{pmatrix}.$$

Now, we note that

$$s_{\tilde{X}}^2 = \frac{1}{7}(-3,-2,1,0,1,2,3)\begin{pmatrix} -3 \\ -2 \\ \vdots \\ 3 \end{pmatrix}$$

$$s_{\tilde{Y}}^2 = \frac{1}{7}(-15,-10,-5,0,5,10,15)\begin{pmatrix} -15 \\ -10 \\ \vdots \\ 15 \end{pmatrix}$$

$$s_{\tilde{X}\tilde{Y}} = \frac{1}{7}(-3,-2,1,0,1,2,3)\begin{pmatrix} -15 \\ -10 \\ \vdots \\ 15 \end{pmatrix}.$$

Therefore,

$$\Sigma = \frac{1}{7}\begin{pmatrix} -3 & -2 & \cdots & 3 \\ -15 & -10 & \cdots & 15 \end{pmatrix}\begin{pmatrix} -3 & -15 \\ -2 & -10 \\ \vdots & \vdots \\ 3 & 15 \end{pmatrix}.$$

In general, given a data set A of n d-dimensional mean-adjusted data, the covariance of the d random variables is given by the expression:

$$\Sigma = (1/n)A^T A.$$

Before proceeding with the description of the PCA algorithm, we derive two more useful expressions. Let Y be a mean-adjusted random variable expressed as a linear combination of d mean-adjusted random variables $X_1, X_2, \ldots, X_d$; i.e., $Y = e_1X_1 + e_2X_2 + \cdots + e_dX_d$, where $e_1, e_2, \ldots, e_d$ are weights. Then, using the expression of the variance of the sum of random variables (see Section 2.5), we have

$$\mathrm{Var}(Y) = \mathrm{Var}(e_1X_1 + e_2X_2 + \cdots + e_dX_d)$$

$$= e_1^2\mathrm{Var}(X_1) + e_2^2\mathrm{Var}(X_2) + \cdots + e_d^2\mathrm{Var}(X_d)$$

$$+ \sum_{k=1}^{d}\sum_{l=1}^{d}\mathrm{Cov}(e_kX_k, e_lX_l).$$

Since $\mathrm{Cov}(e_kX_k, e_lX_l) = e_ke_l\mathrm{Cov}(X_k, X_l)$, the above expression can be written as follows:

$$\mathrm{Var}(Y) = \sum_{k=1}^{d}\sum_{l=1}^{d}e_ke_l\sigma_{kl},$$

where $\sigma_{kk} = \mathrm{Var}(X_x)$. Therefore,

$$\mathrm{Var}(Y) = e^T\Sigma e,$$

where $e = (e_1, e_2, \ldots, e_d)^T$. If the covariance matrix can be estimated from a data set A of n d-dimensional mean-adjusted data, then, we have

$$\mathrm{Var}(Y) = (1/n)e^T A^T Ae.$$

Now, let $Y_1 = e_{11}X_1 + e_{12}X_2 + \cdots + e_{1d}X_d$ and $Y_2 = e_{21}X_1 + e_{22}X_2 + \cdots + e_{2d}X_d$ be two mean-adjusted random variables obtained as a linear combination of the mean-adjusted random variables $X_1, X_2, \ldots, X_d$. Then, it can be shown that

$$\mathrm{Cov}(Y_1, Y_2) = \sum_{k=1}^{d}\sum_{l=1}^{d}e_{1k}e_{2l}\sigma_{kl},$$

$$= e_1^T\Sigma e_2,$$

where $e_1 = (e_{11}, e_{12}, \ldots e_{1d})^T$ and $e_2 = (e_{21}, e_{22}, \ldots e_{2d})^T$. As above, if the covariance matrix can be estimated from a data set A of n mean-adjusted d-dimensional data, then

$$\mathrm{Cov}(Y_1, Y_2) = (1/n)e_1^T A^T Ae_2.$$

PCA transforms a mean-adjusted set of d-dimensional data points $x_i = (x_{i1}, x_{i2}, ..., x_{id})$, $i = 1, 2, ..., n$, into a different d-dimensional space using a set of weights e_{ij}, $i = 1, 2, ..., d$, $j = 1, 2, ..., d$. The axes of the new space are the principal components. Let $y_i = (y_{i1}, y_{i2}, ..., y_{id})$, $i = 1, 2, ..., n$, be the transformed data points, which are basically the projection of the original data points onto the new d-dimensional space. The transformed data points are obtained as a linear combination of x_i, $i = 1, 2, ..., n$; i.e.,

$$y_{i1} = e_{11}x_{i1} + e_{12}x_{i2} + \cdots e_{1d}x_{id}$$

$$y_{i2} = e_{21}x_{i1} + e_{22}x_{i2} + \cdots e_{2d}x_{id}$$

$$\vdots$$

$$y_{id} = e_{d1}x_{i1} + e_{d2}x_{i2} + \cdots e_{dd}x_{id}$$

or

$$\begin{pmatrix} y_{i1} \\ y_{i2} \\ \vdots \\ y_{id} \end{pmatrix} = \begin{pmatrix} e_{11} & e_{12} & \cdots & e_{1d} \\ e_{21} & e_{22} & \cdots & e_{2d} \\ \vdots & \vdots & \cdots & \vdots \\ e_{d1} & e_{d2} & \cdots & e_{dd} \end{pmatrix} \begin{pmatrix} x_{i1} \\ x_{i2} \\ \vdots \\ x_{id} \end{pmatrix}.$$

The first set of weights $e_1 = (e_{11}, e_{12}, ..., e_{1d})^T$ is chosen so that the projections onto the first pc has a maximum variance; i.e., let Y_1 be a random variable representing the projections of the data points onto the first pc. Then, we have the following maximization problem:

$$\underset{e_1}{\operatorname{argmax}} \operatorname{Var}(Y_1) = \underset{e_1}{\operatorname{argmax}} \sum_{i=1}^{n} y_{i1}^2$$

$$= \underset{e_1}{\operatorname{argmax}} \sum_{i=1}^{n} \left(e_{11}x_{i1} + e_{12}x_{i2} + \cdots e_{1d}x_{id}e_1x_1 \right)^2.$$

Using the expression derived above for the variance of Y, the maximization problem can be re-written as follows:

$$\underset{e_1}{\operatorname{argmax}} \operatorname{Var}(Y_1) = \underset{e_1}{\operatorname{argmax}} e_1^T (1/n) A^T A e_1$$

or

$$\underset{e_1}{\operatorname{argmax}} \operatorname{Var}(Y_1) = \underset{e_1}{\operatorname{argmax}} e_1^T A^T A e_1$$

since n is a constant. The above maximization is subject to the constraint $e_1^T e_1 = 1$ that guarantees that e_1 is a unit vector.

The second set of weights $e_2 = (e_{21}, e_{22}, \ldots, e_{2d})^T$ is selected so that the projections Y_2 on to the second pc has a maximum variance, i.e.,

$$\underset{e_2}{\operatorname{argmax}} \operatorname{Var}(Y_1) = \underset{e_2}{\operatorname{argmax}} \; e_2^T A^T A e_2$$

subject to

$$e_2^T e_2 = 1,$$

and

$$\operatorname{Cov}(Y_1, Y_2) = 0.$$

The first constraint guarantees that e_1 is a unit vector, and the second one guarantees that the second pc is orthogonal to first pc. Using the above-derived expression for the covariance of two variables, we can re-write the second constraint as follows:

$$\operatorname{Cov}(Y_1, Y_2) = e_1^T (1/n) A^T A e_2 = 0.$$

In general, the set of weights $e_i = (e_{i1}, e_{i2}, \ldots, e_{id})^T$ is selected in the same way; i.e., let Y_i be a random variable representing the projections of the data points onto the ith pc. Then, we have

$$\underset{e_i}{\operatorname{argmax}} \operatorname{Var}(Y_i) = \underset{e_i}{\operatorname{argmax}} \; e_i^T A^T A e_i$$

subject to

$$e_i^T e_i = 1,$$

$$\operatorname{Cov}(Y_1, Y_2) = e_1^T (1/n) A^T A e_2 = 0$$

$$\operatorname{Cov}(Y_2, Y_3) = e_2^T (1/n) A^T A e_3 = 0$$

$$\vdots$$

$$\operatorname{Cov}(Y_{i-1}, Y_i) = e_{i-1}^T (1/n) A^T A e_i = 0.$$

The first constraint guarantees that e_i is a unit vector, and the remaining constraints guarantee that the ith principal component is orthogonal to the $i - 1$ previous principal components.

It turns out that the above maximization is achieved when e_i, $i = 1, 2, \ldots, d$, are set equal to the eigenvectors of the covariance matrix. Let $\lambda_1, \lambda_2, \ldots, \lambda_d$ and $\varepsilon_1, \varepsilon_2, \ldots, \varepsilon_d$, be its eigenvalues and eigenvectors. Then, the first set of weights is assigned the values of the eigenvector with the largest eigenvalue, the second set of weights is assigned the values of the eigenvector with the second largest eigenvalue, and so on. We only retain the first k principal components, where $k < d$, which capture most of the variance, i.e.,

$$\frac{\lambda_1 + \lambda_2 + \cdots + \lambda_k}{\lambda_1 + \lambda_2 + \cdots + \lambda_d} \approx 1;$$

i.e., we reduce the d-dimensional space to a k-dimensional space, where $k < d$.

For example, let us consider a two-dimensional data set. The scatter diagram of the data set is given in Figure 7.3. We note that the range of the X_1 values is from 4.3 to 7.9, and for the X_2 values is from 2 to 4.2. The covariance matrix is

$$\begin{pmatrix} 0.686 & 1.274 \\ 1.274 & 3.113 \end{pmatrix}$$

and the eigenvalues and eigenvectors are as follows:

$$\lambda_1 = 3.659, \; e_1 = (0.394, 0.919)$$

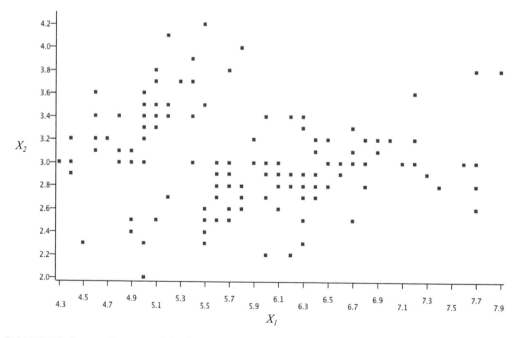

FIGURE 7.3 Scatter diagram of the data set.

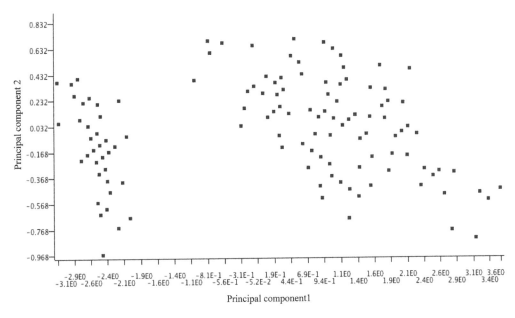

FIGURE 7.4 Scatter diagram of the data set along the two principal components.

$$\lambda_2 = 0.14, \, e_2 = \left(-0.919, 0.394\right).$$

Vector e_1 defines the first principal component (pc1), and e_2 the second one (pc2) since $\lambda_1 > \lambda_2$. The scatter diagram of the transformed data is shown in Figure 7.4. We note that the projected values on pc1 range from −3.1 to 3.6, and on pc2 from −0.968 to 0.73. We see that the range of the projected values on pc1 is larger than the projected values on X_1. Also, the projected values on pc2 has a very small range compared to that on pc1. Thus, we can see clearly the effects of the PCA algorithm. In addition, we have that

$$\frac{\lambda_1}{\lambda_1 + \lambda_2} = \frac{3.659}{3.659 + 0.14} \approx 1.$$

This means that the data set can be reduced to a single dimension defined by pc1.

We note that the transformed data in Figure 7.4 is grouped into two distinct clusters. This may lead us to think that PCA can also be used for clustering data, i.e., grouping data into a number of clusters of similar data. PCA is not useful for clustering data since it maximizes the variance of the projections along a small number of principal components, which may not be necessarily helpful for clustering.

7.3 LINEAR AND MULTIPLE DISCRIMINANT ANALYSIS

Discriminant analysis is a classification technique, whereby an unlabeled data point is assigned a label given a labeled data set. Specifically, we have a data set that is already grouped into different classes, i.e., each data point has been assigned a class id or a label. Discriminant analysis can be used to assign a label to a new unlabeled data point.

In this section, we present the LDA technique which is based on Fisher's discriminant analysis. LDA applies to two-class data. Its generalization to more than two classes is known as multiple discriminant analysis (MDA) or multiclass LDA. However, the term LDA is used often to refer to both techniques. The data set is assumed to be represented by continuous variables and the label is a nominal (or categorical) variable, i.e., it takes a small number of discrete values. In Machine Learning, this technique is typically used only for dimension reduction with a view using a different classification algorithm subsequently.

The main idea of LDA is to find a line so that the projection of the data points onto it are separated per their label. For example, let us consider the two-class data set shown in Figure 7.5, where the data points in the first class are represented by red dots and the data points in the second class by blue dots. Figure 7.5(a) shows an example of a line where the projected data onto it are not separated by class. On the other hand, Figure 7.5(b) shows a line with very good separation between the two classes. LDA calculates a line that maximizes the separability of the two classes. We first present LDA for data with two classes, and then we extend the discussion to more than two classes.

7.3.1 Linear Discriminant Analysis (LDA)

In LDA, we are interested in the length of the projection of a data point x onto a line defined by a unit vector v, as shown in Figure 7.6. It can be shown that the length of the projection is the dot product $v^T x$ (see Section 11.1). The idea behind the LDA method is to find an optimum vector v that provides a maximum separation between the projections of two-class data points onto it.

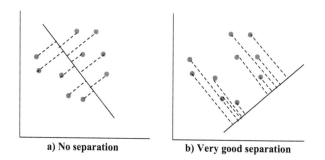

a) No separation   b) Very good separation

FIGURE 7.5 Scatter diagram of data set: (a) No separation; (b) Very good separation.

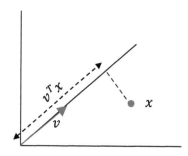

FIGURE 7.6 Length of projection of x.

Let μ_1 and μ_2 be the mean of class 1 and 2 data points, respectively; i.e.

$$\mu_\ell = \frac{1}{n_\ell} \sum_{x \in C_\ell} x, \ell = 1, 2,$$

where ℓ indicates the class and C_ℓ the set of all x that belong to class ℓ. Now, let y be the length of the projection of x onto the projection line. We have $y = v^T x$. Then, the mean of the projections of the class ℓ, $\ell = 1, 2$, data points is

$$\hat{\mu}_\ell = \frac{1}{n_\ell} \sum_{x \in C_\ell} y = \frac{1}{n_\ell} \sum_{x \in C_\ell} v^T x = v^T \left(\frac{1}{n_\ell} \sum_{x \in C_\ell} x \right) = v^T \mu_\ell.$$

Now, if the difference $|\hat{\mu}_1 - \hat{\mu}_2|$ is large, then we can argue that there is good separation of the projected data. In view of this, we could use this difference $|\hat{\mu}_1 - \hat{\mu}_2|$ as the objective function to the problem of determining an optimum projection vector that gives maximum separation between the two classes. However, the problem is that it does not take into account the variance of the projected data within each class. For, if the variance is large, there may be an overlap of the data points between the two classes, even if their respective means are far apart from each other. Fisher suggested maximizing the distance between these two means but normalized by a measure of the variance. Define the quantity:

$$\hat{s}_\ell^2 = \sum_{y \in C_\ell} \left(y - \hat{\mu}_\ell \right)^2, \ell = 1, 2.$$

This is the sum of the squared differences of the projections of the class ℓ data points from their projected class mean. It is in fact the variance of the projected data multiplied by the total number of data points n_ℓ, and it is referred to as the *scatter*. The sum $\hat{s}_1^2 + \hat{s}_2^2$ is referred to as the *within-class scatter*.

The Fisher linear discriminant is the projection vector v that maximizes the criterion:

$$J(v) = \frac{\left| \hat{\mu}_1 - \hat{\mu}_2 \right|^2}{\hat{s}_1^2 + \hat{s}_2^2};$$

i.e., we are looking for a projection line so that the means are as far apart as possible and the projected data within each class are as close as possible.

Before, we proceed to derive the optimum value of v, we recall that the dot product of two vectors $a = (a_1, a_2)^T$ and $b = (b_1, b_2)^T$ is defined as follows: $a \cdot b = a_1 b_1 + a_2 b_2$. It can be easily verified that $a \cdot b = a^T b = b^T a$. Below, for presentation purposes, we assume that the data points are two-dimensional with n_1 and n_2 being the total number of data points in class 1 and 2, respectively, where $n_1 + n_2 = n$.

We first define the quantity:

$$\hat{S}_\ell = \sum_{x \in C_\ell} \left(x - \mu_\ell \right) \left(x - \mu_\ell \right)^T, \ell = 1, 2,$$

which is known as the *scatter matrix* of the original values. In order to understand $\hat{S}_\ell$ better, let us consider the case of $\ell = 1$. Let $x_i = (x_{i1}, x_{i2})^T$ be the ith data point in C_1, $i = 1, 2, \ldots, n_1$, and let $\mu_i = (\mu_{11}, \mu_{12})^T$ be the mean of x_{i1} and x_{i2}, $i = 1, 2, \ldots, n_1$, respectively. Then the above expression can be re-written as follows:

$$\hat{S}_1 = \sum_{i=1}^{n_1} \begin{pmatrix} x_{i1} - \mu_{11} \\ x_{i2} - \mu_{12} \end{pmatrix} \left(x_{i1} - \mu_{11}, x_{i2} - \mu_{12} \right)$$

or,

$$\hat{S}_1 = \sum_{i=1}^{n_1} \begin{pmatrix} \left(x_{i1} - \mu_{11} \right)^2 & \left(x_{i1} - \mu_{11} \right)\left(x_{i2} - \mu_{12} \right) \\ \left(x_{i1} - \mu_{11} \right)\left(x_{i2} - \mu_{12} \right) & \left(x_{i2} - \mu_{12} \right)^2 \end{pmatrix}$$

$$= \begin{pmatrix} \sum_{i=1}^{n_1} \left(x_{i1} - \mu_{11} \right)^2 & \sum_{i=1}^{n_1} \left(x_{i1} - \mu_{11} \right)\left(x_{i2} - \mu_{12} \right) \\ \sum_{i=1}^{n_1} \left(x_{i1} - \mu_{11} \right)\left(x_{i2} - \mu_{12} \right) & \sum_{i=1}^{n_1} \left(x_{i2} - \mu_{12} \right)^2 \end{pmatrix}$$

$$= n_1 \begin{pmatrix} \dfrac{1}{n_1}\sum_{i=1}^{n_1} \left(x_{i1} - \mu_{11} \right)^2 & \dfrac{1}{n_1}\sum_{i=1}^{n_1} \left(x_{i1} - \mu_{11} \right)\left(x_{i2} - \mu_{12} \right) \\ \dfrac{1}{n_1}\sum_{i=1}^{n_1} \left(x_{i1} - \mu_{11} \right)\left(x_{i2} - \mu_{12} \right) & \dfrac{1}{n_1}\sum_{i=1}^{n_1} \left(x_{i2} - \mu_{12} \right)^2 \end{pmatrix}.$$

Now, let X_1 and X_2 be two random variables which represent the data points x_{i1} and x_{i2}, $i = 1, 2, \ldots, n_1$. Then, we observe that the above expression is equal to $n_1\text{Cov}(X_1, X_2)$.

The sum $\hat{S}_W = \hat{S}_1 + \hat{S}_2$ is defined as the *within-class scatter matrix*. Subsequently, we express the within-class scatter of the projected values in terms of the within-class scatter matrix of the original values. We have

$$\hat{s}_\ell^2 = \sum_{y \in C_\ell} (y - \hat{\mu}_\ell)^2 = \sum_{x \in C_\ell} (y - \hat{\mu}_\ell)^T (y - \hat{\mu}_\ell)$$

$$= \sum_{x \in C_\ell} \left(v^T x - v^T \mu_\ell \right)^T \left(v^T x - v^T \mu_\ell \right)$$

$$= \sum_{x \in C_\ell} \left(v^T \left(x - \mu_\ell \right) \right)^T \left(v^T \left(x - \mu_\ell \right) \right).$$

$$= \sum_{x \in C_\ell} \left(\left(x - \mu_\ell\right)^T v \right)^T \left(v^T \left(x - \mu_\ell\right)\right)$$

$$= \sum_{x \in C_\ell} v^T \left(x - \mu_\ell\right)\left(x - \mu_\ell\right)^T v$$

$$= v^T \left(\sum_{x \in C_\ell} \left(x - \mu_\ell\right)\left(x - \mu_\ell\right)^T \right) v$$

$$= v^T \hat{S}_\ell v.$$

Therefore,

$$\hat{s}_1^2 + \hat{s}_2^2 = v^T \hat{S}_1 v + v^T \hat{S}_2 v = v^T \hat{S}_W v.$$

Similarly, the difference between the projected means $\left|\hat{\mu}_1 - \hat{\mu}_2\right|^2 = (\hat{\mu}_1 - \hat{\mu}_2)^2$ can be also expressed in terms of the means of the original data. We have

$$\left(\hat{\mu}_1 - \hat{\mu}_2\right)^2 = \left(v^T \mu_1 - v^T \mu_2\right)^2$$

$$= \left(v^T \mu_1 - v^T \mu_2\right)^T \left(v^T \mu_1 - v^T \mu_2\right)$$

$$= \left(v^T \left(\mu_1 - \mu_2\right)\right)^T \left(v^T \left(\mu_1 - \mu_2\right)\right)$$

$$= \left(\left(\mu_1 - \mu_2\right)^T v\right)^T \left(v^T \left(\mu_1 - \mu_2\right)\right)$$

$$= v^T \left(\mu_1 - \mu_2\right)\left(\mu_1 - \mu_2\right)^T v$$

$$= v^T \hat{S}_B v,$$

where $\hat{S}_B = \left(\mu_1 - \mu_2\right)\left(\mu_1 - \mu_2\right)^T$. $\hat{S}_B$ is referred to as the between-class scatter.
    Using the above-derived expressions, we can write the objective function as follows:

$$J(v) = \frac{v^T \hat{S}_B v}{v^T \hat{S}_W v},$$

where $\hat{S}_B$ and $\hat{S}_W$ are two quantities calculated from the original data. This expression is known as the Rayleigh quotient, and it can be shown that the vector v that maximizes it is given by the expression:

$$v = \hat{S}_W^{-1}\left(\mu_1 - \mu_2\right).$$

This is known as Fisher's discriminant, although it is not a discriminant, but a vector defining the projection line.

For example, let us consider the two-class data set:

Class 1: {(5,1), (2,5), (3,4), (4,7), (5,5)}
Class 2: {(10,11), (7,9), (10,6), (9,8), (11,9)}

Applying LDA gives the projection line given in Figure 7.7, where class 1 is depicted by red squares and class 2 by blue. We see a clear separation of the two classes.

7.3.2 Multiple Discriminant Analysis (MDA)

MDA is a direct generalization of the LDA. It reduces the dimensionality of a data set to $L - 1$ dimensions, where L is the number of classes. Let us consider a data set A of n d-dimensional data points, where each data point belongs to a class ℓ, $\ell = 1, 2, ..., L$, and let C_ℓ, $\ell = 1, 2, ..., L$, be a subset that contains the class ℓ data.

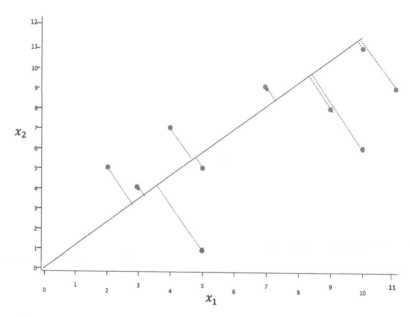

FIGURE 7.7 The LDA projection line.

The single projection line in LDA becomes an $L - 1$ dimensional projection space in MDA, defined by L vectors $v_1, v_2, ..., v_L$. These vectors are grouped into a matrix V, whose ith column is vector v_i. Let $y = (y_1, y_2, ..., y_L)$ be the projection of a data point x onto the projection space. Then, $y_\ell = v_\ell^T x$, $\ell = 1, 2, ..., L$, or $y = V^T x$.

The within-scatter matrix expression generalizes to

$$\hat{S}_W = \sum_{\ell=1}^{L} \hat{S}_\ell,$$

where

$$\hat{S}_\ell = \sum_{x \in C_\ell} \left(x - \mu_\ell \right) \left(x - \mu_\ell \right)^T, \ell = 1, 2, ..., L.$$

$\mu_\ell = (\mu_{\ell 1}, ..., \mu_{\ell d})^T$, where $\mu_{\ell i}$ is the mean of the ith feature in C_ℓ. The between-class scatter becomes

$$\hat{S}_B = \sum_{\ell=1}^{L} n_\ell \left(\mu_\ell - \mu \right) \left(\mu_\ell - \mu_2 \right)^T,$$

where $\mu = (\mu_1, ..., \mu_d)^T$ and μ_i is the super mean of the ith feature.

The objective function to maximize takes the following form:

$$J(V) = \frac{\left| V^T \hat{S}_B V \right|}{\left| V^T \hat{S}_W V \right|}.$$

The optimum projection matrix V is the one whose columns correspond to the eigenvalues of the following generalized eigenvector problem:

$$\hat{S}_B V = \lambda \hat{S}_W V,$$

where λ is the vector of eigenvalues. We note that the projections with a maximum class separation are on the axes that correspond to the largest eigenvalues of $\hat{S}_W^{-1} \hat{S}_B$. Therefore, the projection matrix V is set equal to the first k largest eigenvectors. The set of the projected values Y is then obtained by multiplying A by V, i.e., $Y = AV$.

In summary, LDA reduces the dimensionality of a two-labeled d-dimensional data set, $d \geq 2$, to a single dimension with maximum separation between the two classes. MDA reduces the dimensionality to a number of dimensions up to $L - 1$, depending on the matrix $\hat{S}_W^{-1} \hat{S}_B$. In general, in order for LDA/MDA to work, the discriminatory information should be in the class means and not in the class variance.

EXERCISES

1. Calculate by hand the eigenvectors and eigenvalues of the following matrix:

$$\begin{pmatrix} 2 & 0 \\ -3 & -4 \end{pmatrix}.$$

2. Consider the following data set A:

$$\{(1,2,3), (1.5,3,4.5), (2.5,2.5,5), (3.5,4,7.5), (4.5,3.75,8.25),$$

$$(1.75,1.25,3), (2.5,1.75,4.25), (3.5,1.5,5), (4,2.5,6.5), (5,3,8)\}$$

a. Calculate the covariance matrix and then the eigenvalues and eigenvectors.

b. Determine the number of dimensions to which A can be reduced to.

c. Calculate the projections of the data points in A on the selected principal components.

(Hint: Do not forget to mean adjust the data!)

3. Consider the following data set (x_i, y_i, ℓ_i), $i = 1, \ldots, 10$:

$$A = \{(1,2,1), (1.5,3,1), (2.5,2.5,1), (3.5,4,1), (4.5,3.75,1)$$

$$(1.75,1.25,2), (2.5,1.75,2), (3.5,1.5,2), (4,2.5,2), (5,3,2)\}$$

Color the data per class and plot the scatter diagram of the (x_i, y_i) values. Subsequently, apply the LDA method and plot the scatter diagram of the projected values vs class. Comment on the achieved separation.

REFERENCES

Additional information can be found in the Internet. Note that many authors refer to the LDA and MDA techniques as the LDA technique.

CHAPTER 8

Clustering Techniques

I N THIS CHAPTER, WE ADDRESS THE PROBLEM OF CLUSTERING, I.E., grouping, a data set, which has not been previously clustered, into a number of clusters of similar data points. Clustering a data set is also known as labeling the data set, since each data point is assigned a label, i.e., a cluster id. Clustering a data set is an example of unsupervised learning since the data set is not labeled. In mathematical terms, given a data set D, a clustering provides a partition P of the set into subsets $C_1, C_2, \ldots, C_{1c}$, such that $C_i \subseteq D, i = 1, 2, \ldots, c, C_i \cap C_j = \varnothing$ for any i, j clusters, and $\cup_{i=1}^{c} C_i = D$. Many clustering algorithms have been developed over the years. In this chapter, we examine the following:

- Hierarchical clustering

- k-means clustering

- Fuzzy c-means model

- Gaussian mixture decomposition

- DBSCAN

Another clustering technique that will be examined in Chapter 11 is the support vector machine (SVM). A set of exercises and a clustering project are given at the end of the chapter.

8.1 DISTANCE METRICS

In clustering, the way we determine whether two data points are similar or dissimilar is by measuring their distance. Let $x = (x_1, \ldots, x_d)$ and $y = (y_1, \ldots, y_d)$ be two d-dimensional data points, and let $d(x, y)$ be their distance. There are several ways that can be used to calculate $d(x, y)$, some of which are presented below.

a. The Euclidean Distance
The Euclidean metric is given by the following expression:

$$d_E(x, y) = \sqrt{\sum_{i=1}^{d} (x_i - y_i)^2}$$

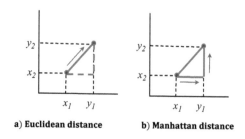

a) Euclidean distance   b) Manhattan distance

FIGURE 8.1 The Euclidean and Manhattan distance: (a) Euclidean distance; (b) Manhattan distance.

For single-dimensional data points x and y, this is equal to their difference, since $d_E(x,y) = \sqrt{(x-y)^2} = x-y$. For two-dimensional data points, this becomes the length of the hypotenuse of the triangle as shown in Figure 8.1(a).

The Euclidean distance is typically notated as $\|x-y\|_2$, or 2-norm, or L_2, where the 2-norm distance $\|x\|_2$ for a d-dimensional vector $x = (x_1, ..., x_d)^T$ is defined as:

$$\|x\|_2 = \sqrt{x_1^2 + \cdots + x_d^2}.$$

We note that

$$\|x\|_2 = \sqrt{x_1^2 + \cdots + x_d^2}$$

$$= \sqrt{x^T x}$$

For simplicity, the Euclidean distance $\|.\|_2$ is notated as $\|.\|$.

b. The Euclidean Squared Distance

This is the same distance as the Euclidean distance, but without the square root. As a result, clustering with this distance is faster than clustering with the regular Euclidean distance.

c. The Manhattan Distance

The Manhattan distance is given by the expression:

$$d_M(x,y) = \sum_{i=1}^{d} |x_i - y_i|$$

For single-dimensional data points x and y, it is equal to their absolute value of their difference. For two-dimensional data points, it is the distance between two points following the shortest path in a Manhattan grid as shown in Figure 8.1(b), i.e., $|x_1 - y_1| + |x_2 - y_2|$. The name Manhattan alludes to the grid layout of most streets in

Manhattan, and the fact that one can only move in an East–West and North–South directions. This distance is the basis of the *taxicab* geometry introduced by Minkowski.

The Manhattan distance is also notated as $\|x - y\|_1$, or 1-norm, or L_1 distance, where the 1-norm $\|x\|_1$ of a d-dimensional vector $x = (x_1, \dots, x_d)^T$ is

$$\|x\|_1 = \sum_{i=1}^{d} |x_i|.$$

d. The Minkowski Distance
   This is a distance that can be considered as a generalization of the Euclidean and Manhattan metrics. It is given by the expression:

$$d_p(x, y) = \sqrt[p]{\sum_{i=1}^{d} |x_i - y_i|^p}$$

The most commonly used values of p are 1, 2, and ∞. The Minkowski distance is also notated as $\|x - y\|_p$ or p-norm, or L_p distance, where for an d-dimensional vector $x = (x_1, \dots, x_d)^T$, $\|x\|_p$ is defined as:

$$\|x\|_p = \sqrt[p]{x_1^p + \cdots + x_d^p}.$$

e. The Chebyshev Distance
   This is the limiting case of the Minkowski distance when $p \to \infty$. We have

$$d_\infty(x, y) = \max\left\{|x_1 - y_1|, \dots, |x_d - y_d|\right\}.$$

f. The Cosine Distance

$$d_C(x, y) = 1 - \frac{x^T}{\|x\|\|y\|},$$

where $\|x\|$ and $\|y\|$ are the 2-norm metrics of x and y.

g. The Mahalanobis Distance
   This is a multi-dimensional generalization of the idea of measuring the distance of a point from its mean in terms of standard deviations. Let us assume that we have a data set of d-dimensional vectors $x = (x_1, \dots, x_d)^T$, and let $\mu = (\mu_1, \dots, \mu_d)^T$ be the vector of the means, where μ_i is the mean of all x_i's on the ith axis. Then, the Mahalanobis distance of the vector x is

$$d_M(x) = \sqrt{(x - \mu)^T \Sigma^{-1} (x - \mu)},$$

where Σ is the covariance matrix of d random variables X_i, $i = 1, 2, \ldots d$, where X_i represents the observations x_i on the ith axis. The covariance matrix is given by the expression:

$$\Sigma = \begin{pmatrix} \sigma_{11}^2 & \sigma_{12}^2 & \ldots & \sigma_{1d}^2 \\ \sigma_{21}^2 & \sigma_{22}^2 & \ldots & \sigma_{2d}^2 \\ \vdots & \vdots & \ldots & \vdots \\ \sigma_{d1}^2 & \sigma_{d2}^2 & \ldots & \sigma_{dd}^2 \end{pmatrix},$$

where $\sigma_{ij}^2 = \mathrm{Cov}\left(X_i, X_j\right)$, and σ_{ii}^2 is the variance of X_i. The matrix is symmetrical about the diagonal since $\sigma_{ij}^2 = \sigma_{ji}^2$.

The Mahalanobis distance between two vectors x and y in the data set is defined as

$$d_M\left(x, y\right) = \sqrt{\left(x - y\right)^T \Sigma^{-1}\left(x - y\right)}.$$

This distance is unitless and scale-invariant and takes into account the correlations of the data set.

If the covariance is the identity matrix, then the Mahalanobis distance becomes the Euclidean distance. This is because the inverse of the identity matrix is the identity matrix, and therefore the term $\left(x - y\right)^T \Sigma^{-1}\left(x - y\right) = \sum_{i=1}^d \left(x_i - y_i\right)^2$.

If the covariance matrix is diagonal (i.e., all the elements off the diagonal are zero), then the inverse of the matrix is

$$\Sigma^{-1} = \begin{pmatrix} 1/\sigma_1^2 & 0 & \ldots & 0 \\ 0 & 1/\sigma_2^2 & \ldots & 0 \\ \vdots & \vdots & \ldots & \vdots \\ 0 & 0 & \ldots & 1/\sigma_d^2 \end{pmatrix},$$

where $\sigma_i^2 = \sigma_{ii}^2$. As a result, we obtain the expression:

$$d_M\left(x, y\right) = \sqrt{\sum_{i=1}^d \frac{\left(x_i - y_i\right)^2}{\sigma_i^2}},$$

which is known as the *standardized Euclidean distance*.

8.2 HIERARCHICAL CLUSTERING

Hierarchical clustering, or hierarchical cluster analysis, is a technique whereby a hierarchy of clusters is built in a bottom-up fashion starting from the initial solution where each data

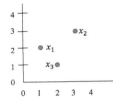

	x_1	x_2	x_3
x_1	0	2.23	1.41
x_2		0	2.23
x_3			0

Euclidean distances

FIGURE 8.2 An example of hierarchical clustering.

point in the set to be clustered belongs to a different cluster. This is known as the *agglomeration hierarchical clustering* technique, where the term *agglomeration* means a bottom-up operation. Alternatively, hierarchical clustering can be used in a top-down fashion, starting from the initial solution where all the points in the data set belong to the same cluster. This is known as the *divisive* approach. Below, we describe the agglomeration hierarchical clustering technique.

Let us consider a data set D consisting of three points x_1, x_2, and x_3 as shown in Figure 8.2. The coordinates of x_1, x_2, and x_3 are (1, 2), (3, 3), and (2, 1), respectively. The Euclidean distances between the three points are given in the table in Figure 8.2. The matrix is symmetric since the distance from x_i to x_j is the same as from x_j to x_i. We note that the distance can be seen as a measure of dissimilarity of two data points; i.e., the higher the distance, the more dissimilar the data points are, and the smaller the distance, the less dissimilar they are.

Using the Euclidean distance, we can determine clusters for a given threshold δ. For instance, in the example given in Figure 8.2, all three data points will be clustered to a single cluster if we set $\delta \geq 2.23$, since the distance between any pair of them is less or equal to 2.23. On the other extreme, if $\delta < 1.41$, then each data point is clustered to a different cluster; i.e., we will get three clusters, each containing a single data point. In general, different values of δ result to different clusters. The problem therefore is to find the value of δ that generates the optimum partition. In the hierarchical clustering technique, a series of partitions is generated by increasing δ from 0 to a large value where all the data points are clustered into a single cluster. Then, the best value of δ is obtained using a heuristic rule (i.e., an intuitive non-theoretical rule).

Hierarchical clustering uses the concept of the *single-linkage* clustering; i.e., a point x_1 joins a cluster C if its distance to the closest point in the cluster is less than δ (see Figure 8.3). Pursuing this further, let us assume that we have two clusters C_1 and C_2 obtained using a threshold value δ, as shown in Figure 8.4. Now, if we increase δ to δ' so that the distance from point x_1 in C_2 to the closest point in C_1 is less or equal to δ', then point x_1 will become part of C_1. Subsequently, due to the single-linkage clustering, all points in C_2 become part of C_1; i.e., the two clusters become a single one C'. Below, we first give an example of hierarchical clustering and then describe the algorithm in detail.

Let us consider the data set $D = \{x_1, x_2, x_3, x_4, x_5, x_6, x_7\}$. The following partitions are obtained as we increase δ from 0 to δ_1 to δ_2 and to δ_3 (see Figure 8.5(a)).

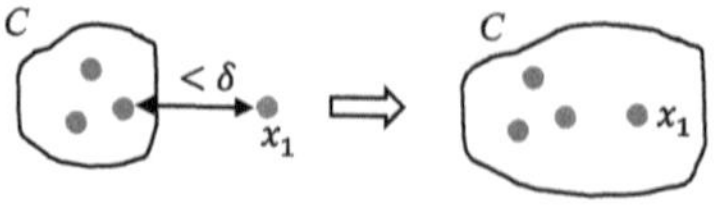

FIGURE 8.3 An example of single-linkage clustering.

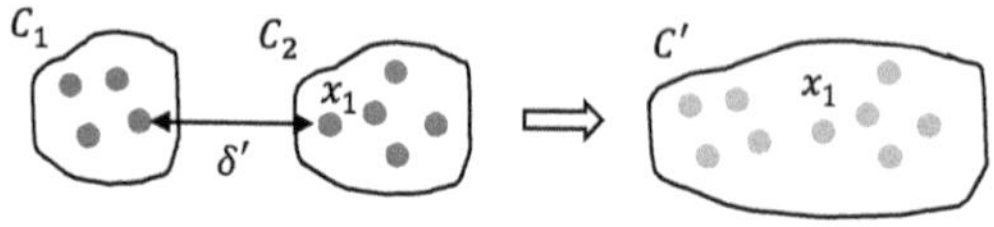

FIGURE 8.4 Joining two clusters as δ increases.

$$\delta = 0: P_0 = \left\{ \{x_1\}, \{x_2\}, \{x_3\}, \{x_4\}, \{x_5\}, \{x_6\}, \{x_7\} \right\}$$

$$\delta = \delta_1: P_{\delta_1} = \left\{ \{x_1, x_3\}, \{x_2\}, \{x_4, x_5, x_6, x_7\} \right\}$$

$$\delta = \delta_2: P_{\delta_2} = \left\{ \{x_1, x_3, x_2\}, \{x_4, x_5, x_6, x_7\} \right\}$$

$$\delta = \delta_3: P_{\delta_3} = \left\{ x_1, x_2, x_3, x_4, x_5, x_6, x_7 \right\}$$

The evolution of the partitions as δ increases is depicted in the diagram shown in Figure 8.5(b). This diagram is known as a *dendrogram* (derived from the Greek dendro which means tree and gramma which means drawing). Clustering software typically display the dendrogram, which can be very dense for large data sets.

We choose an optimum value δ^* such that a slight change in δ should not lead to a completely different partition. One way to do this is to examine the differences between the successive δ values, i.e., $\delta_1 - 0$, $\delta_2 - \delta_1$, and $\delta_3 - \delta_2$, and select the largest one, which in this case is $\delta_3 - \delta_2$. Then, we fix $\delta^* = (\delta_3 + \delta_2)/2$. We note that for this value the best partition is $P_{\delta_2} = \{\{x_1, x_3, x_2\}, \{x_4, x_5, x_6, x_7\}\}$.

8.2.1 The Hierarchical Clustering Algorithm

The input to the algorithm is the data set D, $|D| = n$. The algorithm returns a hierarchically nested partition P_t, $t = 0, 1, ..., n - 1$. The steps are as follows:

1. Let P_0 be the initial partition where each data point belongs to a different cluster.

2. $t = 0$, $\delta_0 = 0$.

3. **while** current partition P_t has more than one cluster, find pair of cluster (C, C') with minimal distance $d(C, C')$. (This is the smallest distance between a point in C and a point in C'.)

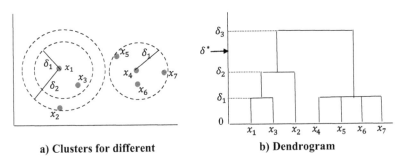

a) Clusters for different   **b) Dendrogram**

FIGURE 8.5 A visual example of the hierarchical clustering algorithm: (a) Clusters for different δ; (b) Dendrogram.

4. $\delta_{t+1} = d(C, C')$

   a. Construct P_{t+1} from P_t by removing C and C' and replacing them with $C \cup C'$.

   b. $t = t + 1$.

5. **end while**

An example of the application of the above algorithm is given in Figure 8.6. Each table in this figure corresponds to one iteration of the algorithm. The highlighted cell corresponds to the minimum distance between two clusters. The dendrogram is given in Figure 8.7. We select the value $\delta^* = (37 + 15)/2$, for which value the best partition is $\{\{x_5, x_6, x_3\}, \{x_1, x_7, x_4\}, \{x_2\}\}$.

In hierarchical clustering, the number of clusters is determined from the dendrogram; i.e., it does not have to be specified in advance as is the case with other clustering techniques described below.

Hierarchical clustering has a complexity of the order of $O(n^2 \log n)$, where n is the number of data points, and therefore it is not suited for large data sets. In addition, the presence of noisy data (i.e., corrupted or erroneous data) tends to chain different clusters into a single one as shown in Figure 8.8. On the left-hand side of this figure we have an example of three different clusters. Now, if we add some noisy data, colored gray, in between these clusters and re-run the algorithm, the original three clusters will be chained into a single one.

8.2.2 Linkage Criteria

There are several other criteria for calculating the distance between two clusters C and C'. Below, we list some of them.

   a. Single-linkage clustering:

$$\min\{d(x, y), x \in C \text{ and } y \in C'\}.$$

This is the method used in the example above. It is the oldest and most well-known method, and it is also known as the minimum or the nearest neighbor method.

	$\{x_1\}$	$\{x_2\}$	$\{x_3\}$	$\{x_4\}$	$\{x_5\}$	$\{x_6\}$	$\{x_7\}$
$\{x_1\}$	0	50	63	17	72	81	12
$\{x_2\}$		0	49	41	42	54	37
$\{x_3\}$			0	52	13	16	61
$\{x_4\}$				0	56	66	15
$\{x_5\}$					0	11	64
$\{x_6\}$						0	74
$\{x_7\}$							0

	$\{x_1\}$	$\{x_2\}$	$\{x_3\}$	$\{x_4\}$	$\{x_5, x_6\}$	$\{x_7\}$
$\{x_1\}$	0	50	63	17	72	12
$\{x_2\}$		0	49	41	42	37
$\{x_3\}$			0	52	13	61
$\{x_4\}$				0	56	15
$\{x_5, x_6\}$					0	64
$\{x_7\}$						0

	$\{x_1, x_7\}$	$\{x_2\}$	$\{x_3\}$	$\{x_4\}$	$\{x_5, x_6\}$
$\{x_1, x_7\}$	0	37	61	15	64
$\{x_2\}$		0	49	41	42
$\{x_3\}$			0	52	13
$\{x_4\}$				0	56
$(x_5, x_6\}$					0

	$\{x_1, x_7\}$	$\{x_2\}$	$\{x_4\}$	$\{x_5, x_6, x_3\}$
$\{x_1, x_7\}$	0	37	15	61
$\{x_2\}$		0	41	42
$\{x_4\}$			0	52
$\{x_5, x_6, x_3\}$				0

	$\{x_1, x_7, x_4\}$	$\{x_2\}$	$\{x_5, x_6, x_3\}$
$\{x_1, x_7, x_4\}$	0	37	52
$\{x_2\}$		0	42
$\{x_5, x_6, x_3\}$			0

	$\{x_1, x_7, x_4, x_2\}$	$\{x_5, x_6, x_3\}$
$\{x_1, x_7, x_4, x_2\}$	0	42
$\{x_5, x_6, x_3\}$		0

FIGURE 8.6 A numerical example of the hierarchical clustering algorithm.

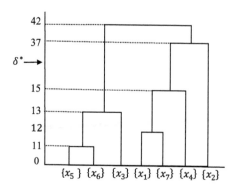

FIGURE 8.7 The dendrogram of the example shown in Figure 8.5.

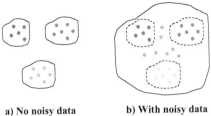

a) No noisy data   **b) With noisy data**

FIGURE 8.8 The hierarchical algorithm is sensitive to noisy data (a) No noisy data; (b) With noisy data.

b. Complete-linkage clustering:

$$\max \left\{ d(x, y), x \in C \text{ and } y \in C' \right\}.$$

This method is also known as the furthest neighbor or maximum method. It defines the distance between two groups as the distance between their two furthest members. This method usually yields clusters that are well separated and compact.

c. Average-linkage clustering:

$$\frac{1}{|C||C'|} \sum_{x \in C} \sum_{y \in C'} d(x, y).$$

It is also called the weighted pair-group method, and it calculates the distance between groups as the average distance of all the pairs of data points between the two clusters, weighted so that the two groups have an equal influence on the final result.

d. Centroid clustering:

$$\| c - c' \|,$$

where c and c' are the centroids of C and C'. This method defines the distance between two groups as the distance between their centroids (the arithmetic mean of all the data points in the cluster.) The method should only be used with Euclidean distances.

e. Weighted pair-group centroid clustering:
This is the weighted version of the centroid clustering. It defines the distance between two groups as the weighted distance between their centroids, where the weights are proportional to the number of individuals in each group. The method should only be used with Euclidean distances.

8.3 THE k-MEANS ALGORITHM

This is a popular algorithm that forms clusters by minimizing the sum of the distance of each data point from its cluster mean. It does not determine the number of clusters as in the hierarchical algorithm. Rather, the number of clusters is provided as input to the algorithm, and subsequently the algorithm determines which data point belongs to each cluster. Specifically, given a set of observations we want to partition the data set into k clusters, so that the sum of the squared distances of the data points from their cluster mean is minimized; i.e.,

$$\underset{P}{\text{argmin}} \sum_{i=1}^{k} \sum_{x \in C_i} \|x - m_i\|^2,$$

where x is a multi-dimensional data point that belongs to cluster C_i, and m_i, $i = 1, 2, \ldots, k$, is the centroid of the ith cluster, i.e., the mean of all the data points in the ith cluster. The inner summation is the sum of the squared Euclidean distances of all points x that belong to the ith cluster C_i from their cluster mean, and the outside summation is the sum over all clusters. We want to find a partition P that minimizes the total sum. The most common algorithm, described below, uses an iterative technique which may converge to a local minimum. (The actual solution of the above optimization problem is NP-hard even for $k = 2$.)

8.3.1 The Algorithm

The k-means algorithm has to be initialized first. Two methods are used: the *Forgy* and the *random partition*. In the Forgy method, we randomly select k observations from the data set and use them as the initial values of the cluster means $m_1^{(1)}$, $m_2^{(1)}$, ..., $m_k^{(1)}$. In the random partition, as the name implies, we partition the data set into k clusters, by allocating each data point to a randomly selected cluster. Then, we calculate the initial values of the cluster means. The iterative steps are as follows:

1. *Assessment step t:* Assign each data point to the cluster j whose mean is the closest, i.e.,

$$C_j^{(t)} = \left\{ x_i : \|x_i - m_j\|^2 \le \|x_i - m_l\|^2, j = 1, 2, \ldots, k \right\}.$$

2. *Update step t + 1:* Calculate the new cluster means, i.e.,

$$m_j^{(t+1)} = \frac{1}{\left|C_j^{(t)}\right|} \sum_{x_i \in C_j^{(t)}} x_i, j = 1, 2, \ldots, k.$$

3. *Stopping rule:* Stop when no more assignments can be done. Otherwise, go to step 1.

As mentioned above, the number of clusters k is specified as input to the algorithm. It is possible that the algorithm may converge to a local minimum, and in view of this, it is a common practice to run the algorithm many times, each time starting with a different initial set of the cluster means $m_1^{(1)}$, $m_2^{(1)}$, ..., $m_k^{(1)}$. The algorithm tends to create spherical clusters of equal size. Other distance metrics than the squared Euclidean distance can be also used depending on the data.

8.3.2 Determining the Number k of Clusters

Various heuristic methods can be used to determine the number of clusters k, one of which is the *elbow method*. This is a simple technique whereby the k-means algorithm is run for different values of k. For each solution, we calculate the sum of squared errors (SSE), which is in fact equal to the double summation we want to minimize; i.e.,

$$SSE = \sum_{j=1}^{k} \sum_{x \in C_j} \left\| x - m_j \right\|^2.$$

Since this is computed in every iteration of the algorithm, the SSE is just the last computed value of the double summation. The SSE values are then plotted against k. If the plot looks like an arm (see Figure 8.9(a)), then the elbow on the arm corresponds to the best value of k. The idea behind it is that the SSE decreases as k increases with SSE becoming zero when k is equal to the number of data points in the data set. This is because in this case, each data point is its own cluster, and therefore there is no error between it and its cluster mean. We want to select a small value of k that has a low SSE, past of which the reduction of the SSE diminishes as k increases. This point is represented by the elbow. For the case in Figure 8.8(a), we will set $k = 3$.

The SSE plot may not always have a clear elbow. Instead, it may be a smooth curve as shown in Figure 8.8(b). In this case, it is difficult to select a good value for k. When this situation arises, one can employ a different method for determining k, such as computing *silhouette scores* and using Akaike's information criterion (AIC) and the Bayesian information criterion (BIC). In addition, the lack of an elbow may be indicative that the data does not cluster, and in this case, one has to pursue a different approach. Below, we describe the silhouette scores and give the AIC and BIC expressions.

a. Silhouette Scores

The technique provides a graphical representation of how well each data point lies within its cluster. The silhouette scores range from −1 to +1, where a high score indicates that the data point is well matched to its own cluster and poorly matched to neighboring clusters. If most

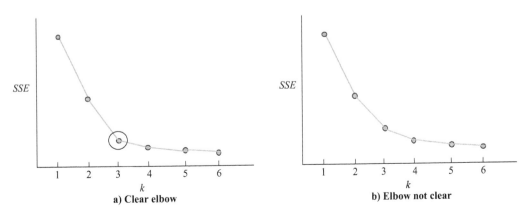

FIGURE 8.9 The elbow method: (a) Clear elbow; (b) Elbow not clear.

data points have a high value, then the cluster partition is deemed appropriate. If many points have a low or negative score, then the partition may have too many or too few clusters.

Assume that the data has been clustered using the k-means algorithm, or any other clustering algorithm, into k clusters. Let $\bar{d}_C(x)$ be the mean distance of a data point x from all the other data points in the same cluster C, measured using any distance metric. This measure of dissimilarity can be interpreted as how well x is assigned to its cluster C, since the smaller the value, the less is the dissimilarity and the better is the assignment.

Also, let $\bar{d}_c(x)$ be the average distance of x from all the data points in cluster c, where $c \neq C$ and $c = 1, 2, ..., k$, and let C' be the cluster among all the clusters excluding C with the minimum average distance $\bar{d}_{C'}(x)$ to x. This cluster is referred to as the *neighboring cluster* because it is the best next fit for the data point x. The silhouette score for x is calculated as follows:

$$s(x) = \frac{\bar{d}_{C'}(x) - \bar{d}_C(x)}{\max\{\bar{d}_C(x), \bar{d}_{C'}(x)\}}$$

which can be written as

$$s(x) = \begin{cases} 1 - \bar{d}_C(x)/\bar{d}_{C'}(x), & \text{if } \bar{d}_C(x) < \bar{d}_{C'}(x) \\ 0, & \text{if } \bar{d}_C(x) = \bar{d}_{C'}(x) \\ \bar{d}_{C'}(x)/\bar{d}_C(x) - 1, & \text{if } \bar{d}_C(x) > \bar{d}_{C'}(x) \end{cases}.$$

From the above-given definition, we see that $-1 \leq s(x) \leq 1$. In addition to the individual scores, we can also calculate the average score $\bar{s}_C$ of all the scores of the data in a cluster C, and the super average score $\bar{\bar{s}}$ computed as the average of the $\bar{s}_C$ values for all clusters. $\bar{s}_C$ is a measure of how tightly grouped are the data in the cluster, and $\bar{\bar{s}}$ is a measure of how well the data has been clustered.

The silhouette scores for each data point in each cluster are plotted out for visual inspection. An example of a silhouette plot is given in Figure 8.10. The scores are given on the x axis, and the data points grouped by cluster on the y axis. A silhouette score near 1 indicates that the data point is far away from the neighboring clusters. A value of 0 indicates that the data point is on or very close to the decision boundary between two neighboring clusters, and a negative value indicates that the data point has been assigned to the wrong cluster. We see that two points in the green cluster have negative scores which means that they have been assigned to the wrong cluster.

Silhouette plots can be obtained for different values of k and subsequently select the best value by inspection. In addition, the average score $\bar{\bar{s}}$ can be plotted for the different values of k and select the value that corresponds to the highest $\bar{\bar{s}}$ value. An example of this plot is given in Figure 8.11. We see that $k = 3$ gives the highest average score value, suggesting that assuming three clusters is the best solution.

We note that the above plots and other similar graphical tools are useful in helping us to determine the number of clusters, and they can be used with any clustering algorithm.

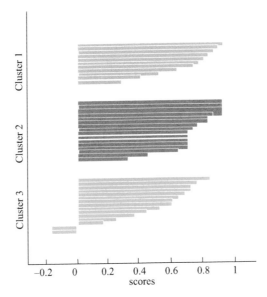

FIGURE 8.10 A silhouette plot.

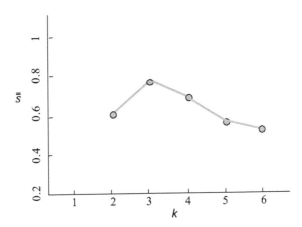

FIGURE 8.11 Average score $\bar{s}$ vs k.

It is imperative that one should not just rely on the output of clustering algorithms and graphical tools, but one should also try to understand the data and try to explain intuitively why the data clusters in a particular way.

b. Akaike's Information Criterion (AIC)

The number of clusters can be determined using Akaike's information criterion (AIC), which is defined as follows for this case:

$$AIC = -2\log(L) + 2k,$$

where L is the minimum achieved SSE value for a given k number of clusters. Another criterion that can be used is the Bayesian information criterion (BIC):

$$BIC = -2\log(L) + k\log(n),$$

where n is the total number of data points. We select the value of k that minimizes both AIC and BIC.

8.4 THE FUZZY c-MEANS ALGORITHM

In the k-means algorithm, a data point is a member of only one cluster (hard clustering). In the fuzzy c-means algorithm, a data point can potentially belong to more than one cluster (soft clustering). As in the k-means algorithm, the number of clusters c is not determined by the algorithm, but it is provided as input to the algorithm. The partitioning of a data set into c fuzzy clusters is similar to the k-means algorithm. For a data set D we want to obtain a partition P that minimizes the weighted sum:

$$\underset{P}{\operatorname{argmin}} \sum_{x \in D} \sum_{i=1}^{c} w_i(x)^m \|x - c_i\|^2,$$

where $w_i(x)$ is a probability indicating the degree of membership of x in the ith cluster, m is a number greater than 1 that controls the level of fuzziness, and c_i, $i = 1, 2, ..., c$, is the centroid of the ith cluster, given by the expression:

$$c_i = \frac{\sum_{x \in D} w_i(x)^m x}{\sum_{x \in D} w_i(x)^m}.$$

The probabilities $w_i(x)$, $x \in D$, $i = 1, 2, ..., c$, where $0 \le w_i(x) \le 1$, are calculated as follows:

$$w_i(x) = \frac{1}{\sum_{j=1}^{c} \left(\frac{\|x - c_i\|}{\|x - c_j\|} \right)^{\frac{2}{m-1}}}.$$

The fuzzy factor m determines the level of cluster fuzziness. A large m results in small values $w_i(x)$, and hence fuzzier clusters. In the absence of additional knowledge about the data, m is set equal to 2.

Given a number of clusters c, the algorithm returns a set of cluster centroids c_i, $i = 1, 2, ..., c$, and a matrix $W = (w_i(x))$ of all the membership probabilities for each data point, where each row in W contains the probabilities of a data point. The algorithm iterates between updating the cluster centroids and the membership probabilities, as follows:

1. *Initialization:* Select an initial value $W^{(0)}$ of the probability matrix W.

2. *Step k:* Calculate the cluster centroids using the expression above for c_i, $i = 1, 2, ..., c$, with the previously updated matrix $W^{(k-1)}$. Update $W^{(k-1)}$ using the expression above for $w_i(x)$ and set it equal to $W^{(k)}$.

3. *Stopping rule:* If $|W^{(k-1)} - W^{(k)}| \leq \varepsilon$, then stop. Else repeat step 2.

8.5 THE GAUSSIAN MIXTURE DECOMPOSITION

In this method, we assume that the data comes from a mixture of k normal (Gaussian) distributions. For instance, let us assume that we have one-dimensional data and let X be a random variable representing the data. If the data comes from a mixture of k normal distributions, then its pdf $f_X(x)$ is given by the expression:

$$f_X(x) = \sum_{i=1}^{k} p_i f_i\left(x; \mu_i, \sigma_i^2\right),$$

where $f_i\left(x; \mu_i, \sigma_i^2\right)$ is the probability that x comes from the ith normal distribution, i.e.,

$$f_i\left(x; \mu_i, \sigma_i^2\right) = \frac{1}{\sigma_i \sqrt{2\pi}} e^{-\frac{1}{2}\frac{(x-\mu_i)^2}{\sigma_i^2}}$$

and p_i is the probability that a data point comes from the ith normal distribution. A good way to understand what is a mixture of normal distributions (or any other distributions) is to see how one can generate random variates from it. For this, we first draw a random number r to determine which normal distribution to select. If $r \leq p_1$, then we select the first normal distribution; if $p_1 < r \leq p_1 + p_2$, then we select the second normal distribution, and so on. Once the normal distribution has been selected, then we draw a random number x from this distribution, which becomes one of our data points. (See Section 3.3.2 for how to generate stochastic variates.)

The same expression for the pdf $f_X(x)$ of the mixture of normal distributions also holds for the case of multivariate data, whereby each data point is d-dimensional, only the expression for $f_i\left(x; \mu_i, \sigma_i^2\right)$ is given by the multivariate normal pdf:

$$f_i\left(x; \mu, \Sigma\right) = \frac{1}{\left(\left(2\pi\right)^d |\Sigma|\right)^{1/2}} \exp\left\{-\frac{1}{2}\left(x - \mu\right)^T \Sigma^{-1}\left(x - \mu\right)\right\},$$

where μ is the vector of means, i.e., $\mu = (\mu_1, \mu_2, ..., \mu_d)^T$, and Σ is the covariance of the d random variables associated with the d-dimensional data.

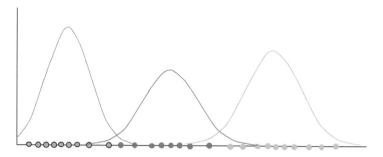

FIGURE 8.12 Data from a mixture of three normal distributions.

For example, let us consider the set of one-dimensional blue, green, and red data points shown in Figure 8.12. We generate this data set using the three different normal distributions blue, red, and green and a set of probabilities p_1, p_2, p_3, where $p_1 + p_2 + p_3 = 1$, as described above. The data points have a pdf defined by the mixture of these three normal distributions. Now, let us forget how these data points were generated and their color, and let us see how we can fit a mixture of normal distributions to the data without knowing the value of k. This can be done by maximizing the likelihood, as described below. We vary k and select the best value that has the highest likelihood. Each estimated normal distribution is then associated with a different cluster, and each data point x is classified to cluster i for which the probability $f_i\left(x; \mu_i, \sigma_i^2\right)$ that it came from that distribution is the highest.

More formally, let θ be the set of all the parameters to be estimated. For instance, in the above example

$$\theta = \left\{ p_1, \mu_1, \sigma_1^2, p_2, \mu_2, \sigma_2^2, p_3, \mu_3, \sigma_3^2 \right\}.$$

The probability that a data point x comes from a mixture of normal distributions with parameters θ is written as $f_X(x; \theta)$. (This is to indicate the dependency of the pdf on θ.) The likelihood L is then defined as

$$L = \prod_{x \in D} f_X\left(x; \theta\right),$$

where D is the data set; i.e., the likelihood is the probability $f_X(x; \theta)$ that the first data point comes from this mixture of normal distributions times the probability that the second data point also comes from the same distribution, and so on. We want to find the set of parameters θ that maximize this total probability; i.e.,

$$\underset{\theta}{\mathrm{argmax}} \prod_{x \in D} f_X\left(x; \theta\right).$$

This is typically done using the expectation-maximization (EM) algorithm (see [1]).

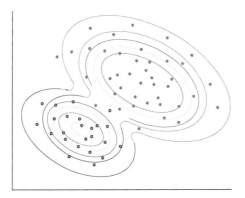

FIGURE 8.13 An example with two-dimensional data and $k = 2$.

An example of fitting a mixture of two normal distributions to two-dimensional data is given in Figure 8.13. The points along each contour have the same pdf value, and the contours are colored from light indicating high pdf values to darker indicating low pdf values. We can distinguish two normal distributions from the set of contours.

As mentioned above, each normal distribution is associated with a different cluster. To determine which cluster a data point belongs to, compute the probability $f_i\left(x; \mu_i, \sigma_i^2\right)$ that the data point x comes from the ith normal distribution for all i's. Then, select the normal distribution for which this probability is maximum. The same procedure is followed to classify a new data point. The number of normal distributions k is determined by trying out different values and select the one that has the maximum likelihood.

8.6 THE DBSCAN ALGORITHM

As we have seen, single-linkage clustering is sensitive to outliers which can act as bridges between clusters forcing them to merge. The clusters may contain a lot of data points, but it only takes a few bridging data points to make them merge. The density-based spatial clustering of applications with noise (DBSCAN) algorithm avoids this problem by ensuring that the density of the bridging data points is the same as that of the clusters. An example of bridging data is shown in Figure 8.14. If we used the hierarchical clustering algorithm, the two clusters would have been merged into a single one. The DBSCAN algorithm will only merge them if the density of the bridging data is the same as that of the clusters. If not, then merging does not take place, and the data is simply considered as outliers.

The DBSAN algorithm uses two thresholds, namely, the radius ε of a hyperspherical neighborhood, and the minimum number *MinPts* of the data points that have to be within an ε-neighborhood. Specifically, the density at a data point x is measured as the number of data points that lie within a hyperspherical neighborhood with radius ε:

$$N_c\left(x\right) = \left\{y \in D: \|x - y\| \le \varepsilon\right\},$$

where D is the data set of the points to be clustered. If the number of data points in $N_c(x)$ is greater or equal to *MinPts*, then the data point x is called a *core object*, and all data points

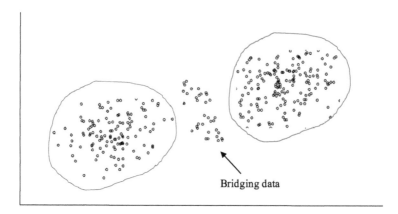

FIGURE 8.14 Density of bridging data.

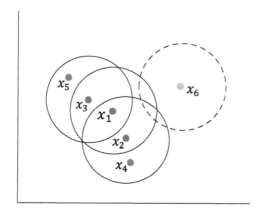

FIGURE 8.15 An example of core objects, not core objects, and noise.

within the ε-neighborhood centered on the core object x belong to the same cluster. Otherwise stated, a data point x is a core object if the ε-neighborhood centered around x contains at least *MinPts* points.

The DBSCN algorithm classifies the data points as *core objects*, *not core objects*, and *noise*. An example of these three types of data points is shown in Figure 8.15, where *MinPts* = 3. Points x_1, x_2, and x_3 in red are core objects, points x_4 and x_5 in blue are not core objects, and point x_6 in green is noise, since its ε-neighborhood only has one point.

The DBSCAN algorithm
Input: Data set D, ε, *MinPts*.
Output: The algorithm labels each data point with a cluster id. Data points not clustered are labeled as noise.
Initialization
Label all data as UNCLASSIFIED, and set cluster counter $cid = 0$.
Main routine
For all $x \in D$
        if x is labelled UNCLASSIFIED
                if expand $(D, x, cid, \varepsilon, MinPts)$
                        increment cluster counter $cid = cid + 1$

Algorithm expand (D, x, cid, ε, MinPts)
Input: Data set D, x, currently unused *cid* number, ε, *MinPts*
Output: Returns true if a new cluster has been found.
    Let $S = \{y \in D: \|x - y\| \leq \varepsilon\}$
    If $|S| < MinPts$, relabel x as NOISE and return false
    For all $x' \in S$, relabel x' with current *cid*
    remove x from S
    For all $x' \in S$ calculate $T = \{y \in D: \|x' - y\| \leq \varepsilon\}$
      If $|T| \geq MinPts$
          for all $y \in T$
               if y does not belong to a cluster
                    if y is labeled UNCLASSIFIED
                        insert y into S
                        relabel y with cluster *cid*
    remove x' from S
    return true

Below, we give an example of the DBSCAN algorithm for $D = \{x_1, x_2, ..., x_7\}$ and *MinPts* = 3. The layout of the data along with their clusters, indicated by circles, is shown in Figure 8.16.

Select x_1 from D: $S = \{x_1, x_2, x_3\}$ with $|S| = 3$. Therefore, label x_1, x_2, x_3 with *cid* = 1, and remove x_1 from S, i.e., $S = \{x_2, x_3\}$.

1. Select x_3 from S: $T = \{x_1, x_3\}$. Since $|T| < 3$, we remove x_3 from S, i.e., $S = \{x_2\}$

2. Select x_2 from S: $T = \{x_1, x_2, x_4, x_5\}$ and $|T| \geq 3$.

    a. x_4 does not belong to a cluster and it is marked UNCLASSIFIED. Therefore, it is added to S, i.e., $S = \{x_2, x_4\}$, and it is labeled with *cid* = 1.

    b. x_5 does not belong to a cluster and it is marked UNCLASSIFIED. Therefore, it is added to S, i.e., $S = \{x_2, x_4, x_5\}$, and it is labeled with *cid* = 1.

    c. Remove x_2 from S, i.e., $S = \{x_4, x_5\}$

    d. Go back to line 2 using the new values in S.

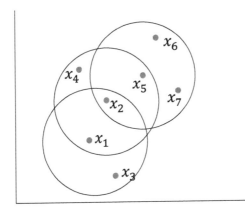

FIGURE 8.16 A DBSCAN example.

Select another value from D and repeat steps 1 and 2. Stop when there are no data points labeled UNCLASSIFIED.

8.6.1 Determining *MinPts* and ε

Increasing the value of *MinPts* makes the clusters denser and more robust. If *MinPts* = 1, then each point is a cluster, which is a trivial solution. If *MinPts* = 2, then the algorithm behaves like the hierarchical clustering algorithm. Consequently, *MinPts* should be set to value equal or greater than 3. In general, for multi-dimensional data, the rule of thumb is that *MinPts* should be greater or equal to the number of dimensions of the data points plus 1.

The radius ε is obtained after fixing *MinPts*, by constructing the following graph. For each data point x_i determine the smallest radius r_i such that the number of data points within the circle with the radius r_i is *MinPts*. If a data point is close to many other data points, then its radius r_i will be small. If the data point is an outlier, then its radius r_i is large. Plot the r_i values for each data point after they are sorted out in a descending order. The value of ε is fixed at the elbow of the graph, as shown in Figure 8.17.

8.6.2 Advantages and Disadvantages of DBSCAN

A main advantage of DBSCAN is that the number of clusters is not specified in advance, as in the k-means and the Gaussian mixture decomposition algorithms described above. Also, the algorithm is robust to outliers, and it can find non-linearly separable clusters, an example of which is shown in Figure 8.18.

A main disadvantage of DBSCAN is that it cannot cluster data with different density. Also, choosing ε may be difficult if the data is not well understood.

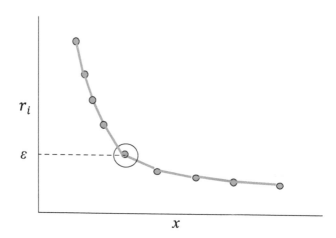

FIGURE 8.17 A plot of the radius r_i values.

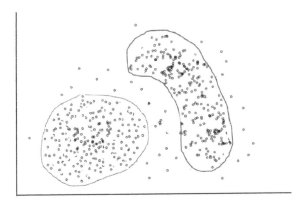

FIGURE 8.18 Non-linearly separable clusters.

TABLE 8.1   Data for exercises 1 and 2

	$\{x_1\}$	$\{x_2\}$	$\{x_3\}$	$\{x_4\}$	$\{x_5\}$	$\{x_6\}$
$\{x_1\}$	0	25	30	28	40	30
$\{x_2\}$		0	26	19	18	28
$\{x_3\}$			0	30	10	15
$\{x_4\}$				0	15	35
$\{x_5\}$					0	9
$\{x_6\}$						0

EXERCISES

1. Apply the hierarchical algorithm with single linkage clustering to the data set $D = \{x_1, x_2, x_3, x_4, x_5\}$ with the Euclidean distance matrix given in Table 8.1. Give each iteration in a different table. Draw the dendrogram, chose δ^*, and give the best partition.

2. Same as in problem 1 but use the complete linkage clustering. Compare the results with the single linkage clustering.

3. Consider the following set of data: (2,2), (3,6), (4,3),(4,8), (7,6), (7,3), (8,4), (8,7), (9,2). Apply the k-means algorithm with $k = 3$. For the initial solution use the cluster center points (2, 2), (7, 6), and (8, 7). Show the clusters after each iteration.

4. Consider the data set given in Figure 8.19. Identify which data points are core objects, not core objects, and noise when using the DBSCAN algorithm with $MinPts = 3$ and for the drawn ε-neighborhoods.

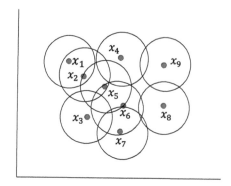

FIGURE 8.19 Data set for Exercise 4.

CLUSTERING PROJECT

The objective of this project is to use various clustering algorithms to determine the best number of clusters in a data set. For this analysis, you can use any software, such as, Python, R, MatLab, and SAS, or a combination of functions from different software.

Data Set Generation

Generate three groups of 3-tuple data, each consisting of 200–300 data points, as follows.

First, generate the number of observations n_i in the ith group, $i = 1, 2, 3$, by drawing a number from the uniform distribution $U[200, 300]$. Subsequently, for each group i generate the 3-tuple data (x_{j1}, x_{j2}, x_{j3}), $j = 1, 2, ..., n_i$. The first data point of a 3-tuple is sampled from a normal distribution, the second data point from an exponential distribution, and the third from a two-stage Erlang distribution (see Section 3.3.2, Chapter 3, for how to generate stochastic variates). The parameters of these distributions depend on the group and they are generated as follows:

For group 1, the parameters of the normal distribution μ and σ^2 are randomly chosen from the uniform distribution $U(10, 20)$ and $U(10, 20)$, respectively. For group 2, μ and σ^2 are randomly chosen from the uniform distribution $U(30, 40)$ and $U(10, 20)$, respectively, and for group 3, μ and σ^2 are randomly chosen from the uniform distribution $U(50, 60)$ and $U(10, 20)$, respectively. For group 1, 2, and 3, the parameter λ of the exponential distribution is randomly chosen from the uniform distribution $U(10, 20)$, $U(40, 50)$, and $U(80, 100)$, respectively. Finally, for group i, $i = 1, 2, 3$, the parameter λ of the Erlang distribution is set equal to $i50$. For each group, first generate the parameters of the normal, exponential, and Erlang distributions. Subsequently, generate the 3-tuples as described above. You can generate different data set by varying the parameters so that to make the three groups come closer or spread out more. You can also generate more than three groups following the same instructions.

Interleave the data, so that the data points within each group are not contiguous. Now, forget (!) how the data set was generated and carry out the following tasks.

Task 1 Hierarchical Clustering

Apply the hierarchical algorithm to the data set to determine the number of clusters.

1.1. Plot the dendrogram.

1.2. Determine the number of clusters.

1.3. Color the data according to their cluster, and do a three-dimensional scatter diagram. Rotate the diagram to identify the clusters.

Task 2 k-Means Algorithm

Apply the k-means algorithm to the data set to determine the number of clusters.

2.1. Apply the algorithm for several values of k starting with $k = 2$.

2.2. Use the elbow method and silhouette scores to determine the best value of k.

2.3. For the best k value, color the data according to their cluster, and do a three-dimensional scatter diagram. Rotate the diagram to visually identify the clusters.

Task 3 DBSCAN Clustering

Apply the DBSCAN algorithm to the data set to determine the number of clusters.

3.1. For $MinPts = 3$, use the elbow method to determine the best values of ε. Run the DBSCAN algorithm for the best value of ε and $MinPts = 3$. Color the data according to their cluster, and do a three-dimensional scatter diagram. Rotate the diagram to visually identify the clusters.

3.2. Repeat the above step for $MinPts = 4,5,6$ (use more values if necessary).

Task 4 Conclusions

For each task, write up your results and conclusions. It is very important that you provide enough results to support your conclusions. Subsequently, compare your results from the three tasks and determine which clustering algorithm gives the best results.

REFERENCES

1. C. Bishop. *Pattern Recognition and Machine Learning*, Springer, 2006.

In addition, numerous references can be found in the Internet on all the clustering topics discussed in this chapter.

CHAPTER 9

Classification Techniques

I N THIS CHAPTER, WE CONSIDER THE PROBLEM OF CLASSIFYING AN unlabeled data point, also known as a *query*, given a set of labeled data points, also known as *examples*. For instance, let us assume that we monitor sensors scattered in a geographical area that produce data that is location dependent. Each data point collected by a sensor is labeled by the sensor id. Now, let us assume that a data point becomes available but for some reason the sensor id is corrupted. This missing information can be restored by using a *classifier* that is a classification algorithm, which will enable us to decide its sensor id. Classification algorithms can also be used for regression.

Classification is an example of supervised learning since we deal with labeled data sets, whereas clustering is an example of unsupervised learning. There are many classification algorithms, such as, the k-nearest neighbor, naive Bayes classifier, decision trees, artificial neural networks, support vector machines, and hidden Markov models. In this chapter, we describe the first three methods. The remaining three are described later in this book in separate chapters. A set of exercises and a classification project is given at the end of the chapter.

9.1 THE k-NEAREST NEIGHBOR (K-NN) METHOD

This method is based on a simple idea, and it can be used for both classification and regression. When used for classification, the labeled data set D is in the form $(x_{i1}, x_{i2}, \ldots, x_{id}, \ell_i)$, $i = 1, 2, \ldots, n$, where $n = |D|$, x_{ij} is the value of the jth feature, and ℓ_i is the label associated with the data point. The label is a nominal variable. A new unlabeled data point is assigned to the label that is the most common among its k nearest neighbors. If $k = 1$, then the data point is simply assigned to the label of the nearest neighbor. When used for regression, the data is in the form $(x_{i1}, x_{i2}, \ldots, x_{id}, y_i)$, $i = 1, 2, \ldots, n$, where y_i is the value of the dependent variable, which is a continuous variable. A new data point is assigned the average of the dependent values of its k nearest neighbors.

We demonstrate the k-NN classification algorithm using the simple example in Figure 9.1. As can be seen, we have a set of data points (x, y) which are labeled + or −. Now, we want to assign a + or a − label to a new data point, indicated by the red dot. The new data point is classified by dissimilarity as measured by a distance metric; i.e., the shorter the distance the less the dissimilarity. We observe that the new data point is closer to a + data point.

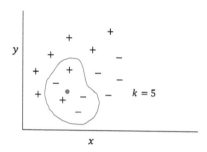

FIGURE 9.1 A k-NN classification example.

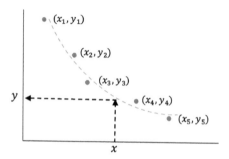

FIGURE 9.2 A k-NN regression example.

Consequently, we predict that the label of the new data point is a +. This is known as 1-nearest neighbor. For $k = 2$, it is closer to a + and also to a – data point. In this case we cannot make a decision, unless we chose to break the tie by randomly selecting one of the two labels. For $k = 3$, the nearest data points are two labeled – and one labeled +. Consequently, by simple majority we predict the label of the new data point to be a –. For $k = 5$, the closest points are those inside the area marked by the red line. There are two + and three – data points, and hence we classify the new data point as a –. Obviously, the choice of k affects the decision.

Now, let us see how the k-NN technique can be used in regression. Let us consider the set of data points (x_i, y_i), $i = 1, 2, 3, 4, 5$, shown in Figure 9.2. If we use a polynomial regression, we will fit a curve such as the blue dotted line shown in Figure 9.2. Then, given a new data point x its y value is obtained by substituting x into the fitted expression, as shown in Figure 9.2. The k-NN technique is a non-parametric method that uses a local approach to estimate the value y of the dependent variable. For $k = 1$, it is set equal to the y value of the nearest data point, i.e., $y = y_4$ since x is closer to x_4. For $k = 2$, y is calculated by averaging the two y values of the points that are the nearest to x, i.e., $y = (y_3 + y_4)/2$. For any k value, y is calculated by averaging the k y values of the points nearest to x.

An example of a 1-NN regression is given in Figure 9.3(b). For comparison purposes, the simple linear regression line is also given in Figure 9.3(a). Smoother regression lines can be generated using kernels as discussed in Section 9.1.2.

In contrast to standard regression, k-NN regression cannot be used to predict a y value outside the range of the existing data points. For instance, in the example in Figure 9.3(b),

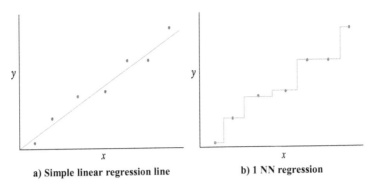

a) Simple linear regression line   b) 1 NN regression

FIGURE 9.3 An 1-NN regression example (a) Simple linear regression line; (b) 1-NN regression.

it is difficult to predict the y value of a point outside the given range of the x values. This is not the case when using the regression line in Figure 9.3(a), assuming of course that the general structure of the data will not change.

Finally, we note that the Euclidean distance metric is typically used to identify the k nearest data points, but other metrics such as those described in Section 8.1 can also be used.

9.1.1 Selection of k

The best value of k is obtained by experimentation; i.e., we measure the classification error for different values of k, and then select the k value that yields the lowest error. Fixing the parameters of a model through experimentation is very common, see for instance Section 6.2.4 that describes how to fix the parameters of the exponential moving average model. Unlike a time series, in this case the order of the classification data is not important. In view of this, we can select the best k value using the following simple scheme. We split randomly the data set D into two disjointed subsets D_1 and D_2, where D_1 contains the bulk of the data, say 80% of the data. This can be done, using a readily available function which assigns each data point to D_1 or D_2 using pseudo-random numbers; that is, for the first data point a pseudo-random number r is drawn and if $r \leq 0.8$ then the data point is assigned to D_1. Otherwise, it is assigned to D_2. This process is repeated for all the data points in D. Data set D_1 becomes the training data set and D_2 the validation data set; i.e., for a given k, we use D_1 to classify each data point in D_2 and since the data points in D_2 are already labeled, we can calculate the number of misclassifications. We repeat this procedure for different values of k and then select k that gives the lowest classification error.

We note that we can also use stratified sampling to generate D_1 and D_2. Let us assume that the data set D consists of 80% of data with label 1 and 20% of data with label 2. Then, stratified sampling ensures that each data set D_1 and D_2 contain approximately the same percentage of the two classes. If the original data set D contains a large number of data points from each class, then stratified sampling amounts to random sampling. However, if one class is not well represented, then stratified sampling can be useful.

A better method than the one described above, which is commonly used and it also allows for more randomization in the calculation of the misclassification error, is the cross-validation method. In this case, the data set D is randomly split into v equal non-overlapping subsets called folds, where v is typically set to 10. The random split is done in the same way as described above, by classifying each data point in D to one of the folds using pseudo-random numbers. For instance, let us assume that $v = 10$. Then, for the first data point, we sample a pseudo-random number r, and if $r \leq 0.1$ then the data point is classified to the first fold. If $0.1 < r \leq 0.2$, then it is classified to the second fold, and so on. We carry out the same procedure for all the other data points. Stratified sampling can also be used to generate the folds.

Subsequently, we fix k to a value, and then we proceed in a similar fashion as in the simple case described above. We combine the first $v - 1$ folds into a single data set, call it D_1, and then we classify each data point in the vth fold using D_1; i.e., D_1 becomes the training data set and the vth fold the validation data set. Since the data points in the vth fold are labeled, we can calculate the number of misclassifications. Subsequently, we construct a new training data set D_1 by combining folds 1, 2, ..., $v - 2$ and v. Then, we classify the data points in the $(v - 1)$th fold using D_1, and count the number of misclassifications. We proceed in this way until we use all the folds as a validation data set and calculate the total number of misclassifications. We repeat this procedure for different values of k and then select the one that yields the lowest number of misclassifications.

A popular rule of thumb for selecting an initial value of k is to set it equal to the square root of the number of data in the data set.

9.1.2 Using Kernels with the k-NN Method

In the k-NN method, the k nearest data points used in the calculation of the label ℓ of a new data point x have the same weight. However, it makes sense that data points nearer x should have more weight than those further away. This can be done by assigning a weight to each data point, which is a function of its distance to x.

A simple method is to set the weight equal to the inverse of the distance $d(x, x_i)$ of x_i to x, i.e., $w_i = 1/d(x, x_i)$. We observe that if x_i coincides with x, the inverse of the distance is infinite, and therefore the label assigned to x is that of x_i. Typically, the weights used in the k-NN method are *kernels*. These are functions of the distance $d(x, x_i)$, and they should not be confused with the kernels used in the support vector machine method described in Chapter 11.

Let $k(d)$ be a kernel, where d is the distance between a data point x_i and the new data point x. The following properties apply: $k(d) \geq 0$, $k(d)$ reaches its maximum when $d = 0$, and $k(d)$ decreases as the absolute value of d increases. Also, it is desirable that a kernel is symmetric, i.e., $k(d) = k(-d)$, and it is also a pdf, i.e., $\int k(u)du = 1$.

Kernels are defined over a range of data points, known as the *support*. If a kernel has infinite support, then all the data points in the data set are used in the decision of the label of a new data point x. If it has a finite support, then only the points within a neighborhood around x with a radius h are used. A popular kernel with infinite support is the *Gaussian kernel:*

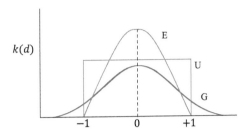

FIGURE. 9.4 Gaussian (G), uniform (U), and Epanechnikov (E) kernels.

$$k(d) = \frac{1}{\sqrt{2\pi}\sigma} \exp\left\{-\frac{d^2}{2\sigma^2}\right\},$$

where σ (not to be confused with the standard deviation) is a parameter that defines how quickly the weights decay as a function of the distance from x.

Finite support kernels are based on a neighborhood defined around the new data point x with a radius h. Within the neighborhood, the distance d of a data point x_i to the new data point x is normalized by setting it equal to $d = (x - x_i)/h$, which means that $|d| \leq 1$. The value d for data points outside the neighborhood is set to zero. The following are some well-known kernels:

- *Uniform kernel:* $k(d) = \dfrac{1}{2}$
- *Triangular kernel:* $k(d) = \dfrac{1}{2}(1-d)$
- *Epanechnikov kernel:* $k(d) = \dfrac{3}{4}(1-d^2)$
- *Quartic* or *biweihgt kernel:* $k(d) = \dfrac{15}{16}(1-d^2)^2$
- *Triweight kernel:* $k(d) = \dfrac{35}{32}(1-d^2)^3$
- *Cosine kernel:* $k(d) = \dfrac{\pi}{4}\cos\left(\dfrac{\pi}{2}d\right)$.

Figure 9.4 gives the plots for the Gaussian, uniform, and Epanechnikov kernels.

Let us now consider the use of kernels in the k-NN classification method. This extension is known as the *weighted k-NN* (*wk-NN*) method. Let $D' = \{(x_i, \ell_i), i = 1, 2, ..., k\}$ be the set of the k nearest data points to x, and let w_i be the weight of the ith data point in D'. Then,

$$\ell = \underset{v}{\operatorname{argmax}} \sum_{(x_i, \ell_i) \in D'} w_i I(v = \ell_i),$$

where $I(v = \ell_i) = 1$ if $v = \ell_i$, and 0 otherwise; i.e., x is assigned the label of the data points whose sum of weights is the highest.

The weight w_i is calculated using a kernel. For the Gaussian kernel, we set $w_i = \left(1/\sqrt{2\pi}\sigma\right)\exp\left\{-d(x, x_i)^2/2\sigma^2\right\}$. The value of σ is selected together with the value for

k by constructing a grid of possible values of (k, σ) and then use cross-validation to find the best combination. Grid searching is used in many cases, see for instance Section 11.7.

For a finite support kernel, we set the radius h to the distance $d(x, x_{[k]})$, where $x_{[k]}$ is the k nearest data point to x. For example, for the Epanechnikov kernel, we set $w_i = (3/4)(1 - d^2)$, where $d = (x - x_i)/d(x, x_{[k]})$.

Now, let us consider the use of kernels in the k-NN regression method. This extension is known as *kernel regression*, and it gives to a smoother fitted line than the one obtained using the plain k-NN method. The data points used are not the k nearest ones, but for a finite support kernel, they come from a neighborhood around x with a radius h. Let $\mathcal{N}(x,h)$ be the set of all the data points (x_i, y_i) that are in the neighborhood. Then, the dependent value y of a new data point x_i is calculated using the Nadaraya–Watson estimator:

$$ y = \frac{\sum_{(x_i,y_i)\in\mathcal{N}(x,h)} k(d) y_i}{\sum_{(x_i,y_i)\in\mathcal{N}(x,h)} k(d)}, $$

where $d = (x - x_i)/h$. For the infinite support Gaussian kernel, all the data points in the data set are used, and in this case the summation in the above expression extends over the entire data set. For further information, see [1].

The use of finite support kernels requires to select a value for the radius h of the neighborhood. This can be done using cross-validation as described in the previous section.

9.1.3 Curse of Dimensionality

This is a term to indicate the various problems that arise when dealing with data with high-dimensionality, which typically do not occur with data with lower dimensions. These problems arise in many areas such as numerical analysis, sampling, combinatorics, Machine Learning, data mining, and databases. The common theme of these problems is that when the data dimensionality increases, the volume of the space within which the data lie increases as well so that the data becomes sparse. For example, let us assume that we have 1000 one-dimensional data points uniformly distributed within a unit interval. Now, let us change this data to two-dimensional data also uniformly distributed in the equivalent space which is now a unit square. Intuitively, we can see that the two-dimensional data is not as densely distributed in the unit square as the one-dimensional data in the unit interval. Likewise, if we assume that the data is now three-dimensional within a unit cube.

This sparsity of the data may be problematic for various algorithms. In the k-NN case, the radius of the circle centered on a data point x that encompasses the k nearest neighbors becomes larger and larger as the number of dimensions increase. In the Euclidean space, for very large dimensions all pairs of data points are at about the same distance.

There are different ways of reducing the impact of dimensionality, such as increase the number of training data points. A typical rule of thumb is that there should be at least five

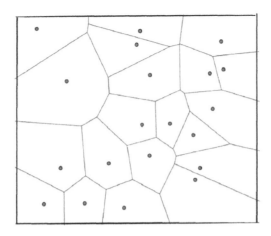

FIGURE 9.5 An example of a Voronoi diagram.

training data points for each dimension. Another approach is to use a dimensionality reduction technique, such as, principal component analysis (PCA) and linear discriminant analysis (LDA) described in Chapter 7, prior to applying the k-NN method.

9.1.4 Voronoi Diagrams

These diagrams were named after G. Voronoi, and they are also known as Voronoi tessellations, Vornoi decompositions, Voronoi partitions, or Dirchlet tessellations. (A tessellation of a flat surface is the tiling of the surface using one or more geometric shapes, called tiles, with no overlaps and no gaps.)

Given a number of points $\{p_1, p_2, \ldots, p_n\}$, the space around these points is divided into convex polygons, known as *cells*, one around each point p_i, so that all points within the ith polygon are closer to p_i than to the other points. The data points on the border line between the ith and jth polygons are equidistant from p_i and p_j. An example of a Voronoi diagram is given in Figure 9.5.

In one dimension, all the data points lie on a single line, and the polygon for a point p_i is a segment which extends half the distance to the point on the left and half the distance to the point on the right. In two dimensions, these polygons are constructed from lines that bisect and are perpendicular to the lines that connect p_i to all its neighbors. An example of this is given in Figure 9.6. The polygon for p_1 is defined by the solid lines. Different metrics can be used, resulting to different size polygons.

The Voronoi diagram provides the complete solution for the 1-NN classification. In this case, points $\{p_1, p_2, \ldots, p_n\}$ are the actual data points of the data set D used for classification. A new data point that falls within the cell associated with p_i is closer to p_i rather than to the other data points. Therefore, for 1-NN classification, its label is the same as that of p_i.

To use the Voronoi diagram for 1-NN, we need to be able to determine which cell a new data point falls in. This is a special case of the point location problem, a fundamental topic of computational geometry. Higher order Voronoi diagrams are used for k-NN classification with $k > 1$. In general, Voronoi diagrams are computationally intensive.

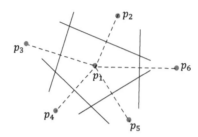

FIGURE 9.6 Polygon construction.

9.1.5 Advantages and Disadvantages of the *k*-NN Method

It is conceptually a simple method and it can be applied to data from any distribution. Also, it is a good classification method if the training data is very large. On the other hand, choosing *k* and an appropriate distance metric may be difficult, and also noise can be problematic. It requires a lot of CPU and memory and as a result it cannot handle more than a few dozen features.

9.2 THE NAIVE BAYES CLASSIFIER

This is a family of simple probabilistic classifiers based on Bayes' theorem (or Bayes' law or Bayes' rule). Bayes' theorem is based on the conditional probability $p(A|B)$ that event *A* will occur conditioned on the occurrence of event *B*:

$$p(A|B) = \frac{p(A,B)}{p(B)}.$$

This expression can be re-written as $p(A, B) = p(A|B)p(B)$. Since, $p(A, B) = p(B, A)$, we also have that $p(B, A) = p(B|A)p(A)$. Consequently, we have

$$p(A|B) = \frac{p(A,B)}{p(B)} = \frac{p(B|A)p(A)}{p(B)}.$$

or,

$$p(A|B) = \frac{p(B|A)}{p(B)} p(A).$$

Probability $p(A)$ is the unconditional probability that event *A* occurs and is known as the *prior probability*. Probability $p(A|B)$ is the probability that *A* occurs given that we know that *B* has occurred, and it is known as the *posterior probability*. The ratio $p(B|A)/p(B)$ is known as the *likelihood ratio*. In Bayesian terms, the above expression can be interpreted as updating our knowledge of the probability that *A* occurs in the light that *B* has occurred, or otherwise stated, $p(A|B)$ is an update of the prior probability $p(A)$ using additional information from event *B*.

Bayes' theorem is used in statistical inference. It is a very popular approach and it is known as *Bayesian inference*. It has applications in many different areas including classification, described below.

9.2.1 The Simple Bayes Classifier

Let us consider a labeled data set of one-dimensional nominal data points; i.e., they take values from a small set of discrete numbers. Now, let x be a new data point whose label we want to determine. The label ℓ of x can be determined as the one that maximizes the probability $p(\ell|x)$ for all labels ℓ, where $p(\ell|x)$ is the probability that given x its label is ℓ. The maximization problem can be written as follows:

$$label(x) = \underset{\ell}{argmax}\, p(\ell|x).$$

Since

$$p(\ell|x) = \frac{p(x|\ell)}{p(x)} p(\ell),$$

we have

$$label(x) = \underset{\ell}{argmax}\, \frac{p(x|\ell)}{p(x)} p(\ell).$$

We note that $p(x)$ is independent of ℓ, and therefore the maximization problem can be reduced to the following:

$$label(x) = \underset{\ell}{argmax}\, p(x|\ell) p(\ell).$$

For example, let us consider two sensors located in two different locations a and b. The data points collected from the sensors, given in Table 9.1, consist of the observation produced by

TABLE 9.1   An example of the simple Bayes classifier

Observation	Location
1	a
2	b
1	b
1	b
3	a
4	b
5	b
6	a

a sensor and a label that indicates the sensor's location. The sensors produce integer obser-vations between 1 and 6. A new observation $x = 1$ becomes available but the sensor's loca-tion is missing. Using the above approach, we can determine the most likely location that it came from as follows.

From Table 9.1, we can calculate the prior probabilities $p(\ell = a)$ and $p(\ell = b)$ are equal to 3/8 and 5/8, respectively. Also, since for location a we have three observations, 1, 3, and 6, and the conditional probability of getting an observation equal to 1 given that it came from location a is 1/3. Likewise, $p(x = 1|\ell = b) = 2/5$. Consequently,

$$p\big(x = 1|\ell = a\big)p\big(\ell = a\big) = \frac{3}{8}\frac{1}{3} = \frac{1}{8}$$

$$p\big(x = 1|\ell = b\big)p\big(\ell = b\big) = \frac{2}{5}\frac{5}{8} = \frac{2}{8}$$

We conclude that the observation most likely came from location b.

9.2.2 The Naive Bayes Classifier

So far, we considered one-dimensional nominal data. In this section, we present an exten-sion of the above approach to d-dimensional nominal data. As mentioned before, each dimension of the data set is referred to as a feature or an attribute, because it represents an aspect of the system under study. Let D be the labeled data set, and let $x = (x_1, x_2, …, x_d)$ be a new data point that we want to classify. As above, we determine the label ℓ of x as the one for which the probability $p(\ell|x)$ for all labels ℓ is maximum; i.e.:

$$label\big(x\big) = \underset{\ell}{argmax}\ p\big(x|\ell\big)p\big(\ell\big).$$

Now, we recall that $p(x|\ell)p(\ell) = p(x, \ell)$, where

$$p\big(x,\ \ell\big) = p\big(x_1,\ x_2,…,x_d,\ \ell\big)$$

$$= p\big(x_1|x_2,…,x_d,\ell\big)p\big(x_2,…,x_d,\ell\big)$$

$$= p\big(x_1|x_2,…,x_d,\ell\big)p\big(x_2|x_3,…,x_d,\ell\big)p\big(x_3,…,x_d,\ell\big)$$

$$\vdots$$

$$= p\big(x_1|x_2,…,x_d,\ell\big)p\big(x_2|x_3,…,x_d,\ell\big)…p\big(x_{d-1}|x_d,\ell\big)p\big(x_d|\ell\big)p\big(\ell\big).$$

Now we make the assumption that each feature is independent of the other features. This is an unrealistic assumption since features are typically correlated, but it has the advantage that it simplifies the above expression. Assuming independence, for any ith feature, we have that

$$p\left(x_i | x_{i+1}, \ldots, x_d, \ell\right) = p\left(x_i | \ell\right),$$

and consequently

$$p\left(x, \ell\right) = \prod_{i=1}^{d} p\left(x_i | \ell\right).$$

Thus, the maximization problem can be re-written as follows:

$$label\left(x\right) = \underset{\ell}{argmax}\, p\left(\ell\right) \prod_{i=1}^{d} p\left(x_i | \ell\right).$$

This classification scheme is known as the *naive Bayes classifier*.

Let n_ℓ be the total number of data points in D with label ℓ, and $n_{x_i, \ell}$ be the total number of data points labeled ℓ for which the ith feature is equal to x_i. Then we have

$$p\left(\ell\right) = \frac{n_\ell}{|D|}$$

$$p\left(x_i | \ell\right) = \frac{n_{x_i, \ell}}{n_\ell},$$

where $|D|$ is the total number of data points in D. We assign the label ℓ to x for which the following term is maximum:

$$p\left(\ell\right) \prod_{i=1}^{d} p\left(x_i | \ell\right) = \frac{n_\ell}{|D|} \prod_{i=1}^{d} \frac{n_{x_i, \ell}}{n_\ell};$$

i.e., the maximization problem can be stated as follows:

$$label\left(x\right) = \underset{\ell}{argmax}\, \frac{n_\ell}{|D|} \prod_{i=1}^{d} \frac{n_{x_i, \ell}}{n_\ell}.$$

It is possible that we may not have any observations for a given value x_i of the ith feature and a label ℓ; i.e., $n_{xi, \ell} = 0$, and consequently $p(x_i | \ell) = 0$, which will zero all the terms that contain $p(x_i | \ell)$. In order to avoid this, $p(\ell)$ and $p(x_i | \ell)$ are calculated by adding a constant γ as follows:

$$p(\ell) = \frac{\gamma + n_\ell}{\gamma |L| + |D|}$$

$$p(x_i|\ell) = \frac{\gamma + n_{x_i,\ell}}{\gamma |L| + n_\ell},$$

where L is the set of all labels and $|L|$ is the total number of labels. This is known as the Laplace correction. For $\gamma = 0$, we obtain the simple estimates given above. For small data sets or small number of data points with the same label, it is better to set $\gamma > 0$. Common choices for γ are $\gamma = 1$, $1/2$, and $\gamma = \gamma_0/$(number of features) $\cdot |L|$, where γ_0 is a user specified number.

Let us consider the example of the two sensors given in Section 9.2.1, but now let us assume that each sensor produces a 2-tuple as opposed to a single observation; i.e., we have a data set with two features. For simplicity, we assume that each feature takes two values 1 and 2. The labelled data set is given in Table 9.2. The prior probabilities $p(\ell = a)$ and $p(\ell = b)$ are equal to 3/8 and 5/8, respectively, and $n_a = 3$ and $n_b = 5$. Now, let us consider a new data point $x = (1, 2)$. Then we have (assuming $\gamma = 0$).

$$p(\ell = a)\left(p(x_1 = 1|\ell = a)p(x_2 = 2|\ell = a)\right) = \frac{3}{8}\frac{1}{3}\frac{2}{3} = \frac{1}{12}$$

$$p(\ell = b)\left(p(x_1 = 1|\ell = b)p(x_2 = 2|\ell = b)\right) = \frac{5}{8}\frac{4}{5}\frac{2}{5} = \frac{1}{5}.$$

Since $1/5 > 1/12$, we conclude that the observation came from location b.

9.2.3 The Gaussian Naive Bayes Classifier

So far, we assumed that the features are nominal; i.e., they take values from a small set of discrete numbers. In this section, we consider the case where some of the features are continuous variables rather than nominal.

TABLE 9.2   An example of the naive Bayes classifier

Feature 1	Feature 2	Location
1	1	a
2	1	b
1	2	b
1	2	b
2	2	a
1	1	b
1	1	b
2	2	a

In the construction of the classifier, it is typically assumed that the continuous values of the ith feature for a given label ℓ have a Gaussian distribution with mean $\mu_{i,\ell}$ and variance $\sigma_{i,\ell}^2$, where $\mu_{i,\ell}$ is the average of all the values of the ith feature which are associated with label ℓ and $\sigma_{i,\ell}^2$ is their variance. The conditional probability $f(x_i|\ell)$ is given by the expression:

$$f(x_i|\ell) = \frac{1}{\sqrt{2\pi}\sigma_{i,\ell}} \exp\left\{ -\frac{\left(x_i - \mu_{i,\ell}\right)^2}{2\sigma_{i,\ell}^2} \right\}.$$

(We use the notation $f(.)$ to describe the equivalent of the probability $p(x_i|\ell)$ for continuous features.)

Likewise, the joint distribution of all the continuous features associated with label ℓ is assumed to follow a multi-normal distribution. Let us assume that we have m continuous features, where $m \leq d$, and $d - m$ nominal features. For presentation purposes, and without loss of generality, we assume that the first m features are continuous and the remaining $d - m$ are nominal. Then, the joint multi-normal distribution of the m features that take the value $x' = (x_1, x_2, \ldots, x_m)$ is

$$f(x'|\ell) = \frac{1}{(2\pi)^{\frac{m}{2}}\sqrt{|\Sigma|}} \exp\left\{ -\frac{1}{2}\left(x' - \mu_{m,\ell}\right)^T \Sigma^{-1}\left(x' - \mu_{m,\ell}\right) \right\},$$

where $\mu_{m,\ell} = (\mu_{1,\ell}, \ldots, \mu_{m,\ell})$ is a vector of all the means for the m continuous features, and Σ is the covariance matrix of the m features with $|\Sigma|$ being its determinant. Assuming independence of features, as in the naive Bayes classifier, the above joint distribution can be simply written out as a product of the individual distributions, i.e.,

$$f(x'|\ell) = f(x_1|\ell)f(x_2|\ell)\ldots f(x_m|\ell),$$

but obviously it is more accurate to use the joint distribution. Now, let $x = (x_1, x_2, \ldots, x_d)$ be a new data point that needs to be labelled. Then, as before, the maximization problem can be stated as follows:

$$label(x) = \underset{\ell}{argmax}\, p(\ell)f(x'|\ell)\prod_{i=m+1}^{d} p(x_i|\ell).$$

This is known as the *full Bayes classifier*. The maximization problems associated with the Bayes classifier can be solved using maximum likelihood estimation (MLE) (see Appendix B).

An alternative, and quite popular way, to dealing with a continuous feature is to discretize it. There are several approaches for doing this. A simple way is to introduce a threshold and divide the values into two sets, one for the values below the threshold and the other one above it. For instance, if the feature is temperature, we can classify it into low and high. We can also use multiple thresholds to create more subsets, each associated with a different label. For instance, using two thresholds, we can classify temperature into low, medium, and high.

9.2.4 Advantages and Disadvantages

Despite the simplicity of the Bayes' classifier and the independence assumption of features, it performs quite well on many classification problems. It tends to produce inaccurate probabilities, but it often assigns the maximum probability to the correct label. In general, it performs well in the case of nominal features.

9.2.5 The k-NN Method Using Bayes' Theorem

In Section 9.1, we presented the k-NN classification method, where the label of a new data point is determined by simple majority of the labels of the k nearest data points. In this section, we present a different method for determining the label based on Bayes' theorem. Let us consider a labeled data set consisting of 20 data points labeled red and of 40 data points labeled green. Now let us assume that a new data point x becomes available, indicated by the black dot in Figure 9.7. The circle drawn around it indicates the $k = 8$ data points that are the closest to the new data point. Using the majority rule, we conclude that the new data point has a green label, since there are three red and six green data points within the circle.

Now let us classify the new data point by applying Bayes' theorem. For this, we want to calculate the following two probabilities:

$$p(red|x) = p(x|red)p(red)$$

$$p(green|x) = p(x|green)p(green).$$

The priors $p(red)$ and $p(green)$ are calculated using the entire data set. We have $p(red) = 20/60 = 1/3$, and $p(green) = 40/60 = 2/3$. (The latter probability can also be

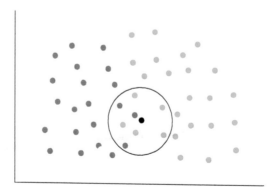

FIGURE 9.7 An example of the k-NN method using the Bayes theorem.

computed as the compliment of $p(red)$, since $p(green) = 1 - p(red)$.) Counting the points inside the circle, we see that we have three red and five green data points. Therefore, $p(x|red) = 3/8$ and $p(x|green) = 5/8$. Consequently,

$$p(red|x) = \frac{3}{8}\frac{1}{3} = \frac{1}{8}$$

$$p(green|x) = \frac{5}{8}\frac{2}{3} = \frac{10}{24}.$$

We note that label green has the largest probability and therefore we assign the green label to x. This is the same conclusion using the simple majority rule. Let us now switch the number of red and green data points in the data set, i.e., let us assume that we have 40 red data points and 20 green data points. Then, we have

$$p(red|x) = \frac{3}{8}\frac{2}{3} = \frac{1}{4}$$

$$p(green|x) = \frac{5}{8}\frac{1}{3} = \frac{5}{24}.$$

Since $1/4 > 5/24$, we conclude that the label of the new data point is red! This is a good example of why one should not rely on a single classification method, but rather use different ones. Above all, one should always develop an intuitive understanding of the data and the results obtained.

9.3 DECISION TREES

Decision trees are simple to understand and interpret since they mirror the way humans make decisions. There are many algorithms for constructing a decision tree from a labeled data set, such as, CART (classification and regression tree), ID3 (iteration dichotomizer 3), C4.5 (successor of ID3), and CHAID (chi-squared automatic interaction detector).

Decision trees are constructed using d-dimensional labeled data, where a feature may be a continuous or a nominal variable. The label is known as the *target variable*. There are two types of decision trees: regression trees and classification trees. In a regression tree, the label is continuous, and in a classification tree it is nominal.

A decision tree consists of a root, decision nodes, and terminal nodes, as shown in Figure 9.8. The decision nodes are the internal nodes of the tree, and the terminal nodes are the leaves of the tree. For simplicity, we shall refer to a decision node as a node and to a terminal node as a leaf. At each node, a number of branches (splits) is generated based on the values of a single feature. The leaves of the tree contain the value of the label to be assigned. When we want to classify a new data point, we follow the branches of the tree according to the values of the features of the new data point and eventually we end up in one of the leaves, which provides the label.

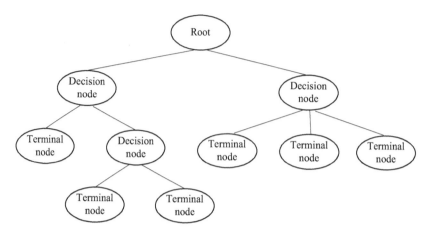

FIGURE 9.8 An example of a decision tree.

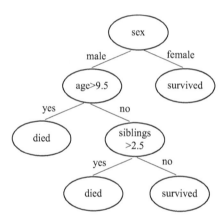

FIGURE 9.9 The decision tree for the Titanic.

A classic example of a decision tree is to decide whether a passenger on the Titanic survived or not. The decision tree is shown in Figure 9.9. The features are {sex, age, number of siblings} and the label takes the values {died, survived}.

A decision tree is constructed top-down, by splitting a feature at each node. Multiway or binary splits can be used. For instance, a continuous variable may be split in multiple ways so that to create a child node for each percentile, such as, 10%, 20%, 30%, etc. Likewise, a nominal variable may be split in multiple ways so that to create multiple child nodes, up to one per nominal value. Typically, binary splits are preferred over multiway splits, since multiway splits can be achieved by a series of binary splits. Also, multiway splits fragment the data quickly leaving insufficient data for subsequent splits.

A decision tree creates a partition of the data set into high-dimensional rectangles, where each rectangle is associated with a label. An example of a decision tree and the associated data set partition is given in Figure 9.10. The data set D consists of two features, X_1 and X_2, which are split as shown in Figure 9.10(a). The resulting leaves are nodes 3, 4, 5, 7,

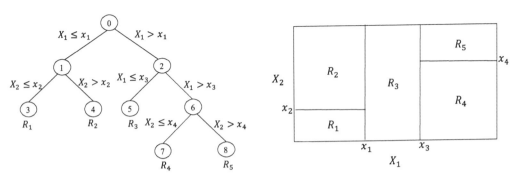

9.10a: The decision tree **9.10b: The data set partition**

FIGURE 9.10 An example of data set partition (a) The decision tree; (b) The data set partition.

and 8, and each leaf i is associated with a partition R_i which is a subset of the data set D as shown in Figure 9.10(b). Each partition R_i is associated with a label. Following the branches of the tree, a new data point will end up in one of the rectangles and will be assigned the label associated with the rectangle.

We now proceed to examine the regression and classification trees in detail.

9.3.1 Regression Trees

A linear multivariable regression is given by the expression

$$Y = a_0 + a_1 X_1 + a_2 X_2 + \cdots + a_d X_d + \varepsilon,$$

where a_0 is the intercept, a_i is the coefficient of the ith random variable X_i, and the random error $\varepsilon \sim N(0, \sigma^2)$ (see Section 5.2). We recall that variables X_i, $i = 1, 2, \ldots, d$ are the independent variables and Y is the dependent variable.

The underlying assumption of this model is that every independent variable contributes separately in an additive manner independently of what the other independent variables are doing. Interactions between independent variables can be represented by introducing additional terms, but the number of coefficients to be estimated becomes very large.

A regression tree provides an alternative model to multivariable regression since it allows interactions between the independent variables. It partitions the data set into high-dimensional rectangles, and the label associated with each rectangle is the average of all the values of the dependent variable in the rectangle. For instance, let $(y_i, x_{i1}, \ldots, x_{id})$, $i = 1, 2, \ldots, n_R$ be the set of all the data in a rectangle R, where $n_R = |R|$. Then, the label associated with R is the average $\bar{y}_R$ of all the y_i, $i = 1, 2, \ldots, n_R$. values. That is, $\bar{y}_R = (1/n_R) \sum_{i=1}^{n_R} y_i$.

A regression tree is grown using binary splits, as in Figure 9.10(a). At each node, we determine the feature to split and how to split it, by calculating the error at the resulting two daughter nodes. The split with the lowest error is selected. The error in a node is computed as the sum of the squared errors; i.e., let $(y_i, x_{i1}, \ldots, x_{id})$, $i = 1, 2, \ldots, n_t$, be the set of data points in node t. Then,

$$E = \sum_{i=1}^{n_t} \left(y_i - \bar{y}_t \right)^2,$$

where $\bar{y}_t$ is the mean value of all the values of the dependent variable in the node.

The above method for growing a regression tree is locally optimal, since we only make an optimal decision at each node. In view of this, we need an additional method so that to determine when to stop growing a tree. There are two techniques for growing a tree, *pre-pruning* and *post-pruning*. In pre-pruning the construction of a tree stops using various stopping criteria. For instance, a node is not split further if the decrease in the error across all the leaves is less than a threshold or when one of its daughter nodes contains less than a pre-determined number of data points. In post-pruning, a decision tree is first grown using loose stopping rules, and then it is pruned by removing parts of it. This topic is discussed in detail in Section 9.3.3.

9.3.2 Classification Trees

Classification trees are used when the label is nominal. They are similar to regression trees except in the way a node is split. In a classification tree, a leaf is assigned a label determined by majority vote; i.e., if a leaf contains data points with different labels, the label that occurs the most is the one assigned to the leaf.

As in a regression tree, a classification tree is grown by splitting a node using a single feature. Different metrics are used to determine which feature to split at each node, such as, the misclassification error, the *information gain* and *Gini impurity*. The decision as to which feature to split is locally optimal, and therefore pre-pruning and post-pruning, described in Section 9.3.3 is used to determine the size of a tree.

The misclassification error is the percent of misclassified data points in a child. The information gain and Gini impurity are preferable metrics over the misclassification error, because they are differentiable and therefore more amenable to numerical optimization. These two metrics are described below.

a. Information Gain

The information gain metric is used in ID3 and C4.5 algorithms, and it is based on the entropy. Let X be a discrete random variable, and let $p_i = p(X = i)$, $i = 1, 2, .., n$, be its probability distribution. Then the entropy of X is defined as follows:

$$H(X) = -\sum_{i=1}^{n} p_i \log_2 \left(p_i \right).$$

Different values for the logarithmic base can be used other than 2, such as e and 10.

Let us now assume that we have selected feature X to split a node. Let S be the data set associated with a child corresponding to $X = i$. Let p_ℓ^i, be the percentage of the data points in S that are labeled ℓ, $\ell = 1, 2, ..., L$, where L is the total number of labels. We have, $p_\ell^i = n_\ell / |S|$, where n_ℓ is the number of data points in subset S labeled ℓ. The set of these percentages $\{p_1^i, p_2^i, ..., p_L^i\}$ is a probability distribution for which we can calculate its entropy using the expression above. The entropy takes values between 0 and 1. It becomes 0 when all the data points in S have the same label, and it achieves its maximum value when all labels in S are equiprobable, i.e., $p_1^i = p_2^i = \cdots = p_L^i = 1/L$.

For each feature X, we calculate the information gain as follows:

$$IG(X) = -\sum_{\ell=1}^{L} p_\ell \log_2(p_\ell) - \sum_{\iota} p(X = \iota) \left(-\sum_{\ell=1}^{L} p_\ell^i \log_2(p_\ell^i) \right).$$

The first term in the right-hand side of the above expression is the *parent* entropy. This is the entropy calculated on the original training data set D using the percentages $\{p_1, p_2, ..., p_L\}$, where p_ℓ is the percent of data points in D labeled ℓ. For instance, let us assume that the label takes the values 0, 1, and 3, and let the data set consist of three data points labeled 0, two data points labeled 1, and five data points labeled 3. Then, $\{p_1, p_2, p_3\} = \{0.3, 0.2, 0.5\}$, and the parent entropy is

$$-\left(0.3\log_2(0.3) + 0.2\log_2(0.2) + 0.5\log_2(0.5)\right)$$

$$= -\left[0.3 \times (-1.7369) + 0.2 \times (-2.3219) + 0.5 \times (-1)\right] = 1.4855.$$

The second term of the information gain expression consists of the weighted entropies of the subsets of the children obtained after splitting the node using feature X. In Figure 9.11, we give an example of how this term is calculated. For simplicity, we assume that X = 1, 2, which results to two splits. Let S_1 and S_2 be the two subsets created from the two splits, and let $\{p_1^1, p_2^1, ..., p_L^1\}$ and $\{p_1^2, p_2^2, ..., p_L^2\}$ be the percentages associated with each subset. Also, let $p(X = 1)$ and $p(X = 2)$ be the probability that X takes the value 1 and 2, respectively. Then, the weighted entropy is

$$p(X = 1) \left(-\sum_{\ell=1}^{L} p_\ell^1 \log_2(p_\ell^1) \right) + p(X = 2) \left(-\sum_{\ell=1}^{L} p_\ell^2 \log_2(p_\ell^2) \right),$$

which is subtracted from the parent entropy to obtain the information gain IG(X) for X.

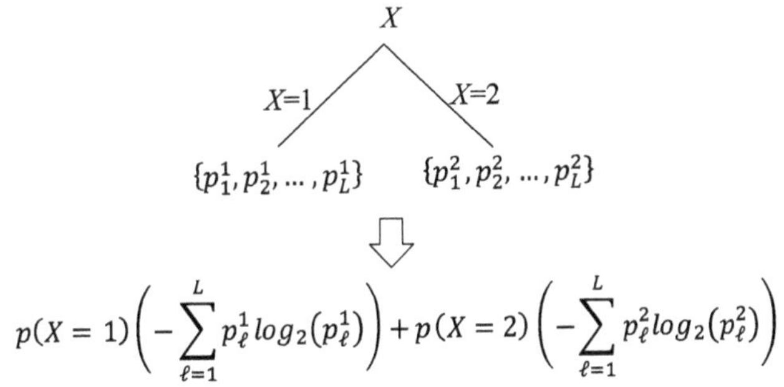

FIGURE 9.11 An example of the weighted entropy.

We calculate the information gain for each candidate feature and select the feature to split with the largest information gain. This causes the biggest reduction in the entropy of the original training data set D.

We demonstrate the use of the information gain though a simple example. Let us assume that we have three features, X, Y, and Z, with each feature taking the values 0, 1, and each data point is labeled 1 or 2. Let the training data set consist of the data points $D = \{(1,1,1,1), (1,0,1,2), (0,1,1,1), (1,0,0,2)\}$. We have: $p(X = 0) = 0.25$, $p(X = 1) = 0.75$, $p(Y = 0) = 0.5$, $p(Y = 1) = 0.5$, and $p(Z = 0) = 0.25$, $p(Z = 1) = 0.75$.

The percentage of the data points in D labeled 1 and 2 are $\{p_1, p_2\} = \{0.5, 0.5\}$, and the parent entropy is

$$-p_1 \log_2(p_1) - p_2 \log_2(p_2)$$

$$= -0.5 \times \log_2(0.5) - 0.5 \times \log_2(0.5)$$

$$= -\log_2(0.5) = -(-1) = 1.$$

a) Information gain for X:

We have two splits, one for $X = 0$ and one for $X = 1$, and therefore $S_1 = \{(0, 1, 1, 1)\}$ and $S_2 = \{(1, 1, 1, 1), (1, 0, 1, 2), (1, 0, 0, 2)\}$. The percentages of data points in S_1 labeled 1 and 2 are $\{p_1^1, p_2^1\} = \{1, 0\}$, and in S_2: $\{p_1^2, p_2^2\} = \{1/3, 2/3\}$. The weighted entropy for S_1 is:

$$0.25\left(-1 \times \log_2(1) - 0 \times \log_2(0)\right) = 0$$

and for and S_2:

$$0.75\left(-\frac{1}{3} \times \log_2\left(\frac{1}{3}\right) - \frac{2}{3} \times \log_2\left(\frac{2}{3}\right)\right)$$

$$= 0.75 \times \left(\frac{1}{3} \times 1.5864 + \frac{2}{3} \times 0.5864\right) = 0.6898.$$

Consequently, $IG(X) = 1 - 0.6898 = 0.3102$.

b) Information gain for Y:

We have two splits corresponding to $Y = 0$ and $Y = 1$, and therefore $S_1 = \{(1, 0, 1, 2), (1, 0, 0, 2)\}$ and $S_2 = \{(1, 1, 1, 1), (0, 1, 1, 1)\}$. The percentages of data points in S_1 labeled 1 and 2 are $\{p_1^1, p_2^1\} = \{0, 1\}$, and in S_2: $\{p_1^2, p_2^2\} = \{1, 0\}$. The weighted entropy for S_1 is

$$0.5\left(-0 \times \log_2(0) - 1 \times \log_2(1)\right) = 0$$

and for S_2

$$0.5\left(-1 \times \log_2(1) - 0 \times \log_2(0)\right) = 0.$$

Consequently, $IG(X) = 1$.

c) Information gain for Z:

We have two splits corresponding to $Z = 0$ and $Z = 1$, and therefore $S_1 = \{(1, 0, 0, 2)\}$ and $S_2 = \{(1, 1, 1, 1), (1, 0, 1, 2), (0, 1, 1, 1)\}$. The percentages of data points in S_1 labeled 1 and 2 are $\{p_1^1, p_2^1\} = \{0, 1\}$ and in S_2: $\{p_1^2, p_2^2\} = \{2/3, 1/3\}$. The weighted entropy for S_1 is

$$0.25\left(-0 \times \log_2(0) - 1 \times \log_2(1)\right) = 0$$

and for S_2:

$$0.75\left(-\frac{2}{3} \times \log_2\left(\frac{2}{3}\right) - \frac{1}{3} \times \log_2\left(\frac{1}{3}\right)\right) = 0.6898.$$

Consequently, $IG(X) = 1 - 0.6898 = 0.3102$.

Therefore, we use Y to split the decision tree, see Figure 9.12. We observe that both the resulting subsets $S_1 = \{(1, 0, 1, 2), (1, 0, 0, 2)\}$ and $S_2 = \{(1, 1, 1, 1), (0, 1, 1, 1)\}$ contain data with only one label. Therefore, both resulting nodes are leaves, and the tree construction stops. A new point is classified solely by looking at the value of Y. For instance, the data point $(0, 0, 0)$ will be labeled 2, since $Y = 0$.

b. Gini Impurity

A node is said to be *pure* if all the data points associated with the node have the same label. The Gini impurity is used in the CART algorithm and it is a measure of node impurity. It should not be confused with the Gini coefficient, which is a measure of statistical dispersion intended to represent the income or wealth distribution of a nation's residents.

Let S be the data set associated with a node before it is split, and let L be the total number of labels in the original training data set. From the data set S, we can calculate the fraction of data points for which the label takes the value ℓ, $\ell = 1, 2, \ldots, L$. These fractions are used as estimates of the probability p_l that the label takes the value ℓ, $\ell = 1, 2, \ldots, L$. Now, let us

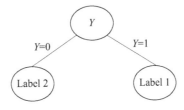

FIGURE 9.12 The decision tree for the information gain example.

select randomly a data point from S, and let us ignore the fact that we know its label. Then the probability that it is labeled $\ell = 1$ is equal to p_1 times $1 - p_1$, which is the probability of $\ell \neq 1$. Likewise, the probability that it is labeled $\ell = 2$ is $p_2(1 - p_2)$, and so on. Summing over these probabilities gives

$$Gini(S) = \sum_{\ell=1}^{L} p_l (1 - p_l) = 1 - \sum_{\ell=1}^{L} p_\ell^2.$$

The above expression is known as the *Gini impurity* and it can be interpreted as the probability of incorrectly labeling a randomly chosen data point in S according to the distribution of labels $\{p_1, p_2, \ldots, p_L\}$.

If S consists of data points with the same label, then $Gini(S) = 0$, since $1 - \sum_{\ell=1}^{L} p_\ell^2 = 1 - p_1^2 = 1 - 1 = 0$. The Gini impurity reaches its maximum value when all labels in S have the same probability. For instance, for two labels and $p_1 = p_2 = 0.5$, we get a value of $1/2$.

In order to determine which feature to split at a node, we calculate the Gini gain:

$$GiniGain(X, S) = Gini(S) - Gini(X, S)$$

$$= Gini(S) - \sum_{i=1}^{k} \frac{|S_i|}{|S|} Gini(S_i),$$

where S is the data set associated with the node, X is the feature, S_i is the subset data of S for which $X = i$, and k is the total number of different values that X takes. We note that the expression for the Gini gain is similar in concept to the information gain. We select the feature with the highest Gini gain.

For example, let us assume that a labeled data set S consists of 9 tuples labeled 1 or 2, as shown in Table 9.3. For simplicity, we have assumed that the two features are binary. Then, using the third column, the probability that the label is 1 or 2 is 4/9 and 5/9, respectively, and hence we have

$$Gini(S) = 1 - \left(\frac{4}{9}\right)^2 - \left(\frac{5}{9}\right)^2 = \frac{40}{81}.$$

The Gini gain for the two features is calculated as follows. For X we have two sets S_1 and S_2 corresponding to $X = 1$ and $X = 2$, where

$$S_1 = \{(1,1,1), (1,1,1), (1,2,2), (1,2,1)\}$$

$$S_2 = \{(2,2,1), (2,1,2), (2,1,2), (2,2,2), (2,1,2).$$

TABLE 9.3   An example for calculating the Gini impurity

X	Y	Label
1	1	1
1	1	1
1	2	2
2	2	1
2	1	2
2	1	2
2	2	2
1	2	1
2	1	2

Therefore,

$$Gini\left(X=1,\,S\right)=1-\left(\frac{3}{4}\right)^{2}-\left(\frac{1}{4}\right)^{2}=\frac{3}{8}$$

$$Gini\left(X=2,\,S\right)=1-\left(\frac{1}{5}\right)^{2}-\left(\frac{4}{5}\right)^{2}=\frac{8}{25}.$$

For Y we also have two sets S_1 and S_2 corresponding to $Y=1$ and $Y=2$, where

$$S_{1}=\left\{\left(1,1,1\right),\left(1,1,1\right),\left(2,1,2\right),\left(2,1,2\right),\left(2,1,2\right)\right\}$$

$$S_{2}=\left\{\left(1,2,2\right),\left(2,2,1\right),\left(2,2,2\right),\left(1,2,1\right).$$

Therefore,

$$Gini\left(Y=1,\,S\right)=1-\left(\frac{2}{5}\right)^{2}-\left(\frac{3}{5}\right)^{2}=\frac{12}{25}$$

$$Gini\left(Y=2,\,S\right)=1-\left(\frac{2}{4}\right)^{2}-\left(\frac{2}{4}\right)^{2}=\frac{1}{2}.$$

Consequently,

$$GiniGain\left(X,\,S\right)=Gini\left(S\right)-\frac{\left|S_{1}\right|}{\left|S\right|}Gini\left(S_{1}\right)-\frac{\left|S_{1}\right|}{\left|S\right|}Gini\left(S_{2}\right)$$

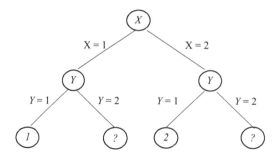

FIGURE 9.13 The decision tree for the example in Table 9.3.

$$= \frac{40}{81} - \frac{4}{9}\frac{3}{8} - \frac{5}{9}\frac{8}{25} = 0.1494.$$

$$GiniGain(Y, S) = Gini(S) - \frac{|S_1|}{|S|} Gini(S_1) - \frac{|S_2|}{|S|} Gini(S_2)$$

$$= \frac{40}{81} - \frac{4}{9}\frac{12}{25} - \frac{5}{9}\frac{1}{2} = 0.005.$$

Therefore, we select X to split, since its Gini gain is higher than that of Y. For the second split, we only have one feature left, Y, and therefore we obtain the decision tree shown in Figure 9.13. The leaves provide the label. We see that in two cases the label cannot be determined since the resulting data sets in both cases contain one data point labeled 1 and another one labeled 2.

9.3.3 Pre-Pruning and Post-Pruning

As we have seen above, the growth of a regression or a classification tree is done using a locally optimal decision at each node. Consequently, there is a need for an additional mechanism that will determine how large a tree should be grown to. To this effect, pre-pruning and post-pruning is used. In pre-pruning, the construction of a tree is stopped using various stopping criteria. In post-pruning, a decision tree is first fully grown, and then it is reduced (pruned) by removing subtrees.

Large decision trees are prone to overfitting; i.e., a data point in the training data set may be classified accurately, but a new data point may not. On the other hand, a small tree may under-fit the data, i.e., it may not capture adequately the structural information of the training data set. Pre-pruning and post-pruning is used to control the size of a tree so that it does not overfit.

a. Pre-Pruning

The following are some of the rules used to stop the construction of a tree in pre-pruning.

- *All data points have the same label:* A node is not split further if all the data points associated with the node have the same label. This of course makes sense, since nothing more can be gained by further splitting the node.

- *Minimum number of data points for a node split:* Before splitting a node, the number of data points should be greater than a given threshold. A high threshold value prevents a model from learning relations which may be too specific to the particular data set, but it may also lead to under-fitting as well.

- *Minimum number of data points for a child node:* This is similar to the above stopping rule, but now we require that before a child node is created, it should contain a minimum number of data points. A low value should be chosen for data sets where the number of data points per label is not well balanced.

- *Maximum depth of the tree:* This parameter controls the depth of the tree. High values are likely to lead to overfitting.

- *Maximum number of leaves:* This parameter can be used instead of the maximum depth parameter.

- *Maximum number of features to consider for a split:* In case of a large number of features, the rule of thumb is to consider a smaller number of features equal to about the square root of the total number of features but maintain it to about 30% to 40% percent of the total number of features.

b. Post-Pruning

In post-pruning, a full tree is grown first using lose stopping criteria that allow the tree to overfit. Then, the tree is cut back by removing subtrees that do not contribute much to the accuracy of the tree. Various methods have been proposed to determine how to prune a tree. Below, we present the *cost complexity pruning* method.

We recall that for a regression tree, the error of a node t is $E_t = \sum_{i=1}^{n_t} (y_i - \bar{y}_t)^2$, where $\bar{y}_t$ is the mean of all the values y_i, $i = 1, 2, \ldots, n_t$, of the dependent variable, and n_t is the number of data points in the node. For a classification tree, the error of a node t is the misclassification rate, i.e., $E_t = (1/n)\sum_{i=1}^{n_t} I(\ell_i \neq \ell)$, where ℓ_i is the label of the ith data point in the node, ℓ is the label determined by majority vote of the labels in the node, $I(x) = 1$ if x is true, and zero otherwise, n_t is the number of data points in the node, and n is the total number of data points in the data set.

Let $R(T)$ be the sum of the node errors of all the leaves of a tree T. In the cost complexity pruning method, we prune a tree T by removing its subtree T' that minimizes the cost function $R_a(T')$:

$$R_a(T') = R(T') + a|T'|,$$

for a given a. $|T'|$ is the number of leaves of T', and a is the *regularization* parameter. The first term $R(T')$ selects a large tree, since a large tree is typically overfitting and therefore it has a lower error. The second term $a|T|$, known as the regularization penalty, balances out the first term since it selects a small tree.

The subtree T' that minimizes the above cost function $R_a(T')$ for a given a is said to be an optimally pruned subtree of T. It is possible that more than one subtree is

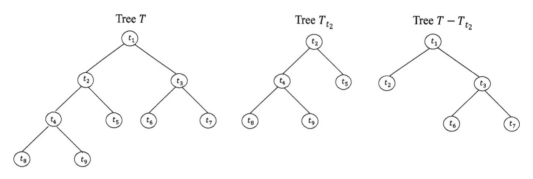

FIGURE 9.14 An example of a pruned tree.

optimal, in which case we chose the smallest one, i.e., the one with the smallest number of leaves.

Let us now consider the example in Figure 9.14. Let T be the original tree, T_{t_2} the subtree which has node t_2 as its root, and $T - T_{t_2}$ the pruned tree after removing T_{t_2}. As can be seen, the removed subtree T_{t_2} is replaced by its root t_2. In other words, when we remove a subtree, its root stays in the tree.

The cost $R(T - T_{t_2})$ of $T - T_{t_2}$ can be calculated by adding the node costs across its leaves, expressed as follows:

$$R(T - T_{t_2}) = R(T) - R(T_{t_2}) + R(t_2),$$

where $R(t_2)$ is the cost of node t_2. Therefore, the cost function $R_a(T - T_{t_2})$ of $T - T_{t_2}$ is

$$R_a(T - T_{t_2}) = R(T - T_{t_2}) + a|T - T_{t_2}|$$

$$= R(T) - R(T_{t_2}) + R(t_2) + a|T - T_{t_2}|.$$

The cost function $R_a(T_{t_2})$ of the removed subtree T_{t_2} can be written as

$$R_a(T_{t_2}) = R_a(T - T_{t_2}) - R_a(T)$$

$$= R(T) - R(T_{t_2}) + R(t_2) + a|T - T_{t_2}| - R(T) - a|T|$$

$$= -R(T_{t_2}) + R(t_2) + a(|T - T_{t_2}| - |T|).$$

It can be easily verified that $|T - T_{t_2}| - |T| = |T_{t_2}| - 1$ by counting the leaves in each tree. Therefore,

$$R_a(T_{t_2}) = -R(T_{t_2}) + R(t_2) + a(|T_{t_2}| - 1).$$

We note that $R_a(T_{t_2})$ becomes zero when

$$a = \frac{R(t_2) - R(T_{t_2})}{|T_{t_2}| - 1}.$$

The cost complexity method consists of two phases. In the first phase, a sequence of trees with a corresponding sequence of a values is genetated using the following algorithm.

Set $a^{(1)} = 0$ and let $T^{(0)}$ be the original tree.

1. *Recursion $i = 0, 1, 2, \ldots$*
Select a node $t \in T^{(i)}$ that minimizes

$$g(t) = \frac{R(t) - R(T_t^{(i)})}{|T_t^{(i)}| - 1}.$$

Let t_{i+1} be this node. Set $T^{(i+1)} = T^{(i)} - T_{t_{i+1}}^{(i)}$ and

$$a^{(i+1)} = \frac{R(t_{i+1}) - R(T_{t_{i+1}}^{(i)})}{|T_{t_{i+1}}^{(i)}| - 1}.$$

2. *Stopping rule*
Stop if the reduced tree $T^{(i+1)}$ consists of only the root of the original tree. Else go back to step 2.

The algorithm generates a sequence of trees $T^{(1)} \supseteq T^{(2)} \supseteq \ldots \supseteq T^{(k)}$, where $T^{(k)}$ is the tree that consists of only the root of the original tree, along with a sequence of a values for which it can be shown that $a^{(1)} \leq a^{(2)} \leq \ldots \leq a^{(k)}$.

We demonstrate the above algorithm using the decision tree given in Figure 9.15. The data set consists of 20 data points, labeled red or blue. The number of red and blue data points in each node t_i, $i = 1, 2, \ldots, 11$ is indicated below the node. Table 9.4 gives the results from the first iteration of the algorithm. We note that all the leaves are pure, and therefore the cost $R(T)$ of the tree and of all its subtrees is zero. From the last column, we see that either T_{t_7} or T_{t_9} can be pruned. T_{t_9} is chosen because it is the smallest tree. Therefore, $T^{(1)} = T - T_{t_9}^{(1)}$ and $a^{(1)} = 0$. The remaining steps are left as an exercise (see Exercise 6) at the end of the chapter.

The second phase of the cost complexity method is to choose one of the k-generated trees $T^{(1)}, T^{(2)}, \ldots, T^{(k)}$. This is done using the cross-validation method as follows. The data set is first divided into 10 folds, i.e., 10 non-overlapping subsets. Then, for each fold i, $i = 1, 2, \ldots, 10$:

TABLE 9.4 First iteration of the cost complexity pruning method

Node t	Label	$R(t)$	$R(T_{t_i})$	$|T_{t_i}|$	$g(t)$
t_1	blue	9/20	0	6	(9/20)/5 = 0.09
t_2	red or blue	4/20	0	2	(4/20)/1 = 0/2
t_3	blue	5/20	0	4	(5/20)/3 = 0.083
t_7	blue	2/20	0	3	(2/20)/2 = 0.05
t_9	red	1/20	0	2	(1/20)/1 = 0.05

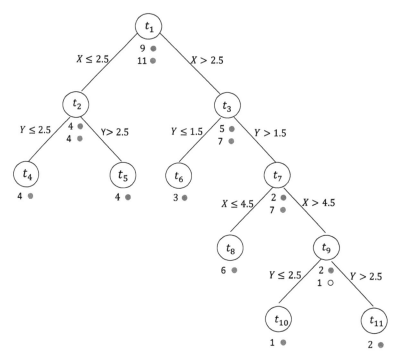

FIGURE 9.15 An example of the cost complexity method example.

1. Grow a full tree using the data set consisting of all the folds except the ith one.

2. Apply the above algorithm to generate a sequence of trees $T^{(1)} \supseteq T^{(2)} \supseteq \ldots \supseteq T^{(k)}$ and corresponding values $a^{(1)}, a^{(2)}, \ldots, a^{(k)}$.

3. Calculate the cost of each tree using the ith fold and select the tree whose a value is the lowest. Let T_i and a_i be the selected tree and a value.

The final result of the cross-validation will be a set of trees $T_1, \ldots, T_{10}$ and corresponding values $a_1, \ldots, a_{10}$. Select the tree with the lowest a value.

9.3.4 Advantages and Disadvantages of Decision Trees

Decision trees are simple to understand and interpret since they mirror the way humans make decisions. They can handle both continuous and nominal variables, and they perform well with large data sets. They do not require normalization, and missing values do not affect the development of a decision tree.

However, they are not as accurate as other techniques, and they are not very robust, since a small change in the training data can result to a big change in the decision tree (see exercise 3 at the end of the chapter).

9.3.5 Decision Trees Ensemble Methods

An ensemble method combines several decision trees to produce a better prediction than using a single decision tree. *Bagging* (also called *bootstrap aggregation*) and *boosting* are two such techniques described in this section.

a. Bagging

This technique is not limited to decision trees and it can be applied to any learning algorithm. Several subsets of the training data set D are first generated randomly with replacement, with a sample size of around $(1 - (1/e))n \approx 0.623n$, where n is the total number of data points in the data set D. Then, for each subset, a separate decision tree is grown, and as a result, we end up with an ensemble of decision trees. To classify a new data point, we use all the decision trees and then combine the results. In regression, the final prediction is averaged from the predictions of all the decision trees. In classification, we use majority vote, i.e., the predicted label is the one that is predicted the most from all the decision trees.

The *random forests* technique is an extension of bagging. In addition to generating random subsets, it also randomizes the selection of features; i.e., at each node, we randomly select a subset of the features from which we decide which feature to split. The random forests technique works well for data sets with high dimensionality and also for missing values.

b. Boosting

This method is primarily used with decision trees, but it can be applied to other learning algorithms. Unlike bagging, in boosting, the entire training data set D is used over many iterations, with each data point weighted by a weight which is updated at each iteration. At each iteration, a one-level decision tree, known as a *decision stump*, is learnt using a single feature. In the case of a nominal feature, the decision stump consists of as many branches as the number of values that the nominal variable takes. In the case of a continuous feature, a threshold value of the feature is selected, and the stump consists of two branches, one per region defined by the threshold. A decision stump has a substantial error, but its accuracy is better than random guessing. The classification of a new data point is done by combining the classification outcome from all the decision stumps.

The best known boosting algorithm is AdaBoost (adaptive boosting) [2], which was developed for binary classification; i.e., the labeled data set has two labels, indicated as +1 and −1. Initially, all data points x_i, $i = 1, 2, ..., n$, in the data set are allocated the weight $w_{i,1} = 1/n$. Then, the decision stump with the lowest error ε_1 is selected and a new set of weights $w_{i,2}$, $i = 1, 2, ..., n$ is computed. This is repeated over many iterations until a stopping rule is satisfied.

More specifically, let T_t be the decision stump learnt at iteration t, $T_t(x_i)$ the label assigned to data point x_i using T_t, and ℓ_i the original label of x_i, $i = 1, 2, ..., n$. Also, let $w_{i, t}$, $i = 1, 2, ..., n$, be the weights calculated at iteration t, and ε_t the error of the decision tree T_t. Then, the new weights $w_{i,t+1}$, $i = 1, 2, ..., n$, in iteration $t + 1$ are computed as follows:

$$w_{i,t+1} = cw_{i,t}e^{-\alpha_t \ell_i T_t(x_i)},$$

where

$$\alpha_t = \frac{1}{2}\log_e\left(\frac{1-\varepsilon_t}{\varepsilon_t}\right),$$

and c is the normalizing constant that guarantees that the sum of all the weights $w_{i,t+1}$ is equal to 1. The next decision stump T_{t+1} at iteration $t + 1$ is selected as the one with the lowest weighted error

$$\varepsilon_{t+1} = \frac{\sum_{i=1}^{n} w_{i,t+1} I\left(T_{t+1}(x_i) \neq \ell_i\right)}{\sum_{i=1}^{n} w_{i,t+1}},$$

where $I(x) = 1$, if x is true, and zero otherwise. The number of iterations is determined by plotting ε_t and then stopping at the inflection point of the curve.

To classify a new data point x, we use each tree stump to predict a label +1 or −1. Then, the prediction of the ensemble is obtained using the weighted sum $\sum_t \alpha_t T_t(x)$. If it is positive, then +1 is predicted. Otherwise, −1 is predicted.

Boosting outperforms bagging and random forests, when the data set does not contain noisy data. However, if the data contains noisy data points, the performance of the boosting technique can degrade quickly because it tends to focus on noisy data points which are typically hard to classify, and as a result boosting overfits the data. For noisy data, bagging and random forests yield better results.

9.4 LOGISTIC REGRESSION

Logistic regression, or *logit* regression, is a classification technique. Unlike other forms of regression, the independent variables in a logistic regression may be continuous or nominal. The dependent variable is a nominal variable and it takes a small number of values. The assumption in multivariable regression that residuals are normally distributed does not hold here. For instance, in binary logistics, the dependent variable is Bernoulli distributed.

There are three types of logistic regression, namely, *binary* or *binomial*, *multinomial*, and *ordinal* as discussed below. Before, we proceed, we review briefly the logistic distribution and the logistic function used in this section.

The *logistic distribution* $f(x)$ of a random variable X is given by the expression

$$f(x) = \frac{e^{-(x-x_0)/s}}{s\left(1+e^{-(x-x_0)/s}\right)^2},$$

where x_0 is the mean of the distribution and s is a scale parameter. Its shape is similar to the normal distribution, except that its tails are fatter. The *standard logistic distribution* is obtained by setting $x_0 = 0$ and $s = 1$, i.e.,

$$f(x) = \frac{e^{-x}}{\left(1+e^{-x}\right)^2}.$$

The cumulative distribution of the standard logistic distribution, referred to as the *standard logistic function*, is given by the expression

$$F(x) = \frac{1}{1+e^{-x}}$$

and shown in Figure 9.16. The logistic function dates back to the 19th century and it has applications in many different fields, including Machine Learning.

9.4.1 The Binary Logistic Regression

In the binary logistic regression model, the dependent variable Y takes the labels "0" notated as 0 and "1" notated as 1. Let $p = p[Y = 1]$. Then, the odds for 1 is defined as $p/(1 - p)$. For instance, if $p = q$ then the odds for 1 is 0.5, if $p = 0.75$ then the odds is 3, and if $p = 0.25$ then the odds is 1/3. The log-based e of the odds $\log(p/(1 - p))$ is referred to as *logit* or *log-odds*. The logistic model is based on the idea that the logit can be expressed as a linear function of the independent variables. For instance, in the case where we have a single independent variable X, $\log(p/(1 - p))$ is expressed as follows:

$$\log\left(\frac{p}{1-p}\right) = a_0 + a_1 X.$$

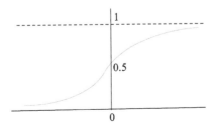

FIGURE 9.16 The standard logistic function.

234 ■ An Introduction to IoT Analytics

The coefficients a_0 and a_1 can be estimated using the MLE procedure. Let $\hat{a}_0$ and $\hat{a}_1$ be the estimated coefficients, then we can obtain p as follows:

$$\frac{p}{1-p} = e^{\hat{a}_0 + \hat{a}_1 X}$$

or

$$p = \frac{1}{1 + e^{-(\hat{a}_0 + \hat{a}_1 X)}}.$$

We note that the above expression for p is a logistic function. Given a new data point x, p is obtained by substituting x into the above expression. If $p \geq 0.5$, then x is assigned label 1, i.e., $y = 1$. Otherwise, it is assigned label 0. If $p \geq 0.5$, then we have that

$$\frac{1}{1 + e^{-(\hat{a}_0 + \hat{a}_1 x)}} \geq \frac{1}{2}$$

or

$$e^{-(\hat{a}_0 + \hat{a}_1 x)} \leq 1.$$

For this to happen, $-(\hat{a}_0 + \hat{a}_1 x) \leq 0$ or $\hat{a}_0 + \hat{a}_1 x \geq 0$; i.e., for a new data point x to be assigned the label 1, we should have $\hat{a}_0 + \hat{a}_1 x \geq 0$ or $x \geq -\hat{a}_0 / \hat{a}_1$. Otherwise, it is assigned label 0; i.e., the decision boundary is point $-\hat{a}_0 / \hat{a}_1$.

In the case of two independent variables, the decision boundary becomes a line. In this case, we have

$$\log\left(\frac{p}{1-p}\right) = a_0 + a_1 X_1 + a_2 X_2$$

and the value of p is given by the expression:

$$p = \frac{1}{1 + e^{-(\hat{a}_0 + \hat{a}_1 X_1 + \hat{a}_2 X_2)}}.$$

A new data point (x_1, x_2) is assigned the label 1 if $\hat{a}_0 + \hat{a}_1 x_1 + \hat{a}_2 x_2 \geq 0$, and the label 0 if $\hat{a}_0 + \hat{a}_1 x_1 + \hat{a}_2 x_2 < 0$. The decision boundary separating the two classes is the line $\hat{a}_0 + \hat{a}_1 x_1 + \hat{a}_2 x_2 = 0$, an example of which is given in Figure 9.17.

From the above expression for p, we observe that when $\hat{a}_0 + \hat{a}_1 x_1 + \hat{a}_2 x_2 \geq 0$, $p \to 1$ rapidly as a new data point (x_1, x_2) moves away from the boundary line. Likewise, when $\hat{a}_0 + \hat{a}_1 x_1 + \hat{a}_2 x_2 < 0$, $p \to 0$ rapidly as the new data point moves away from the boundary line. Therefore, this classifier not only determines the label of a new data point, but it also

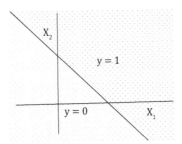

FIGURE 9.17 The decision boundary.

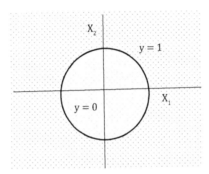

FIGURE 9.18 Higher polynomial boundary.

makes a stronger statement that as the distance of the data point increases from the boundary line, the probability p goes to 1 or 0 more rapidly.

Different boundaries can be created using higher polynomials. For instance, the polynomial $-1 + x_1^2 + x_2^2 = 0$ generates a unit circle boundary shown in Figure 9.18. If $x_1^2 + x_2^2 \geq 1$ then $y = 1$, otherwise $y = 0$.

In the general case where we have d-independent variables, the logit is expressed as

$$\log\left(\frac{p}{1-p}\right) = a_0 + a_1 X_1 + a_2 X_2 + \cdots + a_d X_d,$$

and the probability p is given by the expression:

$$p = \frac{1}{1 + e^{-(\hat{a}_0 + \hat{a}_1 X_1 + \cdots + \hat{a}_d X_d)}},$$

where $\hat{a}_0, \hat{a}_1, \ldots, \hat{a}_d$ are the estimated coefficients. For a given new data point $(x_1, \ldots, x_d)$, $y = 1$ if $\hat{a}_0 + \hat{a}_1 x_1 + \cdots + \hat{a}_d x_d \geq 0$, and $y = 0$ otherwise.

For example, we consider two sensors located in two different places, which are labeled 0 and 1. The sensors transmit measurements of a single variable. The data set D, which is in the form of (x, y) where x is a measurement and y is the sensor's location, is as follows:

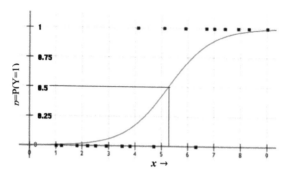

FIGURE 9.19 Plot of p vs x.

$$D = \{(1,0), (1.2,0), (1.8,0), (2.2,0), (2.5,0), (2.9,0), (3.5,0),$$

$$(3.8,0), (4.1,1), (4.7,0), (5.1,1), (5.5,0), (5.9,1),$$

$$(6.3,0), (6.7,1), (7,1), (7.4,1), (7.9,1), (8.3,1), (9,1)\}.$$

The following results were obtained after running the binary logistic regression in one independent variable:

$$\log\left(p/(1-p)\right) = 6.3689 - 1.2097x.$$

The p-value for the constant and for the coefficient of x is 0.0227 and 0.0186, respectively. This means that in both cases, we reject the null hypothesis at 90% and 95% but not at 99%. From the above expression, we have

$$p = p(Y = 1) = \frac{1}{1 + e^{-(6.3689 + 1.2097X)}}.$$

A plot of the above expression is given in Figure 9.19. The boundary point is 6.3689/1.2097 = 5.26. If a sensor transmits a measurement x without its location, then x comes from location 1 if $x \geq 5.26$. Otherwise, it comes from location 0.

We conclude this section by giving an equivalent formulation using the *latent-variable model*. We define the *latent*, i.e., unobserved, variable Y' as a linear function of the independent variables plus a random error ε distributed as the standard logistic distribution; i.e.

$$Y' = a_0 + a_1 X_1 + \cdots + a_d X_d + \varepsilon.$$

If $Y' > 0$, i.e., $a_0 + a_1 X_1 + \cdots + a_d X_d + \varepsilon > 0$, then we set $Y = 1$. Otherwise, $Y = 0$. Therefore,

$$p = p(Y = 1) = p(Y' > 0)$$

$$= p\left(\varepsilon > -\left(a_0 + a_1 X_1 + \cdots + a_d X_d\right)\right)$$

$$= p\left(\varepsilon < a_0 + a_1 X_1 + \cdots + a_d X_d\right)$$

since the standard logistic function is symmetric. We recall that $p(\varepsilon < a) = 1/(1 + e^{-a})$, and therefore, from the above expression we obtain

$$p = \frac{1}{1 + e^{-(a_0 + a_1 X_1 + \cdots + a_d X_d)}}$$

or

$$\log\left(\frac{p}{1-p}\right) = a_0 + a_1 X_1 + \cdots + a_d X_d$$

which is the expression we started with in the first formulation of the logistic regression model. The coefficients can be estimated using MLE.

Let $\hat{a}_0, \hat{a}_1, \ldots, \hat{a}_d$ be the estimated coefficients. Then, for a given new data point $(x_1, \ldots, x_d)$, $y = 1$ if $\hat{a}_0 + \hat{a}_1 x_1 + \cdots + \hat{a}_d x_d \geq 0$, and $y = 0$ otherwise.

We note that if we assume in the above formulation that ε is distributed according to the standard normal distribution, then we obtain the *probit* model, an older model that pre-dates the logistic model.

9.4.2 Multinomial Logistics Regression

This is a generalization of the binary logistic regression to the case where the dependent variable takes more than two labels.

We assume that the data is described by d-independent variables $X_1, \ldots, X_d$, which can be continuous or nominal. The dependent variable Y is a nominal variable that takes L different labels. The multinomial logistic regression is expressed in terms of a set of binary logistic regressions. Specifically, we select a label, say label L, as a pivot and then run $L - 1$ binary logistic regressions by assuming that the log $(p(Y = \ell)/p(Y = L))$, $\ell = 1, 2, \ldots, L$ is a linear function of the independent variables; i.e.

$$\log\left(\frac{p(Y = 1)}{p(Y = L)}\right) = a_{1,0} + a_{1,1} X_1 + \cdots + a_{1,d} X_d$$

$$\vdots$$

$$\log\left(\frac{p(Y = L-1)}{p(Y = L)}\right) = a_{L-1,0} + a_{L-1,1} X_1 + \cdots + a_{L-1,d} X_d.$$

From the above expressions, we have

$$p(Y=1)=p(Y=L)e^{a_1 X}$$

$$\vdots$$

$$p(Y=L-1)=p(Y=L)e^{a_{L-1}X},$$

where $a_\ell = (a_{\ell 0}, a_{\ell 1}, ..., a_{\ell d})$ and $X = (1, X_1, ..., X_d)^T$. Since all the above probabilities have to add up to 1, we have

$$p(Y=L)=1-\sum_{\ell=1}^{L-1}p(Y=\ell)=1-\sum_{\ell=1}^{L-1}p(Y=L)e^{a_\ell X}$$

or

$$p(Y=L)=\frac{1}{1+\sum_{\ell=1}^{L-1}e^{a_\ell X}}$$

and therefore

$$p(Y=\ell)=\frac{e^{a_\ell X}}{1+\sum_{\ell=1}^{L-1}e^{a_\ell X}},\ell=1,2,...L-1.$$

Given new data point $(x_1, ..., x_d)$, we compute all the probabilities $p(Y=\ell)$, $\ell = 1, 2, ...L$ and assign the label with the largest probability.

The unknown parameters in each vector a_ℓ are typically jointly estimated using maximum likelihood with regularization in order to prevent pathological solutions.

Multinomial logistic regression makes the assumption of independence among the classes of the dependent variable; i.e., the choice of one class is not related to the choice of another.

9.4.3 Ordinal Logistic Regression

In the above two models, the dependent variable is assumed to be a nominal variable. In ordinal logistic regression, the dependent variable is assumed to be *ordinal*, i.e., it takes a set of discrete values which are ordered. For instance, let us consider that a variable takes the values {red, green, blue}. These values are not ordered in the sense that we cannot say that green is better than red, and blue is better than green. However, if the dependent variable takes the values {very cold, mild, very hot} then these values are ordered. Ordinal logistic regression is typically used if we have a few values. For more than five values, it is better to use multivariable linear regression.

We assume that the data is described by d-independent variables $X_1, \ldots, X_d$, which may be continuous or nominal. The dependent variable Y is an ordinal variable that takes values from the ordered set of labels {"1", "2", …, "L"}, where "1" ≤ "2", "2" ≤ "3", …, "L − 1" ≤ "L". For simplicity, we notate label "i" as i.

As in the multinomial logistic regression, we express the ordinal logistic regression problem in terms of a number of binary logistic regressions. However, instead of using the logit of the individual probabilities, we use the logit of the cumulative probabilities so that to take advantage of the fact that the labels are ordered. We have

$$\log \frac{p(Y \leq 1)}{1 - p(Y \leq 1)} = \theta_1 - \left(a_0 + a_1 X_1 + \cdots + a_d X_d\right)$$

$$\vdots$$

$$\log \frac{p(Y \leq L-1)}{1 - p(Y \leq L-1)} = \theta_{L-1} - \left(a_0 + a_1 X_1 + \cdots + a_d X_d\right).$$

The case of $p(Y \leq L)$ is not considered, since $p(Y \leq L) = 1$. We note that the coefficients of the independent variables for each logit expression are the same. This means that the contribution of an independent variable is the same for different logits. This is known as the *proportional odds* assumption, and in view of this, this formulation is also known as the *proportional odds model*. The coefficients $\theta_1, \theta_2, \ldots, \theta_{L-1}$ are called the threshold values and they are like the intercept in a linear regression. The unknown coefficients $a_1, \ldots, a_d$ and $\theta_1, \theta_2, \ldots, \theta_{L-1}$ are estimated using maximum likelihood.

From the above expressions, we have

$$p(Y \leq 1) = \frac{1}{1 + e^{-(\theta_1 - aX)}}$$

$$\vdots$$

$$p(Y \leq L-1) = \frac{1}{1 + e^{-(\theta_{L-1} - aX)}},$$

where $a = (a_1, \ldots, a_d)$ and $X = (X_1, \ldots, X_d)^T$. Consequently,

$$p(Y = 1) = p(Y \leq 1)$$

$$p(Y = 2) = p(Y \leq 2) - p(Y \leq 1)$$

$$\vdots$$

$$p(Y = L) = 1 - p(Y \leq L-1)$$

Given a new data point $(x_1, ..., x_d)$, we compute all the probabilities $p(Y = \ell)$, $\ell = 1, 2, ...L$, and assign the label with the largest probability.

EXERCISES

1. Consider the 1-NN regression example given in Figure. 9.3. Redraw Figure 9.3(b) assuming that $k = 2$. Same for $k = 3$. (Since the actual coordinates of the data points are not available, redraw the regression curves approximately.)

2. Consider the example of the two sensors given in Sections 9.2.1 and 9.2.2, but now the labeled data is given in Table 9.5.

   a. Assume that the sensors produce only a single observation given in the first column (feature 1) of Table 9.5. Use the naive Bayes classifier to determine the label of the new data points: 1,2,3,4,5.

   b. Now assume that the sensors produce 2-tuples given by the two columns (feature 1 and feature 2) of Table 9.5. Use the naive Bayes classifier to determine the label of the new data points: (1,1), (2,2), (3,2), (5,1).

3. A small change in the training data set can result to a different decision tree. Consider the example given in Section 9.3.1, but now assume that the training data set D used in the example consists of the additional point: (0,0,1,1); i.e., $D = \{(1,1,1,1), (1,0,1,2), (0,1,1,1), (1,0,0,2), (0,0,1,1)\}$. Use the information gain criterion to build a decision tree. Contrast this decision tree with the one in the example in Section 9.3.1.

4. The Gini impurity reaches its maximum value when all classes in the data set D have equal probability.

   a. Consider a data set with two classes, and let p_1 and p_2 be the percent of data points that are labeled 1 and 2, respectively ($p_1 + p_2 = 1$). Plot the Gini index by varying p_1 from 0.1 to 1. Observe that the peak of the curve occurs when $p_1 = p_2 = 0.5$.

   b. Consider a data set with L classes. Plot the Gini index by varying L from 1 to 200, while keeping $p_1 = \cdots = p_L = 1/L$.

TABLE 9.5   Data for problem 2

Feature 1	Feature 2	Location
1	1	a
2	1	b
3	2	b
4	2	b
2	2	a
5	1	b
1	1	b
5	2	a

TABLE 9.6   Data for problem 5

Temperature	Humidity	Wind	Action
High	High	No	No
High	High	Yes	No
High	High	No	Yes
Mild	High	No	Yes
Cool	Normal	No	Yes
Cool	Normal	Yes	No
Cool	Normal	Yes	Yes
Mild	High	No	No
Cool	Normal	No	Yes
Mild	Normal	No	Yes
Mild	Normal	Yes	Yes
Mild	High	Yes	Yes
High	Normal	No	Yes

5. A multi-sensor station collects weather related information which is used to make an unspecified decision. The training data set is given in Table 9.6. There are three features: temperature {high, mild, cool}, humidity {high, normal}, and wind {yes, no}, and the label is binary {yes, no}.

   a. Use the information gain criterion to determine which feature to split first. Then, for each child determine again which feature to split.

   b. Same as above, but now use the Gini impurity criterion.

   c. Compare and discuss the two sets of results

6. Consider the tree is Figure 9.15. The first iteration of the cost complexity pruning method is given in Table 9.4. Execute the remaining steps of the algorithm and give the sequence of trees $T^{(1)}$, $T^{(2)}$, ..., $T^{(k)}$ and corresponding a values.

CLASSIFICATION PROJECT

The objective of this project is to apply and compare the three classification techniques presented in this chapter, i.e., the k-nearest neighbor, the naive Bayes classifier, and decision trees. For this analysis you can use any software, such as, Python, R, MatLab, and SAS, or a combination of functions from different software.

Generate a labeled training data set following the instructions in the clustering project described in Chapter 8. In addition, following the same instructions, generate a testing data set of 100 data points per group. Use the training data set to develop the models. Then, compare the models by calculating the percent of misclassifications using the testing data set. Discuss your results.

REFERENCES

1. T. Hastie, R. Tibshirani, J. Friedman *The Elements of Statistical Learning*, Second Edition, Springer, 2017. Chapter 6.
2. Y. Freund and R.E. Schapire. "A decision-theoretic generalization of on-line learning and an application to boosting". *Journal of Computer and System Sciences*, 55: 119 (1997).

In addition, there are numerous references in the Internet on the topics discussed in this chapter, and the reader is strongly advised to read some of this material. We note that a good discussion on the logistic regression can be found in Wikipedia.

CHAPTER 10

Artificial Neural Networks

A N ARTIFICIAL NEURAL NETWORK (ANN) IS A MODEL THAT IMITATES the way biological neurons communicate with each other. A neuron consists of a host of fine structures, called dendrites, through which it collects signals from other neurons. As shown in Figure 10.1, the nucleus processes these signals and sends out spikes of electrical activity through a long thin strand, known as the axon. The axon splits into thousands of branches, and each branch connects to the dendrites of another neuron through a synapse.

The basic building block of an artificial neural network is the *artificial neuron* which vaguely imitates a biological neuron. An artificial neural network consists of many artificial neurons which are interconnected so that to form a network. Artificial neural networks can be used for supervised and unsupervised learning. In this book, we only consider supervised learning. An artificial neural network, hereafter referred to as a neural network, can be used for classification and regression. The output is a label when used for classification, and a numerical value when used for regression.

In this chapter, we present the most commonly used neural network model, known as the *feedforward* neural network and describe the *backpropagation* algorithm used to train a neural network. We then describe three different ways that the backpropagation method can be applied to the training data, namely, the *stochastic gradient descent*, the *batch gradient descent*, and the *mini-batch gradient descent*. Finally, a discussion on how to avoid overfitting and also how to select the hyper-parameters of a neural network is given. A set of exercises and a neural network project is given at the end of the chapter.

10.1 THE FEEDFORWARD ARTIFICIAL NEURAL NETWORK

The structure of an artificial neuron used in a neural network is shown in Figure 10.2. It consists of n inputs $X_1, X_2, ..., X_n$, which are analogous to the dendrites, a processing unit which is analogous to the nucleus of a biological neuron, and an output which is analogous to that of the axon. Each input X_i is weighted by a weight w_i, $i = 1, 2, ..., n$, and the total weighted sum $\sum_{i=1}^{n} w_i X_i$ is fed to the processing unit, which produces an output using a non-linear function, known as the *activation function*. The output value may be the final answer or it may become input to other artificial neurons.

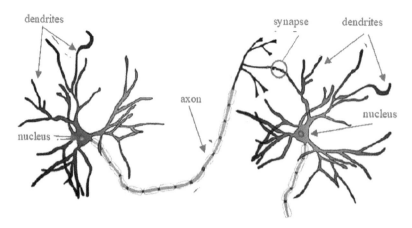

FIGURE 10.1 Neuron connectivity.

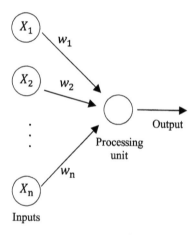

FIGURE 10.2 An artificial neuron.

Artificial neurons can be combined to form a neural network. The simplest neural network model is the feedforward network, an example of which is given in Figure 10.3. It is derived by combining k artificial neurons, and it consists of an input layer, k processing units, and an output layer. The set of all the processing units is referred to as the *hidden* layer, because it is invisible from the outside. The nodes in the input, hidden, and output layers are also referred to as neurons or units. The model in Figure 10.3 consisting of a single hidden layer is known as a *perceptron*.

The feedforward model derives its name from the fact that the information flows forward from left to right without any back loops. Each connection between a node in a layer and a node in the next layer is associated with a weight, as in the connections of the artificial neuron. The weighted sum of the n input nodes is fed to each of the k nodes in the hidden layer, which transforms the information using an activation function (see next section). The weighted sum of the resulting outputs from the k hidden nodes are fed to the m output nodes, which may also apply an activation function before the final answer is calculated.

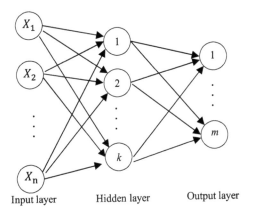

Input layer   Hidden layer   Output layer

FIGURE 10.3 A perceptron.

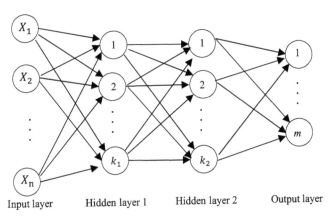

Input layer   Hidden layer 1   Hidden layer 2   Output layer

FIGURE 10.4 A feedforward neural network with two hidden layers.

Feedforward neural networks always consist of a single input layer and a single output layer, and one or more hidden layers. In Figure 10.4, we show a neural network consisting of two hidden layers. A feedforward network with more than one hidden layer is referred to as *multi-layer perceptron* (MLP).

There is some disagreement as to how the layers in a multi-layer perceptron are counted. A multi-layer perceptron with three hidden layers, i.e., a total of five layers when we include the input and output layers, is often referred to as four-layer neural network; i.e., the input layer is excluded from the count because the input nodes are not active; i.e., they do not do any calculations since they only represent the input variables. Consequently, according to this definition, an r-layer neural network consists of an input layer, r-1 hidden layers, and an output layer.

Often a *bias node* is added in each layer of a neural network. An example of a feedforward neural network with added bias nodes is given in Figure 10.5. A bias node provides the value 1 and its connections to the nodes in the next layer are associated with weights, as in the case of the regular nodes. An explanation of the use of a bias node is given in the next section, where we discuss activation functions. Bias nodes are typically not shown in neural network diagrams.

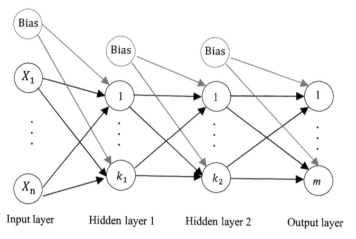

FIGURE 10.5 A feedforward neural network with bias.

10.2 OTHER ARTIFICIAL NEURAL NETWORKS

Other neural network configurations have been extensively studied, such as, *feedforward neural networks with short-cut connections, recurrent neural network* (RNNs), and *convolutional neural networks* (CNNs).

A feedforward neural network with short-cut connections is given in Figure 10.6. The short-cut (or, skip-layer) connections are indicated by the blue dotted lines. We see that the two input nodes are directly connected to the two output nodes, in addition to being connected to the hidden nodes.

RNNs constitute a large class of neural networks used primarily in image/video captioning, word prediction, translation of one language to another, and image processing. They primarily address the issue where the input data set does not consist of independent data points. Typically, a data set D consists of n data points which are independent of each other. However, there are applications where this is not the case. For instance, in order to predict the next word in a sentence we need to know which words came before it. RNNs are trained to predict the next element of a sequence. They are called recurrent because they perform the same task for every element of a sequence of data points, with the output being depended on the previous computations.

CNNs are a class of neural networks used primarily in image recognition and language processing. They have been used successfully in identifying faces and different objects and are used for vision in robots and self-driving cars. They are designed to take advantage of the hierarchical structure of an input image or a speech signal. A CNN consists of one or more hidden convolutional layers, followed by hidden layers as in a multi-layer perceptron. The convolutional layers are used to detect features of an image and the multi-layer perceptron hidden layers are used for classification.

The current processing abilities of computers have permitted the training of very large neural networks using a large amount of data. The use of such large neural networks is known as *deep learning*. The term *deep* refers to the use of neural networks with a large

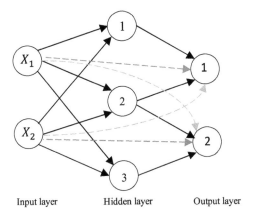

FIGURE 10.6 A feedforward neural network with short-cut connections.

number of hidden layers. The main architectures of deep learning are: CNNs, RNNs, generative adversarial networks, and recursive neural networks.

10.3 ACTIVATION FUNCTIONS

An activation function is used in a node in order to calculate its output. This is a non-linear function applied on the weighted sum of observations fed to the node. Let us consider the perceptron shown in Figure 10.3. A hidden node j receives n inputs $X_1, X_2, ..., X_n$, one from each input node i, and each input is weighted by a weight w_{ij}, $i = 1, 2, ..., n$. The total weighted sum $\sum_{i=1}^{n} w_{ij} X_i$ is used as input to an activation function, the output of which is fed to the nodes of the next layer.

Different activation functions are used. The simplest one is the *binary threshold* function. For a given threshold T, the output Y is set to 1 if $\sum_{i=1}^{n} w_i X_i \geq T$ and to 0 if $\sum_{i=1}^{n} w_i X_i < T$. This step function is not differentiable, a requirement for the backpropagation algorithm, described later on in this chapter, used to train a neural network.

The *sigmoid* function is a popular differentiable non-linear activation function. As shown in Figure 10.7, it is an S-shaped curve that takes values in (0, 1). It is the same as the standard logistic function defined by the expression:

$$s(x) = \frac{1}{1 + e^{-x}},$$

where $x = \sum_{i=1}^{n} w_i X_i$.

Another popular differentiable non-linear activation function is the *hyperbolic tangent* (tan*h*). As shown in Figure 10.8, it is also an S-shaped curve but it maps an input x to a value in the range (−1, 1). It is defined as follows:

$$\tan h(x) = \frac{e^x - e^{-x}}{e^x + e^{-x}}.$$

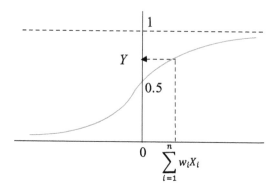

FIGURE 10.7 The sigmoid function.

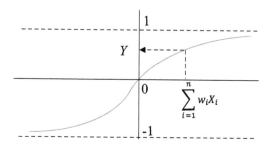

FIGURE 10.8 The tan*h* function.

We note that the sigmoid function and the hyperbolic tangent are related as follows:

$$s(x) = \frac{1}{1+e^{-x}} = \frac{e^x}{e^x+1} = \frac{1}{2} + \frac{1}{2}\tan h\left(\frac{x}{2}\right).$$

The *rectified linear unit* (ReLU) is currently the most widely used activation function in deep learning. This is due to its simplicity and effectiveness. It is given by the expression:

$$f(x) = \max\{0,\, x\}.$$

As shown in Figure 10.9, it returns a zero if it receives a negative input or the actual value it receives if it is positive. For negative x values, the derivative of the ReLU is zero, and for positive x values it is equal to 1.

The above activation functions are used in the hidden nodes of a neural network. In the output layer, different activation functions are used dependent upon the type of problem. These activation functions, known as *output activation functions*, are the identity, the sigmoid, and the *softmax*.

The identity function leaves the input value unchanged. It is used for regression with a continuous dependent variable. The sigmoid function is used in the case of binary

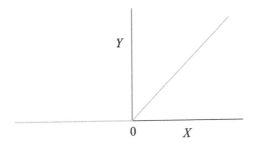

FIGURE 10.9 The ReLU function.

classification; i.e., the label takes only two values. The softmax function is a generalization of the sigmoid function, and it is used in the case of a single label that takes more than two values. It takes as input an un-normalized vector which it normalizes into a probability distribution. It is given by the following expression:

$$Z_i = \frac{e^{Y_i}}{\sum_{j=1}^{k} e^{Y_j}}, \, i = 1, 2, \ldots, k.$$

For instance, let 4, 2, −3 be the offered set of values to the softmax function. These values are transformed to the probability vector $(0.88, 0.119, 0.0008)$.

As mentioned previously, a bias node can be added to a layer. This node always provides the value of 1, which becomes part of the total input to the nodes of the next layer. For instance, let us assume that a bias node is added to the input layer of the perceptron in Figure 10.3, and let w_{0j} be the weight associated with its connection to the jth hidden node, $j = 1, 2, \ldots, k$. Then, the total input to the jth hidden node is $\sum_{i=1}^{n} w_{ij} X_i + w_{0j}$. The effect of this is that the origin point of the activation function of the jth hidden node, i.e., $(0, 1/2)$ for the sigmoid function and $(0, 0)$ for the *tanh*, is moved by w_{0j}, which may be positive or negative. A bias node allows for more flexibility when training a neural network. It can be used in the output layer if a logistic-type of activation function is used.

10.4 CALCULATION OF THE OUTPUT VALUE

In this section, we give an example of the progression of the calculations through a neural network from the input layer to the output layer. Let us assume that we have trained the neural network shown in Figure 10.10, consisting of two input nodes, a single hidden layer with three hidden nodes, and a single output node. Let $w_{i1}^{(1)}, w_{i2}^{(1)}, w_{i3}^{(1)}$ be the weights of the connections between the input node i, $i = 1, 2$, and the three hidden nodes 1, 2, and 3, and let $w_{11}^{(2)}, w_{21}^{(2)}, w_{31}^{(2)}$ be the weights of the connections between the three hidden nodes, 1, 2, and 3, and the output node 1. The value of the weights is known, since we assumed that the neural network has been trained.

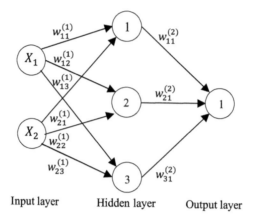

FIGURE 10.10 An example.

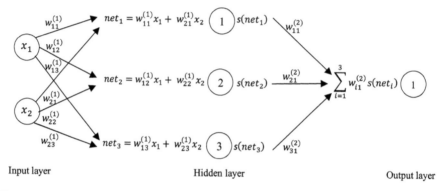

FIGURE 10.11 Progression of calculations.

Now, let us assume that a new data point (x_1, x_2) becomes available. Then, the output value from the neural network is calculated as follows. First, the weighted sum net_i is calculated for each hidden node i, $i = 1, 2, 3$. We have

$$net_1 = w_{11}^{(1)} x_1 + w_{21}^{(1)} x_2.$$

$$net_2 = w_{12}^{(1)} x_1 + w_{22}^{(1)} x_2.$$

$$net_3 = w_{13}^{(1)} x_1 + w_{23}^{(1)} x_2.$$

Subsequently, each hidden node i calculates an output value using the sigmoid activation function $s(x) = 1/(1 + e^{-x})$, where $x = net_i$. Let $s(net_1)$, $s(net_2)$, $s(net_3)$ be the three output values. Then, the weighted sum of these values

$$\sum_{i=1}^{3} w_{i1}^{(2)} s(net_i)$$

is offered to the output node which may apply an activation function to it in order to produce the final output. Figure 10.11 shows the progression of the calculations through the neural network.

10.5 SELECTING THE NUMBER OF LAYERS AND NODES

The number of nodes at the input layer is determined by the number of features in the data set. If there are n features, then the number of input nodes is also n, one per feature.

The number of nodes at the output layer is determined by the type of problem at hand. If we use a neural network for multivariable regression, then the number of input nodes is set equal to the number of independent variables, and the output layer consists of a single node which represents the dependent variable. For example, let us consider the multivariable linear regression example described in the introduction of Chapter 5. In this example, we want to estimate the amount of time Y it takes for a process to run on a server as a function of the CPU time consumed X_1, the number of disk I/Os issued X_2, the number of pages paged in X_3, and the number of Ethernet packets transmitted and received X_4. In this case, we can use a feedforward neural network to estimate Y with four input nodes and a single output node.

In addition to multivariable regression, a neural network can be used for *multivariate* regression. In a multivariate regression, we have one or more independent variables, $X_1, X_2, \ldots, X_n$ as in the case of a multivariable regression, but the dependent variable Y is d-dimensional, i.e., $Y = (Y_1, Y_2, \ldots, Y_d)$. In this case, the input layer consists of n input nodes and the output layer of d nodes.

If we use a neural network for classification, then the data set is in the form $(x_{i1}, x_{i2}, \ldots, x_{id}, \ell_i)$, $i = 1, 2, \ldots, n$, where d is the number of features, x_{ij} is the value of the jth feature of the ith data point and ℓ_i is the label of the ith data point. In this case, the input layer consists of d input nodes, one per feature, and the output layer consists of a single node if the label is binary.

If the label takes more than two values, then it is *one-hot encoded*; i.e., each nominal value that the label takes is converted to a binary string where only a single bit is 1 and the remaining bits are zero. For instance, let us assume that the label takes the values: red, green, and blue. Then, we create three new binary variables, L_1, L_2 and L_3, where $L_i = 0, 1$, $i = 1, 2, 3$. Each nominal value is associated with a different variable, i.e., red with L_1, green with L_2, and blue with L_3. Accordingly, the nominal value red is encoded as $(1, 0, 0)$, green as $(0, 1, 0)$, and blue as $(0, 0, 1)$. An alternative simpler encoding would be to encode the nominal values red, green, and blue as 1, 2, 3, respectively. However, this may be problematic since category 1 appears to be closer to category 2 than category 3!

In the general case, the number of variables used in one-hot encoding is equal to the number of different values that the label takes. Each of these variables is represented by a different output node. The softmax function is used at the output layer to associate a different probability with each output node.

For example, let us consider the classical problem of recognizing hand-written single digit numbers using the MNIST (modified National Institute of Standards and Technology) database [1]. This is a large database of handwritten digits that is commonly used for training various image processing systems. In this database, an image of a hand-written digit consists of a square of 28×28 pixels, i.e., a total of 784 pixels, where each pixel takes the values in the gray scale from 0 to 255. The label indicates the digit, and it is one-hot encoded, represented by the labels $L_1, L_2, \ldots, L_{10}$. Consequently, the ith data point has the form $(x_{i1}, x_{i2}, \ldots, x_{i784}, \ell_1, \ell_2, \ldots, \ell_{10})$, $i = 1, 2, \ldots, n$, where x_{ij} represents the jth pixel of the ith image and takes values from 0 to 255, and $\ell_1 = 0, 1, i = 1, 2, \ldots, 10$. In this example, the input layer of the neural network consists of 784 nodes, and the output layer consists of 10 nodes, each associated with a different digit. The softmax activation function is used at the output layer to determine the digit. When we run an unlabeled single-digit image through the trained neural network, the result from the softmax function will be a vector of 10 probabilities. Each probability is associated with a different output node, which in turn corresponds to one of the 10 digits. The digit with the largest probability is the most likely the digit on the image.

The number of output and input nodes is automatically selected by the software. The number of hidden layers and the number of nodes per hidden layer is selected by the user, and it is not straightforward. The idea behind it is to select the number of hidden layers and nodes per layer so that the neural network has a low error without overfitting. Too few nodes may lead to a high error, and too many nodes may lead to overfitting and also increase the amount of time it takes to train a neural network.

Before the advent of deep learning, at most two hidden layers sufficed. However, additional layers are required in order to train neural networks for complex problems tackled in deep learning, such as, object recognition, handwritten character recognition, and face recognition. For instance, in the problem of identifying hand-written digits, a feedforward neural network was trained with a single hidden layer of 800 nodes. It had an error of 0.7%, i.e., 0.7% of the images in the database were misclassified. In addition, another feedforward neural network was trained with five hidden layers with 2500, 2000, 1500, 1000, and 500 nodes and an error of 0.35%. Different CNNs were also trained, such as one with five hidden layers and with 40, 80, 500, 1000, 2000 nodes and an error of 0.31%. For additional information, see [1].

The issue on how to determine the number of hidden layers, the number of nodes per hidden layer, and other parameters is discussed in Section 10.10.

10.6 THE BACKPROPAGATION ALGORITHM

Training a neural network means calculating the weights of the connections, once the neural network configuration has been determined, so that the predicted labels of the labeled data points have a low error. The *backward propagation of errors* algorithm, commonly referred to as the backpropagation algorithm, is a common method for training neural networks.

Before we present the algorithm, we define a number of variables as shown in Figure 10.12. It is important that the reader becomes familiar with these variables since the derivation of the algorithm is somewhat involved.

Let n be the number of input nodes, k the number of hidden nodes, and m the number of output nodes. The input and output nodes may include a bias node. We will use the subscript $i, j,$ and l to indicate a node in the input, hidden, and output layer, respectively. Let

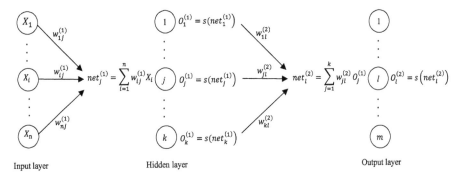

FIGURE 10.12 The variables used in the backpropagation algorithm.

$w_{ij}^{(1)}$, $i = 1, 2, \ldots, n, j = 1, 2, \ldots, k,$ be the weight of the connection between the ith input node and the jth hidden node, and $w_{jl}^{(2)}$, $j = 1, 2, \ldots, k, l = 1, 2, \ldots, m,$ be the weight of the connection between the jth hidden node and the lth output node. Let $X_1, X_2, \ldots, X_n$, be the n inputs from the input layer. Then, the total input $net_j^{(1)}$ into the jth hidden node is

$$net_j^{(1)} = \sum_{i=1}^{n} w_{ij}^{(1)} X_i, j = 1, 2, \ldots, k.$$

Define $O_j^{(1)}$ to be the output from the hidden node j. We have

$$O_j^{(1)} = s\left(net_j^{(1)}\right), j = 1, 2, \ldots, k,$$

where $s(x)$ is the sigmoid activation function; i.e.,

$$s\left(net_j^{(1)}\right) = \frac{1}{1 + e^{-net_j^{(1)}}}, j = 1, 2, \ldots, k.$$

The total input into an output node l is

$$net_l^{(2)} = \sum_{j=1}^{k} w_{jl}^{(2)} O_j^{(1)}, l = 1, 2, \ldots, m$$

We assume that a sigmoid function is used at the output layer. Therefore, the output from the lth output node $O_l^{(2)}$ is

$$O_l^{(2)} = s\left(net_l^{(2)}\right) = s\left(\sum_{j=1}^{k} w_{jl}^{(2)} O_j^{(1)}\right), l = 1, 2, \ldots, m.$$

Now, if we run a data point x through the neural network we will obtain a set of output values $O_{x,l}^{(2)}, l = 1, 2, \ldots m$. Since this data point is already labeled, we also have the correct output values $t_{x,l}$, $l = 1, 2, ..., m$. Then, the sum of the squared differences for all the data points x in the training data set D, i.e.,

$$E_Q = \frac{1}{2}\sum_{x \in D}\sum_{l=1}^{m}\left(O_{x,l}^{(2)} - t_{x,l}\right)^2,$$

is a metric of the cost, i.e., of the error of the estimation. The constant 1/2 is added in order to simplify the derivative of E_Q, as will be seen below. The cost could also be expressed as the average of all the data points n, i.e.,

$$E_Q = \frac{1}{2n}\sum_{x \in D}\sum_{l=1}^{m}\left(O_{x,l}^{(2)} - t_{x,l}\right)^2.$$

These two cost functions are equivalent, and they are referred to as the quadratic cost function or the mean squared error (MSE) cost function. They are typically used in neural networks for regression. For classification, the following cost function based on the cross entropy is used:

$$E_{CE} = \frac{1}{n}\sum_{x \in D}\sum_{l=1}^{m}\left\{t_{x,l}\ln O_{x,l}^{(2)} + \left(1 - t_{x,l}\right)\log_e\left(1 - O_{x,l}^{(2)}\right)\right\},$$

The cross entropy function is also used predominantly in deep learning. In this chapter, we will only make use of the cost function E_Q.

The backpropagation algorithm uses the training data set to find the best values for all the weights so that E_Q is minimum. The algorithm is used in conjunction with an optimization method, such as the gradient descent, which is an iterative optimization algorithm for finding the minimum of a function. Before we proceed to describe the algorithm, we review briefly the gradient descent method.

10.6.1 The Gradient Descent Algorithm

Let $f(x)$ be a single-variable function. The gradient descent algorithm is based on the observation that if $f(x)$ is defined and differentiable in a neighborhood of a point x, then $f(x)$ decreases the fastest if one goes from x in the direction of the negative derivative at x, i.e., $-f'(x)$. Specifically, we start with an arbitrarily selected point x_0, and then we construct an adjacent point x_1 for which $f(x_0) \geq f(x_1)$, by setting $x_1 = x_0 - \gamma f'(x_0)$. The parameter γ is known as the *learning rate*. It is set to a small positive value, and it is used to make sure that we take small steps towards the minimum. For, if we take large steps, then we may overshoot it. Subsequently, we construct a new point $x_2 = x_1 - \gamma f'(x_1)$, for which $f(x_1) \geq f(x_2)$, and so on until we converge to a minimum. This minimum may be a local one and not the global one. On the other hand, if we want to locate the maximum of $f(x)$, then we set $x_1 = x_0 + \gamma f'(x_0)$. Note that γ can change at each step.

The gradient of a multivariable function is a generalization of the derivative of a single-variable function. It is denoted by the symbol $\nabla f(x)$ or $\bar{\nabla} f(x)$, where ∇ is the nabla symbol, and it is a vector computed as follows. Let $f(x, y, z)$ be a multivariable function. Then,

$$\nabla f\left(x, y, z\right) = \left(\frac{\partial f}{\partial x}, \frac{\partial f}{\partial y}, \frac{\partial f}{\partial z}\right)^{T}.$$

The gradient decent algorithm for a multi-variable function is similar to that of the single-variable algorithm. Again, it is based on the observation that if a multivariable function $f(x)$ is defined and differentiable in a neighborhood of a point x, then $f(x)$ decreases the fastest if one goes from x in the direction of the negative gradient of $f(x)$ at x, i.e., $-\nabla f(x)$; i.e., let $x_1 = x_0 - \gamma \nabla f(x_0)$, then $f(x_0) \geq f(x_1)$. For instance, if we assume that x is three-dimensional, then x_1 is obtained as follows:

$$\begin{pmatrix} x_{11} \\ x_{12} \\ x_{13} \end{pmatrix} = \begin{pmatrix} x_{01} \\ x_{02} \\ x_{03} \end{pmatrix} - \gamma \begin{pmatrix} \partial f / \partial x_{01} \\ \partial f / \partial x_{02} \\ \partial f / \partial x_{03} \end{pmatrix}.$$

Subsequently, we construct a new point $x_2 = x_1 - \gamma \nabla f(x_0)$, for which $f(x_1) \geq f(x_2)$, and so on. Iterating in this fashion, we calculate a sequence of points $x_0, x_1, \ldots, x_n$, for which

$$f\left(x_0\right) \geq f\left(x_1\right) \geq \cdots \geq f\left(x_n\right),$$

which converges to a minimum, which may not be the global minimum.

A pictorial view of the iterative procedure is shown in Figure 10.13. The function $f(x, y)$ has a bowl shape, and the concentric contours in the figure are the projection of $f(x, y)$ on the two-dimensional space (x, y). The values of $f(x, y)$ along each contour is constant. We see that the algorithm generates a sequence of points that lead towards the minimum.

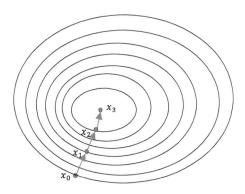

FIGURE 10.13 An example of gradient descent.

10.6.2 Calculation of the Gradients

The backpropagation algorithm calculates the derivative of the cost function for each weight for a given data point, and then updates the weight values. In order to simplify the notation, we will use the term E to indicate the cost of a single data point x; i.e.,

$$E = \frac{1}{2}\sum_{l=1}^{m}\left(O_l^{(2)} - t_l\right)^2,$$

where $O_l^{(2)}$ and t_l, $l = 1, 2, ...m$, is the computed output at the lth node and corresponding label of the data point. Also, let

$$E_l = \frac{1}{2}\left(O_l^{(2)} - t_l\right)^2$$

be the cost of the data point associated with the lth output node. We have that

$$E = \frac{1}{2}\sum_{l=1}^{m}E_l.$$

To help understand the algorithm better, we will make use of the example in Figure 10.14. It consists of two input nodes 1 and 2, two hidden nodes 1 and 2, and two output nodes 1 and 2. A bias node that supplies the value of 1 is included in the input and hidden layers. Let us now assume an initial value of all the weights as shown in Figure 10.14, and let $(1, 2, 0.2, 0.8)$ be a data point from the training data set, i.e., $X_1 = 1$, $X_2 = 2$, $t_1 = 0.2$, and $t_2 = 0.8$. We note that data points also referred to as *patterns*.

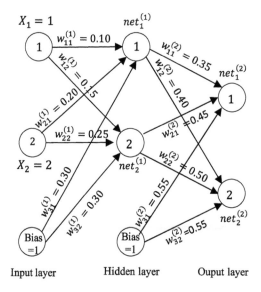

FIGURE 10.14 Example for the backpropagation algorithm.

Given the initial values of the weights, we can calculate the cost E for this data point. We have

$$net_1^{(1)} = 0.10 \times 1 + 0.20 \times 2 + 0.30 \times 1 = 0.8$$

$$net_2^{(1)} = 0.15 \times 1 + 0.25 \times 2 + 0.30 \times 1 = 0.95$$

$$O_1^{(1)} = s\left(net_1^{(1)}\right) = \frac{1}{1+e^{-0.8}} = 0.68997$$

$$O_2^{(1)} = s\left(net_2^{(1)}\right) = \frac{1}{1+e^{-0.95}} = 0.72112.$$

Using the output values from the hidden layer, we have

$$net_1^{(2)} = 0.35 \times 0.68977 + 0.45 \times 0.72112 + 0.55 \times 1 = 1.11592$$

$$net_2^{(2)} = 0.40 \times 0.68977 + 0.50 \times 0.72112 + 0.55 \times 1 = 1.18647$$

$$O_1^{(2)} = s\left(net_1^{(2)}\right) = \frac{1}{1+e^{-1.11592}} = 0.75323$$

$$O_2^{(2)} = s\left(net_2^{(2)}\right) = \frac{1}{1+e^{-1.18647}} = 0.76611$$

Therefore, the total cost is

$$E = \frac{1}{2}\left(O_1^{(2)} - t_1\right)^2 + \frac{1}{2}\left(O_2^{(2)} - t_2\right)^2$$

$$= \frac{1}{2}(0.75323 - 0.2)^2 + \frac{1}{2}(0.76611 - 0.8)^2 = 0.15361.$$

Following the gradient descent method, we calculate a new set of values for all the weights, so that the total cost E is reduced. For this, we need to calculate:

$$\partial E / \partial w_{ij}^{(1)}, i = 1, 2, \ldots, n, j = 1, 2, \ldots, k,$$

and

$$\partial E / \partial w_{jl}^{(2)}, j = 1, 2, \ldots, k, l = 1, 2, \ldots, m.$$

Before we proceed, we obtain the derivative of the sigmoid function $s(x)$. It can be shown that it is equal to $s(x) - s^2(x)$. We have

$$\frac{ds(x)}{dx} = \left(\frac{1}{1+e^{-x}}\right)' = \frac{e^{-x}}{\left(1+e^{-x}\right)^2}$$

$$= \frac{1-1+e^{-x}}{\left(1+e^{-x}\right)^2} = \frac{1+e^{-x}}{\left(1+e^{-x}\right)^2} - \frac{1}{\left(1+e^{-x}\right)^2}$$

$$= \frac{1}{1+e^{-x}} - \frac{1}{\left(1+e^{-x}\right)^2}$$

$$= s(x) - s^2(x) = s(x)(1-s(x)).$$

a. *Calculation of $\partial E / \partial w_{jl}^{(2)}$*

As shown pictorially in Figure 10.15, $w_{jl}^{(2)}$ only affects the cost function E_l. Consequently, we have

$$\frac{\partial E}{\partial w_{jl}^{(2)}} = \frac{\partial}{\partial w_{jl}^{(2)}} \frac{1}{2} \sum_{l=1}^{m} \left(O_l^{(2)} - t_l\right)^2$$

$$= \frac{\partial}{\partial w_{jl}^{(2)}} \left\{ \frac{1}{2}\left(O_1^{(2)} - t_1\right)^2 + \cdots + \frac{1}{2}\left(O_l^{(2)} - t_l\right)^2 + \cdots + \frac{1}{2}\left(O_m^{(2)} - t_m\right)^2 \right\}.$$

$O_l^{(2)}$ is the only term that is a function of $w_{jl}^{(2)}$, while the other terms become zero after the derivative is applied. Therefore, we have

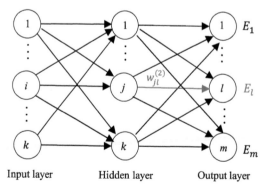

Input layer   Hidden layer   Output layer

FIGURE 10.15 $w_{jl}^{(2)}$ only affects E_l.

$$\frac{\partial E}{\partial w_{jl}^{(2)}} = \frac{\partial}{\partial w_{jl}^{(2)}} \frac{1}{2} \left(O_l^{(2)} - t_l \right)^2 = \frac{\partial E_l}{\partial w_{jl}^{(2)}},$$

where E_l was defined as the cost associated with the lth output node. The above derivative is calculated by applying the chain rule in calculus. We recall that this is a rule for calculating the derivative of the composition of two or more functions. For example, let $y = f(x)$, $z = g(y)$, and $u = h(z)$. Then,

$$\frac{\partial u}{\partial x} = \frac{\partial u}{\partial z} \frac{\partial z}{\partial y} \frac{\partial y}{\partial x}.$$

We have

$$\frac{\partial E_l}{\partial w_{jl}^{(2)}} = \frac{\partial E_l}{\partial O_l^{(2)}} \frac{\partial O_l^{(2)}}{\partial net_l^{(2)}} \frac{\partial net_l^{(2)}}{\partial w_{jl}^{(2)}}$$

where

$$\frac{\partial E_l}{\partial O_l^{(2)}} = \frac{\partial}{\partial O_l^{(2)}} \left(\frac{1}{2} \left(\partial O_l^{(2)} \right)^2 + \frac{1}{2} \left(t_l \right)^2 - \frac{1}{2} 2 O_l^{(2)} t_l \right) = O_l^{(2)} - t_l.$$

$$\frac{\partial O_l^{(2)}}{\partial net_l^{(2)}} = \frac{\partial s \left(net_l^{(2)} \right)}{\partial net_l^{(2)}} = s \left(net_l^{(2)} \right) - s^2 \left(net_l^{(2)} \right).$$

$$= O_l^{(2)} - \left(O_l^{(2)} \right)^2 = O_l^{(2)} \left(1 - O_l^{(2)} \right).$$

$$\frac{\partial net_l^{(2)}}{\partial w_{jl}^{(2)}} = \frac{\partial}{\partial w_{jl}^{(2)}} \sum_{j=1}^{k} w_{jl}^{(2)} O_j^{(1)} = O_l^{(1)}.$$

Hence,

$$\frac{\partial E_l}{\partial w_{jl}^{(2)}} = \left(O_l^{(2)} - t_l \right) O_l^{(2)} \left(1 - O_l^{(2)} \right) O_l^{(1)}.$$

Define

$$\delta_l^{(2)} = \left(O_l^{(2)} - t_l \right) O_l^{(2)} \left(1 - O_l^{(2)} \right).$$

Then

$$\frac{\partial E}{\partial w_{jl}^{(2)}} = \delta_l^{(2)} O_j^{(1)}.$$

Consequently, each weight $w_{jl}^{(2)}, j = 1, 2,\ldots, k, l = 1, 2,\ldots,m,$ is adjusted by subtracting its update $\Delta w_{jl}^{(2)} = \gamma\left(\delta_l^{(2)} O_j^{(1)}\right)$, where γ is the learning rate; i.e.

$$w_{jl}^{(2)} - \Delta w_{jl}^{(2)} = w_{jl}^{(2)} - \gamma\left(\delta_l^{(2)} O_j^{(1)}\right).$$

Let us now go back to our numerical example in Figure 10.14. We have

$$\delta_1^{(2)} = \left(O_1^{(2)} - t_1\right) O_1^{(2)}\left(1 - O_1^{(2)}\right)$$

$$= \left(0.75323 - 0.2\right) \times 0.75323 \times \left(1 - 0.75323\right) = 0.10283.$$

$$\delta_2^{(2)} = \left(O_2^{(2)} - t_2\right) O_2^{(2)}\left(1 - O_2^{(2)}\right)$$

$$= \left(0.76611 - 0.8\right) \times 0.76611 \times \left(1 - 0.76611\right) = -0.00607.$$

Setting $\gamma = 0.5$, we obtain the following new weights:

$$w_{11}^{(2)} - \gamma\left(\delta_1^{(2)} O_1^{(1)}\right)$$

$$= 0.35 - 0.5 \times \left(0.10283 \times 0.68977\right) = 0.31454$$

$$w_{21}^{(2)} - \gamma\left(\delta_1^{(2)} O_2^{(1)}\right)$$

$$= 0.45 - 0.5 \times \left(0.10283 \times 0.72112\right) = 0.41292$$

$$w_{12}^{(2)} - \gamma\left(\delta_2^{(2)} O_1^{(1)}\right)$$

$$= 0.40 - 0.5 \times \left(-0.00607 \times 0.68977\right) = 0.40209$$

$$w_{22}^{(2)} - \gamma\left(\delta_2^{(2)} O_2^{(1)}\right)$$

$$= 0.55 - 0.5 \times \left(-0.00607 \times 0.72112\right) = 0.50219.$$

The weights for the bias node are updated in the same way. We have

$$w_{b1}^{(2)} - \gamma\left(\delta_1^{(2)} \times 1\right) = 0.55 - 0.5 \times 0.10283 = 0.49857$$

$$w_{b2}^{(2)} - \gamma\left(\delta_2^{(2)} \times 1\right) = 0.55 - 0.5 \times (-0.00607) = 0.55304.$$

We now proceed to calculate the derivative of E in terms of the weights between the input nodes and the hidden nodes.

b. *Calculation of* $\partial E / \partial w_{ij}^{(1)}$

As shown pictorially in Figure 10.16, $w_{ij}^{(1)}$ is part of $O_j^{(1)}$ which is part of $O_1^{(2)}, ..., O_l^{(2)}, ..., O_m^{(2)}$, which are part of all the cost terms $E_1, ..., E_l, ..., E_m$. Therefore,

$$\frac{\partial E}{\partial w_{ij}^{(1)}} = \frac{\partial}{\partial w_{ij}^{(1)}} \frac{1}{2} \sum_{l=1}^{m}\left(O_l^{(2)} - t_l\right)^2 = \sum_{l=1}^{m} \frac{\partial}{\partial w_{ij}^{(1)}} \frac{1}{2}\left(O_l^{(2)} - t_l\right)^2.$$

Let us consider the lth term. We have

$$\frac{\partial}{\partial w_{ij}^{(1)}} \frac{1}{2}\left(O_l^{(2)} - t_l\right)^2 = \left(O_l^{(2)} - t_l\right)\frac{\partial O_l^{(2)}}{\partial w_{ij}^{(1)}}.$$

Applying the chain rule we have

$$\left(O_l^{(2)} - t_l\right)\frac{\partial O_l^{(2)}}{\partial w_{ij}^{(1)}} = \left(O_l^{(2)} - t_l\right)\frac{\partial O_l^{(2)}}{\partial net_l^{(2)}}\frac{\partial net_l^{(2)}}{\partial O_j^{(1)}}\frac{\partial O_j^{(1)}}{\partial net_j^{(1)}}\frac{\partial net_j^{(1)}}{\partial w_{ij}^{(1)}}.$$

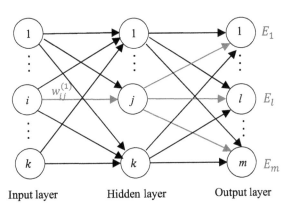

FIGURE 10.16 $w_{ij}^{(1)}$ affects all costs $E_1, ..., E_l, ..., E_m$.

We now calculate each individual derivative. We have

$$\frac{\partial O_l^{(2)}}{\partial net_l^{(2)}} = \frac{\partial s\left(net_l^{(2)}\right)}{\partial net_l^{(2)}} = s\left(net_l^{(2)}\right) - s^2\left(net_l^{(2)}\right)$$

$$= O_l^{(2)} - \left(O_l^{(2)}\right)^2 = O_l^{(2)}\left(1 - O_l^{(2)}\right).$$

$$\frac{\partial net_l^{(2)}}{\partial O_j^{(1)}} = \frac{\partial}{\partial O_j^{(1)}} \sum_{j=1}^{k} w_{jl}^{(2)} O_j^{(1)} = w_{jl}^{(2)}.$$

$$\frac{\partial O_j^{(1)}}{\partial net_j^{(1)}} = \frac{\partial s\left(net_j^{(1)}\right)}{\partial net_j^{(1)}} = s\left(net_1^{(1)}\right) - s^2\left(net_1^{(1)}\right)$$

$$= O_j^{(1)}\left(1 - O_j^{(1)}\right).$$

$$\frac{\partial net_j^{(1)}}{\partial w_{ij}^{(1)}} = \frac{\partial}{\partial w_{ij}^{(1)}} \sum_{k=1}^{n} w_{kj}^{(1)} X_k = X_i.$$

Hence, we have

$$\frac{\partial E}{\partial w_{ij}^{(1)}} = \sum_{l=1}^{m} \left(O_l^{(2)} - t_l\right) O_l^{(2)}\left(1 - O_l^{(2)}\right) w_{jl}^{(2)} O_j^{(1)}\left(1 - O_j^{(1)}\right) X_i$$

$$= O_j^{(1)}\left(1 - O_j^{(1)}\right) X_i \sum_{l=1}^{m} \left(O_l^{(2)} - t_l\right) O_l^{(2)}\left(1 - O_l^{(2)}\right) w_{jl}^{(2)}.$$

We recall that

$$\delta_l^{(2)} = \left(O_l^{(2)} - t_l\right) O_l^{(2)}\left(1 - O_l^{(2)}\right).$$

Therefore,

$$\frac{\partial E}{\partial w_{ij}^{(1)}} = O_j^{(1)}\left(1 - O_j^{(1)}\right) X_i \sum_{l=1}^{m} \delta_l^{(2)} w_{jl}^{(2)}.$$

Analogously to $\delta_i^{(2)}$, we define

$$\delta_j^{(1)} = O_j^{(1)}\left(1 - O_j^{(1)}\right)\sum_{l=1}^{m}\delta_l^{(2)}w_{jl}^{(2)},$$

Therefore, we have

$$\frac{\partial E}{\partial w_{ij}^{(1)}} = \delta_j^{(1)}X_i.$$

Consequently, each weight $w_{ij}^{(1)}$, $i = 1, 2,\ldots, n, j = 1, 2,\ldots, k$ is adjusted by subtracting its update $\Delta w_{ij}^{(1)} = \gamma\left(\delta_j^{(1)}X_i\right)$, where γ is the learning rate; i.e.,

$$w_{ij}^{(1)} - \Delta w_{ij}^{(1)} = w_{ij}^{(1)} - \gamma\left(\delta_j^{(1)}X_i\right).$$

We note that in the calculation of the above quantities, we use the old values for the weights $w_{jl}^{(2)}$, $j = 1, 2,\ldots, k, l = 1, 2,\ldots, m$, and not the updated ones. Let us now go back to our numerical example in Figure 10.14. We recall that $O_1^{(1)} = 0.68977, O_2^{(1)} = 0.72112$ $\delta_1^{(2)} = 0.10283$, and $\delta_2^{(2)} = 0.00607$. We have

$$\delta_1^{(1)} = O_1^{(1)}\left(1 - O_1^{(1)}\right)\left(\delta_1^{(2)}w_{11}^{(2)} + \delta_2^{(2)}w_{12}^{(2)}\right)$$

$$= 0.68977\times\left(1 - 0.68977\right)\left(0.10283\times 0.35 + \left(-0.00607\right)\times 0.40\right)$$

$$= 0.00718.$$

and

$$\delta_2^{(1)} = O_2^{(1)}\left(1 - O_2^{(1)}\right)\left(\delta_1^{(2)}w_{21}^{(2)} + \delta_2^{(2)}w_{22}^{(2)}\right)$$

$$= 0.72112\times\left(1 - 0.72112\right)\left(0.10283\times 0.45 + \left(-0.00607\right)\times 0.50\right)$$

$$= 0.0087.$$

Setting $\gamma = 0.5$, we obtain the following new weights:

$$w_{11}^{(1)} - \gamma\left(\delta_1^{(1)}X_1\right) = 0.10 - 0.5\left(0.00718\times 1\right) = 0.09641$$

$$w_{12}^{(1)} - \gamma\left(\delta_2^{(1)}X_1\right) = 0.15 - 0.5\left(0.0087\times 1\right) = 0.14565$$

$$w_{21}^{(1)} - \gamma\left(\delta_1^{(1)} X_2\right) = 0.20 - 0.5\left(0.00718 \times 2\right) = 0.19282$$

$$w_{22}^{(1)} - \gamma\left(\delta_2^{(1)} X_2\right) = 0.25 - 0.5\left(0.0087 \times 2\right) = 0.2413$$

$$w_{31}^{(1)} - \gamma\left(\delta_1^{(1)} \times 1\right) = 0.30 - 0.5 \times 0.00718 = 0.29641$$

$$w_{32}^{(1)} - \gamma\left(\delta_2^{(1)} \times 1\right) = 0.30 - 0.5 \times 0.0087 = 0.29565.$$

We now use all the revised weights to recalculate E. We have

$$net_1^{(1)} = 0.09641 \times 1 + 0.19282 \times 2 + 0.29641 \times 1 = 0.77846$$

$$net_2^{(1)} = 0.14565 \times 1 + 0.2413 \times 2 + 0.29565 \times 1 = 0.9239$$

$$O_1^{(1)} = s\left(net_1^{(1)}\right) = \frac{1}{1+e^{-0.77846}} = 0.68535$$

$$O_2^{(1)} = s\left(net_2^{(1)}\right) = \frac{1}{1+e^{-0.9239}} = 0.71584.$$

Using the output values from the hidden layer, we have

$$net_1^{(2)} = 0.31454 \times 0.68535 + 0.40209 \times 0.71584 + 0.49859 \times 1$$

$$= 1.00198$$

$$net_2^{(2)} = 0.41292 \times 0.68535 + 0.50219 \times 0.71584 + 0.55304 \times 1$$

$$= 1.19552$$

$$O_1^{(2)} = s\left(net_1^{(2)}\right) = \frac{1}{1+e^{-1.00198}} = 0.73145$$

$$O_2^{(2)} = s\left(net_2^{(2)}\right) = \frac{1}{1+e^{-1.19552}} = 0.76773.$$

Therefore, the cost is

$$E = \frac{1}{2}\left(O_1^{(2)} - t_1\right)^2 + \frac{1}{2}\left(O_2^{(2)} - t_2\right)^2$$

$$= \frac{1}{2}\left(0.73145 - 0.2\right)^2 + \frac{1}{2}\left(0.76773 - 0.8\right)^2 = 0.14174.$$

We note that the revised weights have moved $O_1^{(2)}$ closer to $t_1 = 0.2$, and $O_2^{(2)}$ to $t_2 = 0.8$. The cost was reduced from 0.15361 to 0.14174. This was achieved using a single data point.

c. *Summary of the Algorithm*

The algorithm takes as input a single labeled data point from the training data set D and an estimate of all the weights, and it calculates new values for all the weights. The algorithm consists of a forward and a backward pass, summarized below assuming a perceptron.

Forward pass

Given a labeled data point $(x_1, ..., x_n, t_1, ..., t_m)$ from the training set D, and a set of weights, $w_{ij}^{(1)}, i = 1, 2, ..., n, j = 1, 2, ..., k$, and $w_{jl}^{(2)}, j = 1, 2, ..., k, l = 1, 2, ..., m$, calculate the following:

$$net_j^{(1)} = \sum_{i=1}^{n} w_{ij}^{(1)} x_i, j = 1, 2, ..., k$$

$$O_j^{(1)} = s\left(net_j^{(1)}\right) = \frac{1}{1 + e^{-net_j^{(1)}}}, j = 1, 2, ..., k$$

$$net_l^{(2)} = \sum_{j=1}^{k} w_{jl}^{(2)} O_j^{(1)}, l = 1, 2, ..., m$$

$$O_l^{(2)} = s\left(net_l^{(2)}\right), l = 1, 2, ..., m$$

$$E_Q = \frac{1}{2} \sum_{l=1}^{m} \left(O_l^{(2)} - t_l\right)^2.$$

Backward pass

Select a value for the learning rate γ and calculate the updates:

$$\Delta w_{jl}^{(2)} = \gamma\left(\delta_l^{(2)} O_j^{(1)}\right), j = 1, 2, ..., k, l = 1, 2, ..., m,$$

where

$$\delta_l^{(2)} = \left(O_l^{(2)} - t_l\right) O_l^{(2)} \left(1 - O_l^{(2)}\right), l = 1, 2, ..., m.$$

Set $w_{jl}^{(2)}$ to $w_{jl}^{(2)} - \Delta w_{jl}^{(2)}, j = 1, 2, ..., k, l = 1, 2, ..., m$.

Calculate the updates:

$$\Delta w_{ij}^{(1)} = \gamma \left(\delta_j^{(1)} X_i \right), \, i = 1, 2, \ldots, n, \, j = 1, 2, \ldots, k,$$

where

$$\delta_j^{(1)} = O_j^{(1)} \left(1 - O_j^{(1)} \right) \sum_{l=1}^{m} \delta_l^{(2)} w_{jl}^{(2)}, \, j = 1, 2, \ldots, k,$$

using the input values for the weights $w_{jl}^{(2)}, j = 1, 2, \ldots, k, l = 1, 2, \ldots, m$, and set the weights $w_{ij}^{(1)}$ to $w_{ij}^{(1)} - \gamma \left(\delta_j^{(1)} X_i \right), \, i = 1, 2, \ldots, n, j = 1, 2, \ldots, k.$

10.7 STOCHASTIC, BATCH, MINI-BATCH GRADIENT DESCENT METHODS

The backpropagation method described above is applied to all the training data in order to train a neural network. Three different methods are used for this: the *stochastic gradient descent*, the *batch gradient descent*, and the *mini-batch gradient descent*.

In the stochastic gradient descent training method, we run the backpropagation algorithm for each data point in the training set D using the updated weights obtained from the previous data point; i.e., we start with a randomly generated initial set of the weights, and then we run the algorithm to obtain new values of the weights. Then, using the new weights, we run the algorithm for another data point in D. We iterate in this fashion until we use all the data points in D. The final cost E_Q is the sum of all the individual costs. This method, also known as the *online* training method, is useful when the training data points are not available at the beginning but they become available one at a time.

In the batch method, we run the backpropagation algorithm for each training data point x in D assuming the same initial estimate of the weights. The weights are adjusted after we have calculated the updates for all the training data points. Let

$$\Delta_x w_{jl}^{(2)} = \gamma \left(\delta_l^{(2)} O_j^{(1)} \right), j = 1, 2, \ldots, k, l = 1, 2, \ldots, m,$$

and

$$\Delta_x w_{ij}^{(1)} = \gamma \left(\delta_j^{(1)} X_i \right), \, i = 1, 2, \ldots, n, \, j = 1, 2, \ldots, k,$$

be the updates calculated for a data point x in D. Then, the final updates are obtained by adding up all the individual updates, i.e.,

$$\Delta w_{jl}^{(2)} = \sum_{x \in D} \Delta_x w_{jl}^{(2)}, j = 1, 2, \ldots, k, l = 1, 2, \ldots, m$$

and

$$\Delta w_{ij}^{(1)} = \sum_{x \in D} \Delta_x w_{ij}^{(1)}, \, i = 1, 2, \ldots, n, \, j = 1, 2, \ldots, k.$$

These final updates are used to adjust the initial values of the weights used in the algorithm. The cost E_Q for the entire set is calculated by simply adding the cost calculated for each point.

This method is called the batch method because we use a group of data points, i.e., a batch, to calculate an update of the weights. The batch size is equal to the size of the training data set. This method, also known as the *offline* method, requires that all the training data be available.

The mini-batch gradient descent method is a combination of the above two techniques. In this case, the training data set is divided into a number of subsets, known as minibatches. Starting with an initial estimate of the weights, we apply the batch method to the first subset of the training data set. Then, we compute the updates to all the weights, which are used as the estimate of the weights in the second subset, and so on. The cost E_Q for the entire set is calculated by simply adding the cost calculated for each point. This method can be used when the training data becomes available in small numbers.

The selection of the training method depends on how the training data becomes available. If the training data is produced in real time by some process and we want to train the neural network at the same time that the process is running, then the stochastic gradient descent or the mini-batch method should be used. If all the training data is available, then any of these three methods can be used.

For large data sets, the stochastic gradient descent method can converge faster than the batch method, because it performs updates continuously. The mini-batch method can be faster than the other two methods because it can take advantage of vectorized operations to process an entire mini-batch at once. It can also hop out of local minima that might otherwise trap the batch method.

Running the algorithm through all the training data points, whether we use the stochastic, batch, or mini-batch gradient descent method, is called an *epoch*.

10.8 FEATURE NORMALIZATION

A good rule of thumb is that the input variables should be small, as this can make the gradient descent method converge much faster. This can be done using the min–max normalization or the standardization method. The min–max normalization rescales the range of a feature to $[0, 1]$ or $[-1, 1]$. Let x_{min} and x_{max} be the smallest and the largest number of the range of values of a feature. Then, each value x is rescaled to $[0, 1]$ as follows:

$$x' = \frac{x - x_{min}}{x_{max} - x_{min}}.$$

Rescaling the values of a feature to $[-1, 1]$ is done as follows:

$$x' = 2 \times \frac{x - x_{min}}{x_{max} - x_{min}} - 1.$$

The standardization method expresses each value x in terms of the number of standard deviations from the mean. Let $\bar{x}$ and s be the mean and the standard deviation, respectively, of the values of a feature. Then,

$$x' = \frac{x - \bar{x}}{s}.$$

Normalization of the output data is necessary when we have two or more output nodes and we use the quadratic cost function, because the cost function is scale-sensitive. For instance, if we have two output variables one in the range of $[0, 1]$ and the other in $[0, 10000]$, then the error due to the latter variable may swamp the cost function. In this case, the neural network will try to learn the second output variable at the expenses of the first one.

10.9 OVERFITTING

Neural networks, and other Machine Learning and statistical models, have to be trained so that to fit the training data and also produce good results with new data. Not sufficient training of a model results to underfitting, where the model represents poorly the training data and does not produce good results with new data. The opposite to underfitting is overfitting, which is basically the main problem when training a neural network. It refers to the problem where a neural network is trained so that to fit the training data very well at the expense of not classifying well new data or predicting future observations.

An example of overfitting is shown in Figure 10.17(a). We see that the overfitted model predicts the y values very well, but obviously it will not predict well the y value of a new data point x. The opposite of overfitting is underfitting, an example of which is shown in Figure 10.17(b). As can be seen, the straight regression line is a simplistic model and it has poor prediction ability. On the other hand, the non-linear regression model in Figure 10.17(c) has a good fit as it captures the overall trend of the data. It does not predict the y values of the existing x values as well as the overfitted model, but it will predict more accurately the y value of a new data point x.

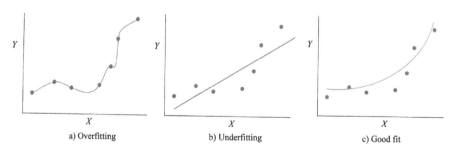

FIGURE 10.17 An example of overfitting, underfitting, and good fit.

Several methods have been proposed in the literature for avoiding overfitting, such as, *early stopping*, *regularization*, and *dropout*. These methods are presented below.

10.9.1 The Early Stopping Method

The early stopping method is a popular method, as it is effective and simple to implement. It involves monitoring the cost E_Q as a function of the number of epochs, as shown in Figure 10.18. The more epochs we train the neural network, the less the cost. However, overfitting is likely to occur as E_Q becomes very low. To prevent this from happening, we divide the data set into two parts, namely, a training data set consisting of about 2/3 of the data and a validation data set consisting of the balance of the data. Each time we train the neural network using the training data for an epoch, we apply it to the validation data set to check the cost. As overfitting sets in, the cost of the validation data begins to increase. This is because the neural network begins to overfit to the training data at the expense of losing its ability to fit other data. As can be seen in Figure 10.18, there is an inflection point passed which the validation cost begins to increase. At that point we stop training the neural network.

We note that in reality the training and validation cost curves in Figure 10.18 are not as smooth because the training of a neural network is stochastic and there may be randomness in the results. Therefore, it is quite likely that the inflection point may not be easy to identify as the validation cost may go up and down for several epochs. In this case, we can continue for a number of epochs until an increase is observed in the cost. There are other stopping rules, such as monitoring the cost averaged over a given number of epochs.

In addition to calculating the validation cost, we can calculate another metric which may be more relevant to the problem under study, such as, the percent of accurately classifying the test data. The plot of this metric will have a similar shape as the validation cost, but it may have a different inflection point, which is the one that should be used to determine the number of epochs.

Splitting the data set D randomly into a training data set and a validation data set may be problematic because the validation test may not turn out to be a representative sample of D. In view of this, cross-validation is used. The data set D is randomly split into v equal folds, where v is typically set to 10. We then combine the first $v - 1$ folds into a single training data

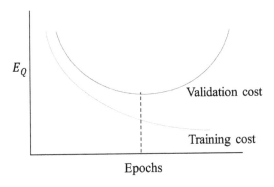

FIGURE 10.18 Inflection point in the early stopping method.

set and the vth fold is used as the validation data set. We train the neural network using the training data set and the resulting model is evaluated using the validation data set. We record the training and validation costs. Subsequently, we construct a new training data set by combining folds 1, 2, ..., $v - 2$, v, and the $v - 1$ fold becomes the validation data set. Again, we train the neural network using the training data set and the resulting model is evaluated using the validation data set. We record the training and validation costs. We repeat this process, until we use all the folds as a validation data set, and average out the weight updates, and the training and validation costs. This concludes one epoch.

10.9.2 Regularization

Regularization is an alternative method to avoiding overfitting, and it is used with the early stopping method. It is based on the observation that overfitting can be reduced by making the model less complex. The complexity of a neural network depends on the number of weights which is indicative of the number of nodes and layers. It also depends on the values of the weights. This is because, small weight values suggest a less complex model because some of them may be close to zero and therefore their connections have no impact on the model. In addition, the resulting model is more stable since it is less sensitive to statistical fluctuations in the input data. For, large weight values tend to cause sharp transitions in the activation functions and thus large changes in the output values for small changes in the input values.

Therefore, we can reduce the complexity of a neural network by either reducing the number of weights or reducing the values of the weights. The most common approach is to ensure that the weights have small values. This is achieved by adding a penalty term, expressed as a function of all the weights, to the cost E; i.e.,

$$E = \frac{1}{2}\sum_{l=1}^{m}\left(O_l^{(2)} - t_l\right)^2 + \lambda\frac{1}{2}\sum_{w\in W}w^2,$$

where λ is the *regularization parameter* and W is the set of all the weights. This penalty is known as *L2 regularization*. Following the derivations given in Section 10.6, we have that

$$\frac{\partial E}{\partial w_{jl}^{(2)}} = \frac{\partial}{\partial w_{jl}^{(2)}}\left\{\frac{1}{2}\sum_{l=1}^{m}\left(O_l^{(2)} - t_l\right)^2 + \lambda\frac{1}{2}\sum_{w\in W}w^2\right\}$$

$$= \frac{\partial}{\partial w_{jl}^{(2)}}\left\{\frac{1}{2}\sum_{l=1}^{m}\left(O_l^{(2)} - t_l\right)^2\right\} + \lambda w_{jl}^{(2)}.$$

This is because all the terms in the summation $\sum_{w\in W}w^2$ except $w_{jl}^{(2)}$ become zero when we take the derivative of E_Q with respect to $w_{jl}^{(2)}$. As a result, we have

$$\frac{\partial E}{\partial w_{jl}^{(2)}} = \delta_l^{(2)} O_j^{(1)} + \lambda w_{jl}^{(2)}$$

and

$$\Delta w_{jl}^{(2)} = \gamma \left(\delta_l^{(2)} O_j^{(1)} \right) + \gamma \lambda \, w_{jl}^{(2)}, j = 1, 2, \ldots, k, l = 1, 2, \ldots, m.$$

Likewise, we obtain that

$$\Delta w_{ij}^{(1)} = \gamma \left(\delta_j^{(1)} X_i \right) + \gamma \lambda \, w_{ij}^{(1)}, i = 1, 2, \ldots, n, j = 1, 2, \ldots, k.$$

Consequently, the new $w_{jl}^{(2)}, j = 1, 2, \ldots, k, l = 1, 2, \ldots, m$ are equal to

$$w_{jl}^{(2)} - \gamma \left(\delta_l^{(2)} O_j^{(1)} \right) - \gamma \lambda \, w_{jl}^{(2)}$$

$$= w_{jl}^{(2)} \left(1 - \gamma \lambda \right) - \gamma \left(\delta_l^{(2)} O_j^{(1)} \right).$$

Likewise, the new $w_{ij}^{(1)}, i = 1, 2, \ldots, n, \, j = 1, 2, \ldots, k$ are equal to

$$w_{ij}^{(1)} \left(1 - \gamma \lambda \right) - \gamma \left(\delta_j^{(1)} X_i \right);$$

i.e., we first rescale a weight by $1 - \gamma \lambda$ and then subtract the update calculated without the L2 regularization. This rescaling is known as *weight decay*. If we assume that $\lambda = 0.03$, then we penalize each weight by 3% of its value.

An alternative penalty to L2 regularization is the *L1 regularization*. In this case, the penalty is the sum of the absolute values of all the weights, i.e.,

$$E = \frac{1}{2} \sum_{l=1}^{m} \left(O_l^{(2)} - t_l \right)^2 + \lambda \frac{1}{2} \sum_{w \in W} |w|.$$

The advantage of L1 regularization is that often some of the weights become zero, which means that some of the connections are eliminated and as a result some of the input nodes may not be used. This is an example of feature selection. In view of this, L1 regularization can be useful when we have too many features. L2 regularization, on the other hand, does not zero weights, but it makes them very small in a balanced way. It also makes the computations more efficient. In general, L2 regularization gives rise to models with better predictive power, and it is more common than L1 regularization.

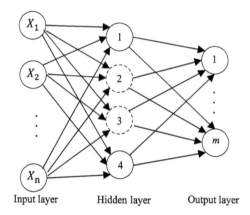

FIGURE 10.19 An example of a thinned neural network.

10.9.3 The Dropout Method

Dropout is a form of regularization, as it constrains adaptation of a neural network model to the data at training time, and it thus helps to avoid overfitting. With this method, we train a "thinned" neural network obtained by dropping some of the hidden layer nodes. Each hidden node is dropped out with a fixed probability p, with p typically set equal to 0.5; i.e., on the average, we drop half the hidden nodes. An example of a thinned network is given in Figure 10.19, where the dropped nodes are indicated by dotted circles. The dropped nodes are not physically removed from the neural network, but they are ignored in the forward and backward passes of the backpropagation algorithm.

This technique can be employed using the mini-batch approach. At the beginning of each mini-batch a new thinned network is created. It is then trained using the backpropagation algorithm on the number of input data points in the mini-batch, and at the end the weights of the thinned neural network are adjusted as described in Section 10.7. We then repeat the process by restoring the dropped nodes and generate a new thinned network. For the validation cost we use the full network. We note that the weights have been learnt under conditions in which half the hidden neurons were dropped. Therefore, when we actually run the full network, twice as many hidden nodes are active. To compensate for that, we multiply the weights outgoing from the hidden neurons by p.

10.10 SELECTING THE HYPER-PARAMETERS

As we have seen above, in order to build a neural network model, we need to decide on the value of various input parameters, such as, the number of hidden nodes and hidden layers, the number of epochs for which we have to train a neural network, the learning rate γ, the momentum, the regularization parameter λ, and the size of the mini-batch. These parameters are collectively referred to as *hyper-parameters*. These parameters are input to a neural network model, as opposed to the weights, which are also parameters but they are determined by the model. Selecting values for the hyper-parameters is time consuming as it involves grid searching. In this section, we describe how to select some of these values manually following the guidelines in [3]. An easier way is to use an automatic grid search procedure that searches systematically through a grid of hyper-parameters and determines the best neural network.

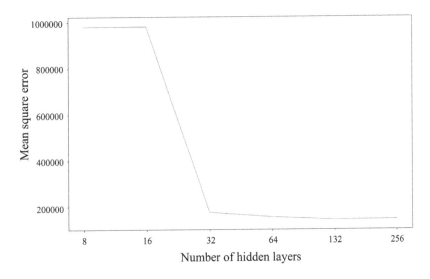

FIGURE 10.20 Number of hidden nodes vs testing cost.

We start by selecting a configuration, i.e., a number of hidden layers and nodes per hidden layer. The safest approach is to start with a simple configuration, such as, a single hidden layer with two or three nodes. Then, we search for the best values of the remaining hyper-parameters. Having determined the best values, we train the neural network as described above. Finally, we run the trained neural network on a data set, referred to as the *test* set, which is separate from the set used for training and validation, in order to obtain the cost of the neural network. Subsequently, we increase the number of nodes, select the hyper-parameters using the previous ones as initial values, and calculate the testing cost. We continue until we observe that the testing cost is not decreasing much (see Figure 10.20). Then, we add another layer with a small number of nodes and repeat the above process. We continue in this manner until overfitting, manifested by the testing cost curve starting to increase (see Figure 10.21).

The number of epochs for which we have to train a neural network is determined using the early stopping method discussed in Section 10.9.1. If the mini-batch method is used, a good value is to set batch size to 32. Several methods have been proposed to select the minimum batch size. The momentum α is a parameter used to accelerate the convergence of the backpropagation algorithm by attenuating the oscillations in the iteration process. Let $\Delta w_k(i)$ be the calculated adjustment of the kth weight in the ith iteration. Then, using the momentum parameter α, $\Delta w_k(i)$ is replaced by the weighted average $\Delta w_k(i) + \alpha \Delta w_k(i - 1)$. The momentum is determined by trial and error.

Below, we focus on how to select a value for the learning rate γ and the regularization parameter λ.

10.10.1 Selecting the Learning Rate γ

We recall that the backpropagation algorithm converges very slowly if the learning rate is very small. On the other hand, if it is very large, the algorithm oscillates since it skips over the minimum. The best value of γ is the largest possible value at which the cost decreases during the first few epochs.

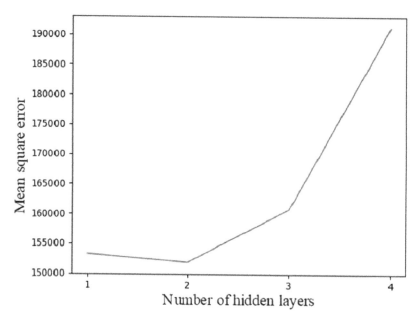

FIGURE 10.21 Number of hidden layers vs testing cost.

The learning rate can be determined using the following simple procedure assuming that $\lambda = 0$. We run a few epochs with $\gamma = 0.01$. If the training cost decreases during the first few epochs, then we change it to 0.1 and check the training cost again for a few epochs. If it decreases, we change the rate to 1 and repeat the above procedure. We continue in this manner until the training cost begins to increase or oscillate in the first few epochs. On the other hand, if we observe that for $\gamma = 0.01$ the training cost increases or oscillates in the first few epochs, then we retry with reduced values, such as, $\gamma = 0.001, 0.0001, 0.00001$, until we find a value for which the training cost for the first few epochs decreases. In either case, we locate approximately the largest value for which the training cost decreases in the first few epochs. We can fine-tune this value by reducing the step by which we increased or decreased γ. To be on the safe side, the final value of γ is the half the value obtained as above.

10.10.2 Selecting the Regularization Parameter λ

Having fixed γ, we select the best value for λ by calculating the validation cost for different values of λ, typically in the range [0.0005,5]. We select the value of λ for which the validation cost is minimum, as shown in Figure 10.22. Note that in this figure, λ decreases from left to right. Once we fix λ we go back and re-adjust γ.

After we select the values for γ and λ, we can apply the early stopping rule to get the final estimates of the weights. The learning rate can be changed according to a schedule as we iterate over the epochs in order to achieve better results. Different learning rate schedules have been proposed. One solution is to use the γ value obtained above until the validation cost begins to go up. At this point, we decrease γ by a factor of 2 (or 10) and continue to iterate until the validation cost begins to go up again. We repeat this until the learning rate is decreased by a factor of 1024 (or 1000).

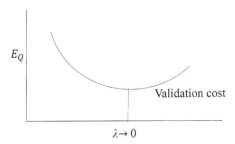

FIGURE 10.22 Selecting the best value of λ.

EXERCISES

The following four problems deal with the implementation of logical gates using neural networks. Note that the activation function used in these neural networks is a binary threshold function.

1. Consider a neural network that simulates the AND gate:

A	B	A AND B
1	1	1
0	1	0
1	0	0
0	0	0

Here, A and B are Boolean variables and 0 indicates true and 1 false. The neural network consists of two input nodes, a hidden node, and an output node. Determine the weights from the input nodes to the hidden node and the threshold of the activation function so that the neural network simulates the AND gate without using the backpropagation algorithm.

2. Consider a neural network that simulates the OR gate:

A	B	A OR B
1	1	1
0	1	1
1	0	1
0	0	0

The neural network consists of two input nodes, a single hidden node, and an output node. Determine the weights from the input nodes to the hidden node and the threshold of the activation function so that the neural network simulates the OR gate.

3. Consider a neural network that simulates the NAND gate:

A	B	A NAND B
1	1	0
0	1	1
1	0	1
0	0	1

The neural network consists of two input nodes, a single hidden node and an output node. Determine the weights from the input nodes to the hidden node and the threshold of the activation function without using the backpropagation algorithm so that the neural network simulates the NAND gate.

4. Consider a neural network that simulates the XOR gate:

A	B	A XOR B
1	1	0
0	1	1
1	0	1
0	0	0

The neural network consists of two input nodes, two single hidden nodes and an output node. The first hidden node simulates a NAND gate and the second node simulates an OR case. The output node simulates an AND gate. Determine the weights from the input nodes to the hidden nodes, and from the hidden nodes to the output node, and the threshold of the activation function at each node so that the neural network simulates the XOR gate without using the backpropagation algorithm.

5. Apply the backpropagation algorithm to the example in Figure 10.14 using the data points $(-0.5, 1, 0.6, 1)$. (You can apply the algorithm manually or write a small program to implement the algorithm.)

NEURAL NETWORK PROJECT

The objective of this project is to determine the configuration of a feedforward neural network and its hyper-parameters using a labeled data set. For this analysis, you can use any software, such as, Python, R, MatLab, and SAS, or a combination of functions from different software.

Data Set Generation

Use the same data generation algorithm you used in the regression project in Chapter 5, only this time generate 3000 points. Generate an additional 500 points to use them as test data.

Task 1 Train a Feedforward Neural Network

1.1. Follow the instructions in Section 10.10 to determine the number of hidden layers and hidden nodes and the hyper-parameters γ and λ. If you use the mini-batch method, set the batch size to 32. Do not use automatic grid search tools.

1.2. Comment on your results.

Task 2 Automatic Grid Search

2.1. Train your neural network using an automatic grid search procedure.

2.2. Compare the resulting neural network model with the one determined in task 1 above. Comment on your results.

Task 3 Compare the Best Trained Neural Network Model with Multivariable Regression

3.1. Re-run task 3 of the linear multivariable regression project, Chapter 5, using the 3000 data points.

3.2. Compare the SSE of the regression model with that of the best neural network model determined above. Comment on which model is the best.

REFERENCES

1. See MNIST database – Wikipedia, for a description of MNIST and a summary of the various classification models that have been used.
2. For a good description of all aspects of neural networks, see http://www.faqs.org/faqs/ai-faq/ neural-nets/part1/preamble.html. *Neural Networks and Deep Learning*, by M. Nielsen, a free on-line book.

CHAPTER 11

Support Vector Machines

A SUPPORT VECTOR MACHINE (SVM) IS A SUPERVISED MACHINE LEARNING CLASSIFI-
CATION technique for two classes. For two-dimensional data, SVM calculates an optimal straight line that separates the two classes, as shown in Figure 11.1(a). For d-dimensional
data, it calculates an optimal hyperplane. Real-world data are not always linearly separated, and the two classes may have a non-linear boundary as shown in Figure 11.1(b).
Non-linear boundaries can also be determined with SVMs after the data is transformed to
a higher-dimensional space using a kernel function, where they actually become linearly
separated.

The two-class SVM classifier can be applied to more than two classes by combining the
multiple classes into just two. For instance, each class can be considered separately and all
the other classes can be lumped into a single one. The SVM algorithm has been adapted so
that it can be used for regression and also for clustering.

As mentioned above, the SVM technique calculates an optimal hyperplane that separates
two classes. For presentation purposes, let us consider the two-dimensional data in
Figure 11.2. Obviously, there is an infinite number of lines that can be drawn, as shown in
Figure 11.2(a). The idea behind SVM is to construct a line so that the nearest points from
the two classes to the line are as far as possible; i.e., the length of the projections of the nearest data points onto the line (see Figure 11.2(b)) is maximized.

In this chapter, we present the SVM algorithm for linearly separable data, the soft-margin
SVM (C-SVM) algorithm, the SVM algorithm for non-linearly separable data, how to select
the hyper-parameters of the SVM algorithm, and the ε-support vector regression (ε-SVR).
A set of exercises and an SVM project is given at the end of the chapter. Before we proceed
to describe these techniques, we review some basic concepts.

11.1 SOME BASIC CONCEPTS

We note that a point $x = (x_1, x_2, ..., x_d)$ is represented by a vector which originates at
$(0, 0, ..., 0)$ and terminates at $(x_1, x_2, ..., x_d)$. We shall use the terms point and corresponding
vector interchangeably. Let us now consider two vectors u and v, where $u = (u_1, u_2)^T$ and
$v = (v_1, v_2)^T$, and let p be the length of the projection of v onto u (see Figure 11.3); i.e., p is the
distance from $(0, 0)$ to where the projection of v meets u. The following relation holds:

$$u^T \cdot v = \|u\| \|v\| \cos\theta,$$

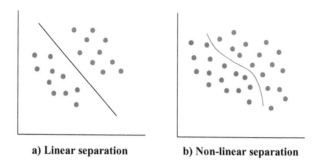

a) Linear separation   **b) Non-linear separation**

FIGURE 11.1 Examples of linear and non-linear separation (a) Linear separation; (b) Non-linear separation.

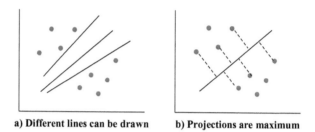

a) Different lines can be drawn   **b) Projections are maximum**

FIGURE 11.2 The basic idea of the SVM algorithm (a) Different lines can be drawn; (b) Projections are maximum.

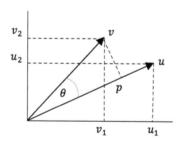

FIGURE 11.3 The projection p of v onto u.

where the dot operation is the inner product of two vectors, defined as follows:

$$u^T \cdot v = \left(u_1, u_2\right)\begin{pmatrix} v_1 \\ v_2 \end{pmatrix} = u_1 v_1 + u_2 v_2.$$

We note that the dot product $u^T \cdot u = u_1^2 + u_2^2 = \|u\|^2$, where $\|u\|$ is the Euclidean distance. Using the above expression, we have

$$\cos\theta = \frac{u^T \cdot v}{\|u\|\|v\|} = \frac{p}{\|v\|},$$

where the second fraction is obtained using the definition of the cosine function, i.e., $\cos\theta = p/\|v\|$. Using the last two terms, we have

$$p = \frac{u^T \cdot v}{\|u\|}.$$

We note that $\cos\theta$ is positive if $\theta < 90°$, zero if $\theta = 90°$, and negative if $\theta > 90°$. Correspondingly, p is positive, zero, or negative.

Let us now consider the projection of a vector on to a hyperplane. The most popular form of representing a line is the slope-intercept form, $y = ax + b$. Thinking of a line as an object and not as the graph of a function, we can express it as $ax + by - c = 0$ with the stipulation that either a or b is not zero. This can be easily converted to the slope-intercept form by solving for y, i.e., $y = (-a/b)x + c/b$. In the general case, a hyperplane in the d-dimensional space can be expressed as

$$w_1 x_1 + w_2 x_2 + \cdots + w_d x_d - b = 0.$$

If the plane passes through the origin, then $b = 0$. Let

$$w = \begin{pmatrix} w_1 \\ w_2 \\ \vdots \\ w_d \end{pmatrix} \text{ and } x = \begin{pmatrix} x_1 \\ x_2 \\ \vdots \\ x_d \end{pmatrix}.$$

Then, the above expression can be written as

$$w^T x - b = 0.$$

It can be easily shown that the vector w is orthogonal to the hyperplane. We demonstrate this using a plane in the three-dimensional space. Let x' and x'' be two vectors starting at the origin $(0,0,0)$ and terminating on the plane, as shown in Figure 11.4. We have $w^T x' - b = 0$ and $w^T x'' - b = 0$. Taking the difference between these two expressions gives $w^T(x' - x'') = 0$, which means that vectors w and $x' - x''$ are orthogonal. Since the vector $x' - x''$ lies on the plane, we conclude that w is orthogonal to the plane.

Vector w is known as the *normal vector* and it is analogous to the slope in the slope-intercept expression of a line. A hyperplane can be defined by the vector w and a point x' on the hyperplane, which is used to calculate b, i.e., $b = w^T x'$. By varying b we can obtain an infinite number of hyperplanes parallel to the original plane.

Now let us consider a vector x that terminates on a plane and a vector x' that terminates at a point above the plane. Both points originate at $(0,0,0)$. The difference $x' - x$ is a vector as shown in Figure 11.5. We are interested in the length of the projection p of $x' - x$ onto w. We have

$$p = \frac{w^T \left(x' - x \right)}{\|w\|} = \frac{w^T x' - w^T x}{\|w\|}.$$

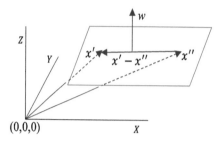

FIGURE 11.4 w is orthogonal to its plane.

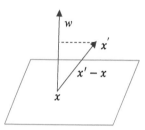

FIGURE 11.5 w is orthogonal to its plane.

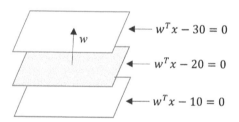

FIGURE 11.6 Parallel planes.

Since $w^T x - b = 0$, we have

$$p = \frac{w^T x' - b}{\|w\|}.$$

We note that point x' was selected to be on the side of the plane where the angle θ between the vectors $x' - x$ and p is less than 90°. In this case $\cos\theta > 0$, and therefore the length of the projection p onto w is a positive number. From the above expression of p, we conclude that in this case $w^T x' - b > 0$. If x' is below the plane where the angle θ is greater than 90°, then p is negative since $\cos\theta < 0$, and therefore $w^T x' - b < 0$. Therefore, given a hyperplane $w^T x - b = 0$ we can determine which side of the plane a point x' lies by computing the sign of $w^T x' + b$.

Given a hyperplane, we can obtain an infinite number of parallel hyperplanes by varying b. If b is increased, then we get hyperplanes in the direction of w. If it is decreased, then the parallel hyperplanes are in the opposite direction (see Figure 11.6).

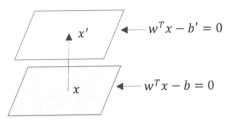

FIGURE 11.7 Calculation of the distance between two planes.

Of interest is the distance between a plane $w^T x - b = 0$ and a parallel one obtained by changing b to b'. Let x' be a point on the plane $w^T x - b' = 0$, and let x be the point of its projection to $w^T x - b = 0$, as shown in Figure 11.7. The vector projection $x' - x$ is a multiple of some number k of w, i.e., $x' - x = kw$, and therefore, $x' = x + kw$. In addition, since x' lies on the plane $w^T x - b' = 0$, we have that $w^T x' - b' = 0$. Substituting $x' = x + kw$ into $w^T x' - b' = 0$ gives $w^T(x + kw) - b' = 0$, or $w^T x + kw^T w - b' = 0$, which can be re-written as $w^T x + k\|w\|^2 - b' = 0$. Solving for k gives

$$k = \frac{b' - w^T x}{\|w\|^2} = \frac{b' - b}{\|w\|^2}.$$

Now the length δ of the vector projection kw is $\delta = \|kw\| = |k|\|w\|$. Substituting the above expression for k into it gives

$$\delta = \frac{|b - b'|}{\|w\|^2} \|w\| = \frac{|b - b'|}{\|w\|}.$$

An easier way to obtain the above expression is to simply observe that the distance δ between the two planes is equal to the absolute value of the projection of point x' on to the vector w; i.e.,

$$\delta = \frac{|w^T x' - b|}{\|w\|} = \frac{|b - b'|}{\|w\|}.$$

Finally, we note that the representation $w^T x - b = 0$ of a hyperplane is not unique; i.e., any non-zero scalar multiple k of the equation of a hyperplane gives another equation for the same plane; i.e., the expressions $kw^T x + kb = 0$ and $w^T x + b = 0$ describe the same hyperplane.

11.2 THE SVM ALGORITHM: LINEARLY SEPARABLE DATA

Let (x_i, ℓ_i), $i = 1, 2, \ldots, n$ be a set of data points where x_i is a d-dimensional data point, and the label ℓ_i takes the values $-1, +1$. We assume that the two classes are linearly separated, and let $w^T x - b = 0$ be the optimum hyperplane. Given a data point x_i, $i = 1, 2, \ldots, n$, we assume that its label is +1 if $w^T x_i - b \geq 0$ and −1 if $w^T x_i - b < 0$. Because of the choice of the labels, we can see that the following holds:

$$\ell_i \left(w^T x_i - b \right) \geq 0.$$

Therefore, the distance of x_i from the optimum hyperplane is

$$\frac{\ell_i \left(w^T x_i - b \right)}{\|w\|}.$$

We want to calculate a hyperplane $w^T x - b = 0$ so that the minimum distance of the data points to the plane is as large as possible. This can be expressed by the following optimization problem:

$$\underset{w,b}{\operatorname{argmax}} \left\{ \underset{i}{\min} \frac{\ell_i \left(w^T x' - b \right)}{\|w\|} \right\}$$

$$\text{s.t. } \ell_i \left(w^T x' - b \right) \geq 0, \; i = 1, 2, \ldots, n.$$

This is a difficult non-linear optimization problem, and in view of this the following alternative optimization problem is solved. We select two parallel hyperplanes $w^T x - (b + k) = 0$ and $w^T x - (b - k) = 0$ that separate the two classes so that the distance between them is as large as possible. The area between these two hyperplanes is called the *margin*. The optimum hyperplane lies in the margin at an equal distance from the two hyperplanes. These two hyperplanes can be written as $w^T x - b = k$ and $w^T x - b = -k$ or $w^T x_i - b = 1$ and $w^T x_i - b = -1$ since the expression of a hyperplane does not change if we multiply it by a non-zero scalar.

In two dimensions, the two hyperplanes are given by the dotted lines in Figure 11.8. The red points are associated with the label +1 and the blue points with the label −1. Note the direction of the normal vector. For the red points we have $w^T x_i - (b - 1) \geq 0$ or $w^T x_i - b \geq 1$; and likewise for the blue points we have $w^T x_i - (b + 1) < 0$ or $w^T x_i - b \leq -1$.

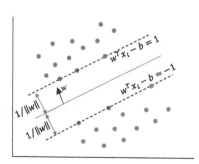

FIGURE 11.8 The two hyperplanes.

The distance of the hyperplane $w^T x_i - b = 1$ from the optimum hyperplane is

$$\frac{\left|b - (b+1)\right|}{\|w\|} = \frac{1}{\|w\|}.$$

Likewise, the distance of the hyperplane $w^T x_i - b = -1$ to the optimum hyperplane is also $1/\|w\|$. Therefore, the total gap between the two hyperplanes is $2/\|w\|$. We note that the bigger the normal vector is, the smaller is the gap. For instance, for $\|w\| = 2$ the gap is 1, and for $\|w\| = 4$ the gap is 1/2.

The above optimization problem can now be reformulated as finding a hyperplane so that the gap is maximum while at the same time the two classes of data are separated by the two hyperplanes, i.e.,

$$\max_{w,b} \frac{1}{\|w\|}, \quad \text{s.t. } \ell_i \left(w^T x_i - b \right) \geq 1, \, i = 1, \, 2, \ldots, n,$$

or, equivalently

$$\min_{w,b} \|w\|, \quad \text{s.t. } \ell_i \left(w^T x_i - b \right) \geq 1, \, i = 1, \, 2, \ldots, n.$$

For mathematical convenience, the following objective function is minimized:

$$\min_{w,b} \frac{1}{2} \|w\|^2, \quad \text{s.t. } \ell_i \left(w^T x_i - b \right) \geq 1, \, i = 1, \, 2, \ldots, n.$$

We recall that $\|w\|^2 = w^T w = w_1^2 + w_2^2 + \cdots + w_d^2$, where d is the number of dimensions or features. Therefore, the above formulation is a convex quadratic programming optimization problem with d variables, and it can be solved using the Lagrange multipliers method.

Let α_i, $i = 1, 2, \ldots, n$ be the Lagrange multipliers. Then, the problem can be reformulated to the following *dual form* problem:

$$\max_{\alpha_1, \ldots, \alpha_n} \left(\sum_{i=1}^{n} \alpha_i - \frac{1}{2} \sum_{i,j=1}^{n} \alpha_i \alpha_j \ell_i \ell_j x_i^T x_j \right)$$

$$\text{s.t.} \quad \alpha_i \geq 0, \, i = 1, \, 2, \ldots, n,$$

$$\sum_{i=1}^{n} \alpha_i \ell_i = 0.$$

We observe that the formulation of the objective function requires the dot product of the data points and not in the individual data points. It can be shown that the vector w is a linear combination of all the data points, i.e.,

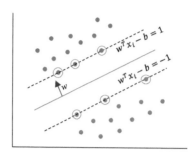

FIGURE 11.9 Support vectors indicated by a circle.

$$w = \sum_{i=1}^{n} \alpha_i \ell_i x_i.$$

An interesting feature of this solution is that each Lagrange multiplier is either zero or positive. The data points whose Lagrange multiplier is positive are known as the *support vectors*, and it is these data points that are used in the above expression for w. All the other points do not affect the solution, since their Lagrange multiplier is zero. The support vectors lie on the boundary hyperplanes of the margin (see Figure 11.9), irrespective of the number of features and the number of data points. The minimum number of support vectors is 2.

Once, we have computed w we can compute b using one of the constraints of the original problem: $\ell_i(w^T x_i - b) \geq 1$. This constraint is still true because the dual formulation is equivalent to the original optimization problem. For a support vector x_i, we have: $\ell_i(w^T x_i - b) = 1$, or by multiplying both sides by ℓ_i we have $w^T x_i - b = \ell_i$, since $\ell_i^2 = 1$. Consequently, we have

$$b = \ell_i - w^T x_i.$$

It is recommended to take the average of the b value calculated for all the support vector.

A new data point x is classified according to the sign of the expression

$$\text{sign}\left(\sum_{i=1}^{s} \alpha_i \ell_i x_i^T x - b \right),$$

where s is the total number of support vectors. The above expression requires the dot products of the new data point x with only the support vectors. If the sign is positive, then the new data point x is classified as +1. Otherwise, it is classified as −1.

This optimization problem is known as the *hard-margin SVM*, because we try to find a hyperplane that has a maximum distance from either class so that all the data points of the same class are on the same side of the hyperplane. This constraint may make the optimization unsolvable if the two classes cannot be separated perfectly.

11.3 SOFT-MARGIN SVM (C-SVM)

The hard-margin SVM optimization described above is sensitive to outliers. As shown in Figure 11.10, an outlier may be close to the data points of the other class, or even within the data points of the other class. In Figure 11.10(a), a red data point is very close to the blue data points, which causes the classifier to have a very small margin. In Figure 11.10(b), a red data point lies within the blue data points which makes the two classes non-linearly separated. Outliers can be handled by the algorithm described above by introducing a penalty in the objective function.

We first introduce the notion of the *slack* variable of a data point. Let us consider the red data points shown in Figure 11.11. We assume that the class of the red data points is above the optimal line and that of the blue data points, not shown, is below. For point A, we have that $w^T x_A - b = 0$ since it lies on the optimal line. B lies on the boundary line of the margin, and therefore $w^T x_B - b = 1$. C lies beyond the boundary line, and hence $w^T x_c - b > 1$. D lies on the margin, and therefore $0 \leq w^T x_D - b < 1$. For E, we have $w^T x_E - b = -1$ since it lies on the boundary line, and for F we have $w^T x_F - b > -1$. Finally, G lies on the margin and therefore $-1 \leq w^T x_G - b \leq 0$.

Let ξ_i be the slack variable of a data point x_i. This variable takes the following values. For points B and C, i.e., for points on the boundary line and beyond, $\xi_i = 0$. Otherwise, $\xi_i = |\ell_i - (w^T x_i - b)|$. For a point on the optimal line, such as point A, $\xi_i = |\ell_i - (w^T x_i - b)| = |1 - 0| = 1$. For a point on the margin such as point D, $\xi_i = |\ell_i - (w^T x_i - b)| = |1 - (w^T x_i - b)| < 1$.

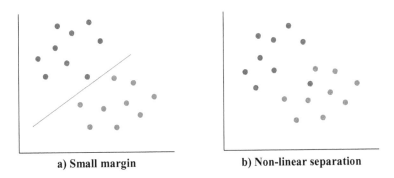

a) Small margin   **b) Non-linear separation**

FIGURE 11.10 Two cases of outliers (a) Small margin; (b) Non-linear separation.

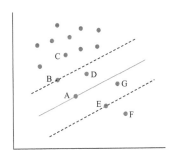

FIGURE 11.11 Red data points A to F.

For a point on the margin below the optimal line, such as point G, $\xi_i = |\ell_i - (w^T x_i - b)| = |1 - (w^T x_i - b)| > 1$, since $w^T x_i - b < -1$ For a point on the lower boundary line, such point E, $\xi_i = |\ell_i - (w^T x_i - b)| = |1 - (w^T x_i - b)| = 2$, since $w^T x_i - b = -1$. For points beyond, such as F, we have $\xi_i > 2$.

The slack variables are used to introduce a regularization penalty in the objective function. The optimization problem is now as follows:

$$\min_{w,b,\xi} \left(\frac{1}{2}\|w\|^2 + C\sum_{i=1}^{n}\xi_i \right),$$

$$\text{s.t. } \ell_i\left(w^T x_i - b\right) \geq 1-\xi_i, \ i=1, 2,\ldots, n,$$

where $\sum_{i=1}^{n}\xi_i$ is the penalty introduced due to the outliers and C is the *penalty factor*. Following the same approach as in the hard-margins SVM, we recast the above optimization problem to its dual form:

$$\max_{\alpha_1,\ldots,\alpha_n} \left(\sum_{i=1}^{n}\alpha_i - \frac{1}{2}\sum_{i,j=1}^{n}\alpha_i\alpha_j\ell_i\ell_j x_i^T x_j \right)$$

$$\text{s.t. } 0\leq\alpha_i\leq C, \ i=1, 2,\ldots, n,$$

$$\sum_{i=1}^{n}\alpha_i\ell_i = 0,$$

where α_i, $i = 1, 2, \ldots, n$ are the Lagrange multipliers. The optimum solution and the classification of a new data point is the same as before.

The above optimization requires to select a value for C. A small C value gives a solution that has a wide margin at the cost of some misclassifications. On the other hand, a large C value gives a hard margin solution with no misclassifications. The best value of C is found by doing a grid search as will be discussed later on in this chapter.

The support vectors lie on the boundary hyperplanes of the margin and within the margin, as shown in Figure 11.12. The number of support vectors depend on the penalty in the objective function. If the penalty is small, then we have a large number of support vectors. If it is large, then we have a small number of support vectors.

This SVM optimization is referred to as *soft-margin SVM*, because it allows some of the data points of one class to be on the wrong side of the hyperplane, as opposed to the hard-margin SVM. In view of this, it is preferable to the hard-margin SVM algorithm. The soft-margins SVM is also referred to as the *C-SVM* because of the use of the penalty factor C.

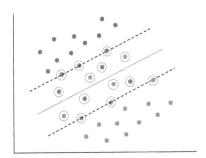

FIGURE 11.12 Support vectors in C-SVM.

11.4 THE SVM ALGORITHM: NON-LINEARLY SEPARABLE DATA

So far, we have assumed that the two classes are linearly separable. In practice, however, this is not always possible. This problem can be solved using the soft-margin C-SVM algorithm described above but after we transform the data into a higher dimension space, referred to as the feature space, where they are linearly separable.

Let us now examine this transformation. In Figure 11.13(a), we give a plot of a one-dimensional data set consisting of two non-linearly separable classes, red and blue. If we apply the transformation $x_i \rightarrow \varphi(x_i) = (x_i, x_i^2)$, then we will obtain the two-dimensional data set shown in Figure 11.13(b), which as can be seen is linearly separable. A second example is given in Figure 11.14, where the data points are two-dimensional, i.e., $x_i = (x_{i1}, x_{i2})$. In this case, if we apply the transformation $x_i \rightarrow \varphi(x_{i1}, x_{i2}) = (x_{i1}^2, \sqrt{2}x_{i1}x_{i2}, x_{i2}^2)$, we will obtain the three-dimensional data set shown in Figure 11.14(b), where the two classes are linearly separable.

Once we have transformed the data set, we can apply the C-SVM method to calculate the optimum hyperplane. For example, for the data set in Figure 11.14(a), we first transform each data point (x_{i1}, x_{i2}) to $(x_{i1}^2, \sqrt{2}x_{i1}x_{i2}, x_{i2}^2)$. Subsequently, since we only need the dot products of the data points, we compute the dot products

$$\left(x_{i1}^2, \sqrt{2}x_{i1}x_{i2}, x_{i2}^2\right)^T \left(x_{j1}^2, \sqrt{2}x_{j1}x_{j2}, x_{j2}^2\right),$$

and then calculate the optimum hyperplane. This is a time-consuming process, and in view of this, instead of transforming each data point and then calculate their dot products,

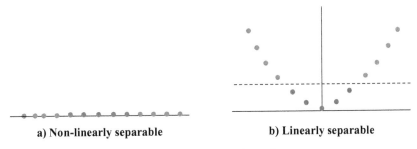

a) Non-linearly separable   b) Linearly separable

FIGURE 11.13 Applying the transformation $x_i \rightarrow \varphi(x_i) = (x_i, x_i^2)$ (a) Non-linearly separable; (b) Linearly separable.

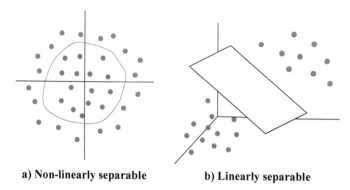

a) Non-linearly separable **b) Linearly separable**

FIGURE 11.14 Applying the transformation $x_i \rightarrow \varphi(x_i) = \left(x_{i1}^2, \sqrt{2}x_{i1}x_{i2}, x_{i2}^2\right)$ (a) Non-linearly separable; (b) Linearly separable.

we use a function to transform their dot products directly. Such a function is known as a *kernel* and the idea of transforming the dot products rather than the individual data points is known as the *kernel trick*. For instance, let us consider again the transformation $x_i \rightarrow \varphi(x_i) = \left(x_{i1}^2, \sqrt{2}x_{i1}x_{i2}, x_{i2}^2\right)$. Then

$$\varphi(x_i)\varphi(x_j)^T = \left(x_{i1}^2, \sqrt{2}x_{i1}x_{i2}, x_{i2}^2\right)\left(x_{j1}^2, \sqrt{2}x_{j1}x_{j2}, x_{j2}^2\right)^T$$

$$= x_{i1}^2 x_{j1}^2 + 2x_{i1}x_{i2}x_{j1}x_{j2} + x_{i2}^2 x_{j2}^2$$

$$= \left(x_{i1}x_{j1} + x_{i2}x_{j2}\right)^2.$$

Therefore, instead of using $\varphi(x_i)$, we use the kernel

$$k(x_i, x_j) = \left(x_{i1}x_{j1} + x_{i2}x_{j2}\right)^2$$

to transform the dot products directly without first having to transform the individual data points. The set of all the kernel values $k(x_i, x_j)$, $i, j = 1, 2, \ldots, n$ forms the kernel matrix K, where

$$K = \begin{pmatrix} k(x_1, x_1) & k(x_1, x_2) & \cdots & k(x_1, x_n) \\ k(x_2, x_1) & k(x_2, x_2) & \cdots & k(x_2, x_n) \\ \vdots & \vdots & \cdots & \vdots \\ k(x_n, x_1) & k(x_n, x_2) & \cdots & k(x_n, x_n) \end{pmatrix}.$$

This matrix contains all the information we need for the calculation of the hyperplane. A matrix of all the inner products of a set of vectors, such as the above matrix K, is known as a Gram or Gramian matrix.

A kernel can be constructed by defining a $\varphi(.)$ function, as we did above. Alternatively, we can construct kernels by combining simpler kernels. For instance, let $k_1(x_i, x_j)$ and $k_2(x_i, x_j)$ be two kernel functions. Then, the following combinations are also kernels.

- $k(x_i, x_j) = ck_1(x_i, x_j)$, where c is a constant
- $k(x_i, x_j) = f(x_i)k_1(x_i, x_j)f(x_j)$, where $f(.)$ is a function
- $k(x_i, x_j) = p(k_1(x_i, x_j))$, where $p(.)$ is a polynomial with non-negative coefficients
- $k(x_i, x_j) = \exp(k_1(x_i, x_j))$
- $k(x_i, x_j) = k_1(x_i, x_j) + k_2(x_i, x_j)$
- $k(x_i, x_j) = k_1(x_i, x_j)k_2(x_i, x_j)$.

Below, we describe some of the commonly used kernels.

Linear kernel: $k\left(x_i, x_j\right) = x_i^T x_j$; i.e., no transformation to a higher dimension space. This is useful when the number of features, i.e., the number of dimensions of the data points is very large. This is the case, for instance, in document classification. Also, when the data set is extremely large, the linear kernel can be used to speed up the execution of the SVM algorithm.

Polynomial kernel:

$$k\left(x_i, x_j\right) = \left(x_i^T x_j\right)^d \text{ or } k\left(x_i, x_j\right) = \left(x_i^T x_j + c\right)^d.$$

This kernel function is the same as the linear one if we do not assume a constant c and set $d = 1$.

Gaussian or radial base function (RBF) kernel: This kernel is derived from the normal distribution and it is given by the expression:

$$k\left(x_i, x_j\right) = e^{-\frac{1}{2}\frac{\|x_i - x_j\|^2}{\sigma^2}}.$$

The above expression is not interpreted as a probability distribution, and therefore, the normalization coefficient $1/(\sigma\sqrt{2\pi})$ in the expression of the normal distribution is omitted. In addition, σ^2 is just a free parameter and not the variance. This kernel is commonly written as

$$k\left(x_i, x_j\right) = \exp\left(-\gamma\|x_i - x_j\|^2\right)$$

where $\gamma = (1/2)\sigma^2$. γ is called the *kernel bandwidth*, and it is determined through a grid search, as described below.

The RBF kernel takes values between 0 and 1. This is because $k(x_i, x_j) = 1$ when $x_i = x_j$. On the other hand, $k(x_i, x_j)$ tends to zero as the distance between the two data points x_i and x_j tends to infinity, since their squared Euclidean distance $\|x_i - x_j\|^2$ tends to inifnity. Because of this behavior, the RBF is a similarity measure. The higher its value, the closer are the two points.

RBF corresponds to an infinite dimensional feature space and $\varphi(x_i)$ cannot be defined. RBF is one of the most commonly used kernel.

Other kernels are the *hybrid kernel*

$$k(x_i, x_j) = (x_i^T x_j + p)^d \exp(-\gamma \|x_i - x_j\|^2),$$

and the sigmoidal kernel

$$k(x_i, x_j) = \tanh(kx_i^T x_j - \delta).$$

Given a kernel, the optimization problem is the same as in C-SVM, except that we substitute $x_i^T x_j$ by k; i.e.,

$$\max_{\alpha_1,...,\alpha_n} \left(\sum_{i=1}^{n} \alpha_i - \frac{1}{2} \sum_{i,j=1}^{n} \alpha_i \alpha_j \ell_i \ell_j k(x_i, x_j) \right)$$

$$\text{s.t.} \quad 0 \le \alpha_i \le C, \, i = 1, 2, ..., n,$$

$$\sum_{i=1}^{n} \alpha_i \ell_i = 0,$$

where α_i, $i = 1, 2, ..., n$ are the Lagrange multipliers. A new data point x is classified according to the sign of

$$\text{sign}\left(\sum_{i=1}^{s} \alpha_i \ell_i k(x, x_j) - b \right),$$

where s is the total number of support vectors.

The number of support vectors can range from very few to every single data point if the model completely overfits the data. An example of a non-linear separation boundary and the support vectors in the original space of the data set is shown in Figure 11.15.

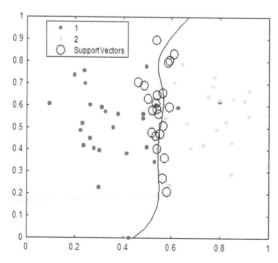

FIGURE 11.15 An example of a non-linear separation boundary.

11.5 OTHER SVM METHODS

The SVM optimization problem described above is typically solved using the *sequential minimal optimization* (SMO) algorithm, which solves the following optimization problem:

$$\min_{\alpha_1,\dots,\alpha_n} \left(\frac{1}{2} \sum_{i,j=1}^{n} \alpha_i \alpha_j \ell_i \ell_j k\left(x_i, x_j\right) - \sum_{i=1}^{n} \alpha_i \right)$$

$$\text{s.t.} \quad 0 \le \alpha_i \le C, i = 1, 2,\dots, n,$$

$$\sum_{i=1}^{n} \alpha_i \ell_i = 0.$$

The SMO algorithm was created to solve the SVM optimization problem faster. It achieves this by finding the optimum value of only a pair of the Lagrange multipliers at a time. This approach is guaranteed to converge to the same optimum value as in the original algorithm where all the Lagrange multipliers are considered at the same time.

In the soft-margin SVM optimization, the value of C depends on the actual data. v-SVM is an alternative formulation where C is replaced by the parameter v, $0 \le v \le 1$, which depends on how noisy the data is rather than the actual values of the data set. The optimization problem is defined as follows.

$$\min_{w,b} \left(\frac{1}{2} \|w\|^2 - v\rho + \frac{1}{n} \sum_{i=1}^{n} \xi_i \right),$$

$$\text{s.t.} \quad \ell_i \left(\kappa\left(w, x_i\right) - b \right) \ge \rho - \xi_i, i = 1, 2,\dots, n,$$

where $\rho \ge 0$ and $0 \le v < 1$.

Another popular SVM method is the *one-class SVM*. This is used in the case where we have a lot of data for one class and none or some data for the other, contrary to the typical case where we have enough data for both classes. This situation occurs in detection of rare events, such as, failure of a sensor. In such a case, it is relatively easy to collect data when the sensor functions correctly, but it may be impossible or very expensive to collect data when it malfunctions. In view of this, the number of data points collected when the sensor malfunctions may be small or none at all. A similar situation also occurs in structural defect detection, machine malfunction, and medical condition monitoring.

The one-class SVM algorithm determines the optimum hyperplane that separates the two classes after the data has been transformed using a kernel. In this case, class 1 is the normal data set, and class 2 consists of only one point which is the origin. The quadratic optimization problem is as follows:

$$\min_{w,\xi,\rho}\left(\frac{1}{2}\|w\|^2 + \frac{1}{vn}\sum_{i=1}^{n}\xi_i - \rho\right),$$

s.t. $w^T x_i - b \geq \rho - \xi_i, \xi_i \geq 0, i = 1, 2, \ldots, n.$

The decision of whether a new data point x is determined by the sign of

$$\text{sign}\left(\sum_{i=1}^{s}\alpha_i k(x, x_j) - \rho\right),$$

where the +1 sign means that the new data point x is not different from the normal data, and the −1 sign indicates otherwise.

11.6 MULTIPLE CLASSES

The SVM method can only be used to generate a binary classifier. Different techniques have been proposed for data sets with more than two classes. The *one-against-all* is probably the simplest method. We select one class which we associate with the +1 label, and we group all the other classes into second class which we associate with the −1 label. We obtain an SVM classifier for these two classes. We repeat this process for all the classes. For instance, if we have three classes 1, 2, and 3, we will compute three classifiers for the following two-class cases: 1 and {2,3}, 2 and {1,3}, and 3 and {1,2}, as shown in Figure 11.16. A new data point is classified by considering all the classifiers. This approach may be inconsistent as a point may be assigned to more than one class or to none. For instance, the point indicated by a star in Figure 11.17 will not be assigned to any class, and the point indicated by a square will be assigned to both the red and blue classes. The suggested solution around this problem is to compute the actual value $w^T x - b$ of the new data point x rather than its sign for each classifier and assign x to the class with the highest value. Another problem is that each class may not have enough data points compared to the aggregate of the remaining classes, which may skew the decision.

An alternative method is to construct a classifier for each pair of classes. This method is known as the *one-against-one* and leads to $c(c - 1)/2$ combinations for c classes. Prediction

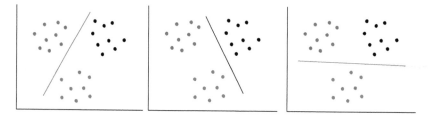

FIGURE 11.16 One-against-all.

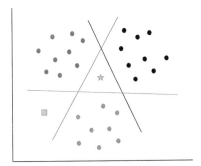

FIGURE 11.17 The data point star will be assigned to no class.

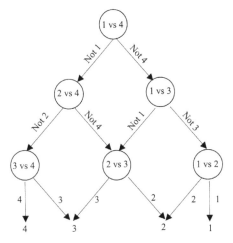

FIGURE 11.18 The DAGSVM method.

is made by predicting the class of a new data point for each classifier, and then use majority voting to determine its class.

An extension of the one-against-one method is the directed acyclic graph SVM (DAGSVM) method. This method uses a directed acyclic graph to decide on the sequence of the one-against-one comparisons so that to speed up the classification process. An example of this method is shown in Figure 11.18. The data set consists of four classes, 1, 2, 3, and 4. We start by running the classifier on classes 1 and 4. If we select class 1, which is equivalent to not selecting class 4, we branch out on the "Not 4" branch, otherwise to the "Not 1". Assuming that we follow the "Not 4" branch, we run the classifier on classes 1 and 3. If we select class 1, we follow the branch "Not 3" and run the classifier on classes 1 and 2, and accordingly, we select class 1 or 2. Following this approach, we eliminate unnecessary comparisons between classes. For instance, if the outcome of the first comparison 1 vs 4 is "Not 4", then we can skip pairs of classes containing class 4.

In addition to the above heuristics, there are optimization algorithms that extend the two-class SVM method to multiple classes, by adding additional constraints, one per class, to the two-class optimization problem. These methods are CPU intensive since the size of the optimization problem depends on the number of classes.

Comparisons among the proposed multi-class methods have shown that the one-against-one and the DAGSVM methods perform better. For further details on these methods (see [1]).

11.7 SELECTING THE BEST VALUES FOR C AND γ

The soft-margin SVM algorithm used with the RBF kernel requires to select values for C and γ. This can be done using a grid search.

First, all the data points have to be numerical. The values of a nominal variable have to be converted to binary strings using one-hot encoding. In these binary strings only one bit is 1 and the remaining bits are zero. Let us assume that a nominal variable takes the values: red, green, and blue. Then, these values can be mapped to integers 1, 2, 3, i.e., red is mapped to 1, green to 2, and blue to 3. This encoding can lead to erroneous predictions, because it implies that category 1 is closer to category 2 than to category 3. One-hot encoding avoids such problems, by creating three new binary variables, X_1, X_2 and X_3, where $X_i = 0, 1, i = 1, 2, 3$. Each nominal value is associated with a different variable, i.e., red with X_1, green with X_2, and blue with X_3. As a result, the nominal value red is encoded as $(1, 0, 0)$, green as $(0, 1, 0)$, and blue as $(0, 0, 1)$. If the number of values of a nominal variable is large, then it is acceptable to map the nominal values to the integers 1, 2, 3, ..., one integer per value.

Subsequently, the data points have to be scaled in the region of $[0, 1]$ or $[-1, 1]$ so that a variable that takes very large values does not dominate other variables that take small values, when calculating the Euclidean distance in the RBF kernel.

The values for C and γ are selected by doing a grid search. As described in [3], we start with a coarse search by setting $C = 2^{-5}, 2^{-3}, ..., 2^{15}$ and $\gamma = 2^{-15}, 2^{-13}, ..., 2^3$. For each combination of C and γ calculate the percentage of accuracy using cross-validation; i.e., the data set D is randomly split into v equal folds, where v is typically set to 10. We then combine the first $v - 1$ folds into a single data set, call it D_1, which becomes the training data set and the vth fold becomes the validation data set. We run the C-SVM algorithm on D_1 and then we use the resulting classifier to classify the data points in the vth fold, and calculate the percent of data points that were accurately classified. Subsequently, we construct a new training data set D_1 by combining folds 1, 2, ..., $v - 2$, and v, and the $v - 1$ fold becomes the validation data set. Again, we run the C-SVM algorithm on D_1 and then we use the resulting classifier to classify the data points in the $v - 1$ fold and calculate the percentage of accuracy. We proceed in this way until all the v folds are used as a validation data set and calculate the average percentage of accuracy.

Following the coarse search, we continue with a finer search by focusing in a narrow area of the values of C and γ where we have observed the highest percentage of accuracy.

An example of a coarse and finer grid search is given in Figures 11.19 and 11.20. Figure 11.19 gives the heat map of the results of the coarse grid search, obtained by varying $C = 2^{-5}, 2^{-3}, ..., 2^{15}$ and $\gamma = 2^{-15}, 2^{-13}, ..., 2^3$. The different colors in the table correspond to different percentages of accuracy, as indicated by the legend on the left of the table. The highest accuracy was obtained for $C = 69.123$ and $\gamma = 0.00195$. Based on this information, a finer search was done by narrowing down the region of values for C and γ as shown in

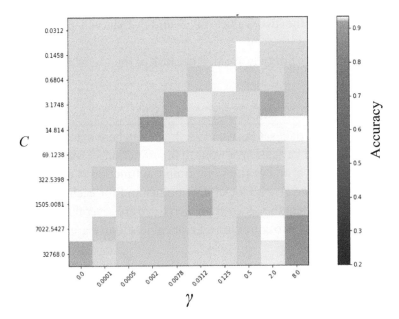

FIGURE 11.19 Heat map of the coarse search.

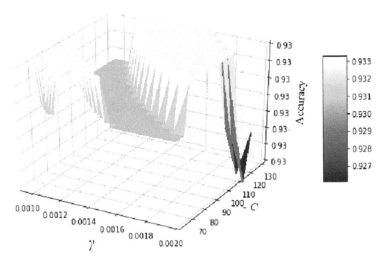

FIGURE 11.20 Finer search.

Figure 11.19. The highest accuracy of 0.93 was achieved for $C = 130.0$ and $\gamma = 0.00145$. Consequently, these are the two values of C and γ that will be used in the SVM classifier.

Finally, we note that overfitting is avoided because of the use of regularization and the grid search for C and γ.

11.8 ε-SUPPORT VECTOR REGRESSION (ε-SVR)

Let us consider a training data set (x_i, y_i), $i = 1, 2, \ldots, n$, where x_i is a d-dimensional data point and y_i is a single real variable. We want to fit a function so that the deviations from the y_i values are less than ε, and at the same time the function is as flat as possible. For example, in Figure 11.21 we give a plot of data points and for each y_i we have drawn an upper and lower bound of length ε. As can be seen there are several lines through the data points that satisfy the above two conditions. Errors of less than ε are acceptable, but errors larger than ε are not.

Using the SVM method, we can fit a function $f(x) = w^T x + b$ so that the errors $|y_i - (w^T x_i + b)| \leq \varepsilon$. Flatness of the function means that w is as small as possible. In view of this, we have the following formulation analogous to the hard-margins SVM:

$$\min_{w,b} \frac{1}{2}\|w\|^2,$$

$$\text{s.t. } y_i - w^T x_i - b \leq \varepsilon,$$

$$w^T x_i + b - y_i \leq \varepsilon, \ i = 1, 2, \ldots, n.$$

In view of the use of the parameter ε, this method is known as the ε-support vector regression (ε-SVR). The value of ε affects the fitted function, as can be seen in Figure 11.22.

The above formulation requires that all the data points have an absolute error of less or equal to ε. Consequently, in the presence of outliers, we will be obliged to increase the value of ε, which may not result to a good fit. Alternatively, we can use an algorithm analogous to the soft-margins SVM, where outliers are penalized through the use of slack variables.

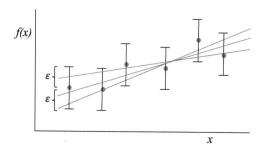

FIGURE 11.21 Many lines satisfy the two conditions.

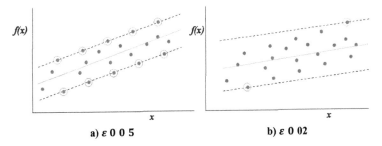

a) ε 0 0 5   b) ε 0 02

FIGURE 11.22 Fitted line for two different values of ε (a) ε = 0.05; (b) ε = 0.02.

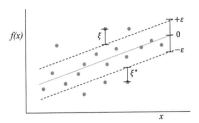

FIGURE 11.23 Definition of slack variables ξ and ξ*.

As can be seen in Figure 11.23, the slack variable ξ is defined as the error over and above ε. For data points above the fitted line, $\xi = y_i - (w^T x_i + b) - \varepsilon$, and for data point below the line $\xi* = (w^T x_i + b) - y_i - \varepsilon$. Therefore, the SVR optimization problem can be stated as follows:

$$\min_{w.b,\xi} \left(\frac{1}{2} \|w\|^2 + C \sum_{i=1}^{n} \left(\xi_i + \xi_i^* \right) \right),$$

$$\text{s.t.}\quad y_i - w^T x_i - b \le \varepsilon + \xi_i,$$

$$w^T x_i + b - y_i \le \varepsilon + \xi_i^*,$$

$$\xi_i, \xi_i^* \ge 0, \quad i = 1, 2, \ldots, n.$$

So far, we assumed that the data can be modeled by a linear function. In the case where the data is non-linear, we can use a kernel function to transform the data into a higher dimension space where it becomes linear, and then apply ε-SVR (see Figure 11.24). As before, the selection of C, ε, and kernel parameters is done using a grid search. The selection of ε depends on the range of the variables in the data set, and in view of this, the data set has to be re-scaled (see Section 11.7).

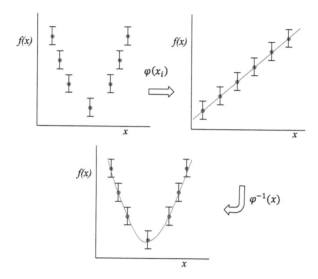

FIGURE 11.24 Use of a kernel for non-linear data.

EXERCISES

1. Suppose the input data set is three-dimensional: $x = (x_1, x_2, x_3)$. Let

$$x \rightarrow \varphi(x) = \left(x_1^2,\ x_2^2,\ x_3^2,\ \sqrt{2}x_1x_2,\ \sqrt{2}x_1x_3,\ \sqrt{2}x_2x_3\right).$$

Calculate the corresponding kernel function $k(x, y)$.

2. Same as (1). Calculate the number of additions and multiplications needed if you (a) transform x and y and then perform their dot product and (b) compute their kernel function directly. Compare the results.

3. Construct an example of how the values of the RBF kernel can be affected if you do not scale the input data.

4. What is the value, or range of values, of the slack variable ξ for points A, B, and C in Figure 11.25?

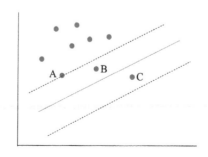

FIGURE 11.25 Problem 4.

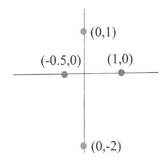

FIGURE 11.26 Problem 5.

5. Consider the data set consisting of two red and two blue data points, as shown in Figure 11.26. The two classes are not linearly separable. Apply a kernel to transform the data into a higher dimension space where they are linearly separable.

SVM PROJECT

The objective is to train an SVM classifier using a data set. For this analysis, you can use any software, such as, Python, R, MatLab, and SAS, or a combination of functions from different software.

Data Set Generation

Generate two clusters of 2-tuple data, each consisting of 400–500 data points, as follows.

1. First, generate the number of observations n_1 in the first cluster by drawing a number from the uniform distribution $U(400, 500)$. Subsequently, generate a stochastic variate from each of the following two normal distributions in order to form a 2-tuple.

   a. $N_1(\mu_1, \sigma_1^2): \mu_1 \sim U(10, 20),\ \sigma_1^2 \sim U(10, 15)$

   b. $N_1(\mu_2, \sigma_2^2): \mu_2 \sim U(30, 40),\ \sigma_2^2 \sim U(10, 15)$

2. Repeat step 3 until you have generated all the n_1 tuples in the first cluster.

3. For the second cluster, repeat steps 1–3, but use the following two distributions:

   a. $N_1(\mu_3, \sigma_3^2): \mu_3 \sim U(50, 60),\ \sigma_3^2 \sim U(10, 15)$

   b. $N_1(\mu_4, \sigma_4^2): \mu_4 \sim U(70, 80),\ \sigma_4^2 \sim U(10, 15)$

4. Compute the average Euclidean distance between the two clusters. Accept the data file if it is greater than 250. Else, reject the data file and create a new one by repeating steps 1–3. Also, you may alter the mean values of the normal distributions to achieve different levels of non-linear separation between the two clusters.

5. Place all the generated tuples and their cluster id number into a three columns table, and interleave the data, so that the data within each group is not contiguous.

Now, forget how you generated the data (!) and carry out the following tasks.

Tasks

1. Color your data set and plot a scatter diagram to get a visual view of how the data is clustered.

2. Scale the values of each feature in the range $[0, 1]$.

3. Apply the C-SVM algorithm with the RBF kernel. For this, you will have to obtain the best values of C and γ using the grid search described above in Section 11.7.

   a. Use a 5-fold cross-validation.

   b. Present the results obtained from the coarse search in either a heat map or in a three-dimensional plot.

   c. Present the results obtained from the finer search in a three-dimensional plot.

4. Discuss your results.

REFERENCES

1. A. Kowalczyk. *Support Vector Machines Succinctly.* Free e-book, https://www.syncfusion.com/ebooks/support_vector_machines_succinctly.
2. A. Statnikov, C.F. Aliferis, D.P. Hardin, and I. Guyon. *A Gentle Introduction to Support Vector Machines in Biomedicine – Volume 1: Theory and Methods,* World Scientific, 2001.
3. C.-W. Hsu, C.-C. Chang, and C.-J. Lin. *A Practical Guide to Support Vector Classification,* https://www.csie.ntu.edu.tw/~cjlin/papers/guide/guide.pdf.

CHAPTER 12

Hidden Markov Models

A HIDDEN MARKOV MODEL (HMM) IS A SPECIAL CLASS OF MARKOV models where the system under study is modeled as a Markov Chain, but the actual states of the system and other parameters are not observable; i.e., they are hidden, and thus the name of these models. What is observable are the values of a parameter or a vector of parameters, which depend on the current state of the Markov Chain. HMMs are very flexible models and they can be used for classification and forecasting. They have been used extensively in many different areas, such as, speech and pattern recognition, biology, and econometrics.

In this chapter, we start by first giving a brief introduction to Markov Chains. Then, we give an example of an HMM, and describe the three basic problems associated with HMMs. Subsequently, we present the algorithms used for solving these three problems, including the forward–backward algorithm and the Viterbi algorithm. We conclude the chapter with a discussion on how to forecast future values of an HMM when it is used as a forecasting tool, followed by a short introduction to autoregressive HMMs. A set of exercises and an HMM project is given at the end of the chapter.

12.1 MARKOV CHAINS

We introduce the concept of a Markov Chain through a simple example. A system of sensors tracks people as they move through a museum, which consists of five rooms, four exhibition rooms (Ancient Greece, Medieval, Renaissance, and Impressionists) and a foyer.

To simplify the model, we assume that a visitor spends a fixed amount of time in each room; i.e., we impose a slotted time axis where a slot is equal to a fixed amount of time. At the end of a time slot, a visitor may choose to stay in the same room for another time slot or move to another room.

We consider a single visitor, and we define the state of the system as the room where the visitor is currently in; i.e., the system can be in one of the five states given in Table 12.1. Now, let p_{ij} be the probability of moving from one room to another; i.e.,

$$p_{ij} = \text{Prob}(\text{visitor is in room } j \text{ at time slot } t \mid \text{visitor is in room } i \text{ at time slot } t-1).$$

These probabilities are known as the *one-step transition probabilities*, because they tell us in which state the system may be in the next time slot. Obviously, all the probabilities for a

TABLE 12.1   The five states

State	Description
1	Foyer
2	Ancient Greece
3	Medieval
4	Renaissance
5	Impressionists

given state i add up to 1; i.e., in the above example, for $i = 1$ we have that $p_{11} + p_{12} + \cdots + p_{15} = 1$. These probabilities can be estimated using the data collected by the sensors. For instance, let us assume that n individuals entered room 1 over a period of time. Of these, in the next time slot, n_1 stayed in the room, n_2 moved to room 2, and n_3 moved to room 3. Then, the branching probabilities $p_{1j}, j = 1, 2, ..., 5$, can be computed as follows: $p_{11} = n_1/n$, $p_{12} = n_2/n$, $p_{13} = n_3/n$, $p_{14} = p_{15} = 0$. The remaining branching probabilities can be computed in the same way. Obviously, the longer the observation period, the better the estimates of the branching probabilities. The one-step probabilities are typically summarized in a matrix, as follows:

$$
P = \begin{pmatrix}
p_{11} & p_{12} & p_{13} & p_{14} & p_{15} \\
p_{21} & p_{22} & p_{23} & p_{24} & p_{25} \\
p_{31} & p_{23} & p_{33} & p_{34} & p_{35} \\
p_{41} & p_{24} & p_{43} & p_{44} & p_{45} \\
p_{51} & p_{25} & p_{53} & p_{54} & p_{55}
\end{pmatrix}.
$$

Matrix P is known as the *one-step transition probability matrix*. For example, let us consider the following matrix:

$$
P = \begin{pmatrix}
0.2 & 0.8 & 0 & 0 & 0 \\
0.2 & 0.2 & 0.6 & 0 & 0 \\
0 & 0.2 & 0.6 & 0.2 & 0 \\
0 & 0 & 0.6 & 0.2 & 0.2 \\
0 & 0 & 0 & 0.8 & 0.2
\end{pmatrix}
$$

Assume that the visitor is in state i during a time slot t. Then, the one-step transition probabilities in the ith row of P give the probable transitions of the visitor from state i to another state at the end of the time slot t. For instance, if the visitor is in state 1 at the end of time slot t, then the visitor may remain in state 1 (foyer) with probability 0.2, or may move to state 2 (Ancient Greece) with probability 0.8. Likewise, if the visitor is in state 2 at the end of a time slot t, then the visitor may remain in state 1 (foyer) with probability 0.2, or move to state 2 (Ancient Greece) with probability 0.2, or move to state 3 (Medieval) with probability 0.6.

In addition to the matrix P, we also use the vector $\pi^{(t)}$ which gives the probability that the visitor is in each state of the system during time slot t. This vector is known as the *state vector*. Of interest is $\pi^{(0)}$, which tells us where the visitor is at time $t = 0$. For instance, if $\pi^{(0)} = (1,0,0,0,0)$, then the visitor is in the foyer (state 1) at time $t=0$. If $\pi^{(0)} = (0.2,0.2,0.2,0.2,0.2)$, then the visitor has the same probability of being in any of the five rooms. The probability vector $\pi^{(0)}$ is known as the *initial state vector*, and it is conceptually similar to the initial conditions we need to assume in order to start a simulation (see Section 3.2.1, Chapter 3).

Now, given the initial state vector $\pi^{(0)} = (1,0,0,0,0)$, we would like to know in which room the visitor may be in the next time slot $t = 1$. By inspecting the one-step transition probabilities in the first row of matrix P, it is evident that $\pi^{(1)} = (0.2,0.8,0,0,0)$. This can also be obtained by multiplying $\pi^{(0)}$ with P; i.e.,

$$\pi^{(1)} = \pi^{(0)}P = \left(1,0,0,0,0\right)P = \left(0.2,0.8,0,0,0\right).$$

Now, if we multiply $\pi^{(1)}$ with P, we will get $\pi^{(2)}$ which gives the probability that the visitor is in room 1, 2, 3, 4, 5 in the next time slot; i.e.,

$$\pi^{(2)} = \pi^{(1)}P = \left(0.2,0.8,0,0,0\right)P = \left(0.2,0.32,0.64,0,0\right).$$

We observe that $\pi^{(2)} = \pi^{(1)}P = \pi^{(0)}P^2$. In general, we have

$$\pi^{(n)} = \pi^{(n-1)}P$$

or

$$\pi^{(n)} = \pi^{(0)}P^n.$$

Therefore, in order to calculate the state vector after n steps, we need to multiply P by itself n times. It turns out that as we keep powering P, eventually P^n will stop changing; i.e., it will converge to a matrix Π where the elements in each column are the same, i.e.,

$$\lim_{n\to\infty} P^n = \Pi.$$

In our example, after multiplying P by itself 100 times, we get

$$P^{100} = \begin{pmatrix} 0.045 & 0.1818 & 0.545 & 0.1818 & 0.045 \\ 0.045 & 0.1818 & 0.545 & 0.1818 & 0.045 \\ 0.045 & 0.1818 & 0.545 & 0.1818 & 0.045 \\ 0.045 & 0.1818 & 0.545 & 0.1818 & 0.045 \\ 0.045 & 0.1818 & 0.545 & 0.1818 & 0.045 \end{pmatrix}.$$

306 ■ An Introduction to IoT Analytics

(We note that the fact that some of the above columns are identical is purely coincidental and it has to do with the values of the transition probabilities. In general, the columns are different from each other.)

This means that $\pi^{(n)}$ will also converge to a vector π known as the *steady-state* or *stationary probability vector*, i.e.,

$$\lim_{n \to \infty} \pi^{(n)} = \pi^{(0)} \lim_{n \to \infty} P^{(n)} = \pi^{(0)} \Pi = \pi.$$

Going back to our numerical example, we have

$$\pi = \pi^{(100)} = \left(1,0,0,0,0\right) P^{100} = \left(0.045, 0.1818, 0.545, 0.1818, 0.045\right).$$

If we use $\pi^{(0)} = (0.2, 0.2, 0.2, 0.2, 0.2)$ as the initial vector, we obtain the same result; i.e.,

$$\pi = \pi^{(100)} = \left(0.2, 0.2, 0.2, 0.2, 0.2\right) P^{100} = \left(0.045, 0.1818, 0.545, 0.1818, 0.045\right).$$

In fact, the choice of the initial vector does not affect the value of π. This is analogous to the observation that the initial conditions in a simulation do not affect the final estimated values (see Section 3.2.1, Chapter 3).

The steady-state probability vector gives the probability that the visitor is in room i. $i = 1, 2, \ldots 5$ in the long run, i.e., assuming that the visitor continues to visit the museum rooms for ever.

There are different methods for calculating π, the simplest of which is to keep powering matrix P until it converges. A more commonly used method is based on the expression:

$$\pi P = \pi.$$

This expression is obtained from the realization that since π is the steady-state probability vector, multiplying it by the one-step transition matrix P will not change it. In addition, we impose the condition $\pi e = 1$, where e is the unit vector, since the sum of the steady-state probabilities should be 1. To guarantee this condition, we replace one of the equations in the system $\pi P = \pi$ by $\pi e = 1$. This system of linear equations is typically solved numerically, and the interested reader is referred to a book on numerical linear algebra.

There are various types of Markov models depending on whether the underlying time is discrete or continuous, and whether the set of the states is discrete or continuous. The above example of the museum is an example of a discrete-time and discrete-space Markovian model. These models are known as Markov Chains.

The objective in a Markov Chain is to calculate the stationary probability vector π, given that we know the number of states, the one-step transition matrix, and the initial probability vector, though the latter is not important in the calculation of π.

In HMMs, the determination of the stationary probability vector π is not the objective. In addition, the one-step transition matrix P and $\pi^{(0)}$ are not known, i.e., they are hidden. What is known, is a set of observations of a parameter, or a vector of parameters, which are generated as a function of the state of the Markov Chain. From this sequence of observations, we estimate the parameters of the HMM.

12.2 HIDDEN MARKOV MODELS – AN EXAMPLE

We introduce the concept of HMMs through an example. Let us consider a closed environment whose temperature takes the following values:

- 1: very cold
- 2: cold
- 3: normal
- 4: warm
- 5: very warm

These five different values constitute the five different states of the system of the closed environment. We assume that the temperature within this environment fluctuates according to a Markov Chain; i.e., it stays constant within a time slot, at the end of which it may switch to another state or stay at the same state. The transitions to other states at the end of a slot are dictated by a Markov Chain.

In the HMM formulation, we assume that the one-step transition matrix of this system is not known. Instead, we can observe a parameter which takes values as a function of the current state of the system. For instance, let us assume that there is a sensor embedded in the system that reports on the temperature using only three values: 1 (low), 2 (medium), 3 (high). The value it produces depends on the current state of the system.

In order to clarify this point further, let us assume that we can observe the actual temperature of the closed environment and that the one-step transition matrix P is

$$P = \begin{pmatrix} 0.2 & 0.8 & 0 & 0 & 0 \\ 0.2 & 0.2 & 0.6 & 0 & 0 \\ 0 & 0.2 & 0.6 & 0.2 & 0 \\ 0 & 0 & 0.6 & 0.2 & 0.2 \\ 0 & 0 & 0 & 0.8 & 0.2 \end{pmatrix}.$$

The value that the sensor reports is a function of the current temperature and it is given by the following matrix B:

$$B = \begin{pmatrix} 0.8 & 0.2 & 0 \\ 0.6 & 0.3 & 0.1 \\ 0.3 & 0.6 & 0.1 \\ 0.1 & 0.4 & 0.5 \\ 0 & 0.1 & 0.9 \end{pmatrix}.$$

State	1	2	3	4	5	4
Time	0	1	2	3	4	5
Observation	1	2	2	3	3	2

FIGURE 12.1 An example of a sequence of states and observations.

This matrix is interpreted as follows. Each row corresponds to a different state of the system. Row i gives the probabilities that the sensor produces the values 1, 2, 3 when the system is in state i. We assume that these probabilities follow an arbitrary empirical distribution. For instance, when the system is in state 1, the sensor produces the value 1 with probability 0.8 and 2 with probability 0.2. Likewise, when the system is in state 2, the sensor produces the value 1 with probability 0.6, 2 with probability 0.3, and 3 with probability 0.1. The sum of the values in each row add up to 1.

Now, let us follow the evolution of this system for a few slots starting at time 0, where we assume that the system is in state 1. Figure 12.1 gives a possible path of the states of the system along with possible observations of the sensor.

At time $t = 0$, the system is in state 1, i.e., the temperature is "very cold", and the sensor shows 1, i.e., "low" temperature. In the next time slot, the system switches to state 2, i.e., the temperature is "cold", and the sensor shows 2, "medium" temperature, and so on.

In the HMM formulation of the above example, we assume that we know the number of the states, i.e., 5, but we do not know P, the initial state vector $\pi^{(0)}$, and B. What we know is just the sequence of the observations produced by the sensor. There are three basic problems associated with HMMs whose solution is necessary in order to use HMMs in real-life applications. These are described below.

12.3 THE THREE BASIC HMM PROBLEMS

Let us consider an HMM, and let $P = (p_{ij})$ be its one-step transition matrix, $\pi^{(0)}$ the initial state vector, notated as just π for simplicity, and $B = (b_i(k))$ the set of the state-dependent probability distributions of the observations, where $b_i(k)$ is the probability of observing a value k given that the system is in state i. Matrix B is known as the *event matrix*.

The probability distribution $b_i(k)$ may be assumed to be an arbitrary empirical discrete distribution, or it may be a theoretical distribution, such as, Poisson, normal, multivariate normal, and mixture of multivariate normal distributions. It could also be defined by an auto-regressive process AR(p). For instance, in the case when it is defined by a normal distribution, the value observed when the system is in state i is normally distributed with parameters μ_i and σ_i^2 which depend on the state i. In the case where the observations come from an AR(p), the observations are generated by an autoregressive process, say an AR(1), whose parameters depend on the current state i, i.e., $X_t = \delta_i + a_i X_{t-1} + \varepsilon_t$, with the variance of the error also depending on the current state i, i.e., $\varepsilon_t \sim N\left(0, \sigma_i^2\right)$. In this chapter, we assume that $b_i(k)$ is an arbitrary empirical discrete distribution. At the end of the chapter, we will discuss the case of continuous distributions and distributions defined by an autoregressive process.

We now proceed to describe the three problems. We note that the symbol λ is used below to indicate the set of the parameters associated with an HMM, i.e., $\lambda = (P, \pi, B)$. Also, a sequence of observations $O_1, O_2, \ldots, O_T$ from $t = 1, 2, \ldots, T$, is notated as $O_1 O_2 \ldots O_T$, i.e. we write the sequence without commas.

12.3.1 Problem 1 – The Evaluation Problem

Given that we know the parameters of an HMM, P, π and B, and given that we have observed a sequence of observations $O = O_1 O_2 \ldots O_T$, calculate the probability $p(O_1 O_2 \ldots O_T | \lambda)$; i.e., calculate the probability that this sequence came from the given HMM. For instance, let us assume that in the above example with the sensor, we have observed the sequence $O = 122332$. Then, we would like to know the probability that this sequence was generated by the HMM $\lambda = (P, \pi, B)$ described in Section 12.2. This problem is referred to as the *evaluation problem* in HMMs.

The solution to this problem is used in the solution of the other two problems, as will be seen below. In addition, it can be used to solve classification problems. For instance, in the above example with the sensor we have assumed that the sensor operates correctly. Now, what if the sensor has gone bad, but we do not know that. How can we determine this? In this case, we can calculate the probability that an observed sequence comes from the given HMM. Given a threshold for this probability, we can determine the health of the sensor; i.e., if this probability is below the threshold, we conclude that the sensor has gone bad.

In speech recognition, there is a finite number of unique spectral coded vectors, and each word is represented by a sequence of such spectral vectors. When we pronounce a word, we create a sequence of spectral vectors $O_1 O_2 \ldots O_T$. Now, each word is represented by a different HMM, and consequently we can calculate the $p(O_1 O_2 \ldots O_T | \lambda)$ for all λ, where λ represents an HMM, and then chose the word for which this probability has the highest value.

12.3.2 Problem 2 – The Decoding Problem

In this problem, referred to as the *decoding problem*, we assume again that we have complete knowledge of the parameters P, π and B of an HMM. Now, given a sequence of observations $O = O_1 O_2 \ldots O_T$ we want to compute the most likely sequence of states that the system went through that best explains O.

Going back to our sensor problem, let us assume again that we have observed the sequence $O = 122332$. Given that we know the parameters P, π, and B, we want to determine the sequence of states (temperatures) that the closed environment went through that best explains O. Since the states are hidden, we cannot observe this sequence. We can only calculate the most likely sequence that gave rise to the observed sequence O.

12.3.3 Problem 3 – The Learning Problem

This is the most difficult problem of the three problems and it has to do with training an HMM. In this case, we assume that that we do not know the HMM parameters P, π and B. We only know that the underlying Markov Chain consists of N states, and we also know the set of values that the observations come from. For instance, in the sensor example, we know that we have 5 states, and that the observations come from the set $\{1, 2, 3\}$. Now, given a sequence of observations $O = O_1O_2 \ldots O_T$, we want to train the HMM, i.e., estimate the HMM parameters P, π, and B. This problem is known as the *learning problem* in HMMs.

We now proceed to obtain the solution to each of the three basic problems. For further details see [1].

12.4 MATHEMATICAL NOTATION

Before, we start, we need to define a number of mathematical symbols that we will use in the remaining of this chapter. The reader is urged to understand and memorize these symbols.

N: The total number of states.

S: The set of all the states of the Markov Chain, numbered from 1 to N, i.e., $S = \{1, 2, \ldots, N\}$.

q_t: The state of the Markov Chain at time t.

Q: A sequence of states that the Markov Chain went through from $t = 1$ to $t = T$, i.e., $Q = q_1q_2 \ldots q_T$.

P: The one-step transition matrix. $P = (p_{ij})$, where p_{ij} is the probability that at the end of a time slot the system will jump to state j given that it was in state i.

M: The total number of observation symbols.

v: The set of the observation symbols, i.e., $v = \{v_1, v_2, \ldots, v_M\}$. For instance, in the sensor example, $M = 3$, and the set of symbols is $v = \{1, 2, 3\}$.

O: A sequence of observations, i.e., $O = O_1O_2 \ldots O_T$ observed from time $t = 1$ to $t = T$, where T is the current time slot. Observation O_i takes a value from the set $\{v_1, v_2, \ldots, v_M\}$.

B: The event matrix. $B = (b_i(k))$, where

$$b_i(k) = p\{O_t = v_k \text{ at time } t | q_t = i\}.$$

For $i = 1, 2, \ldots, N$, the probability distribution $b_i(k)$, $k = 1, \ldots, M$, is assumed to be an arbitrary discrete distribution.

π: The initial state of the Markov Chain. $\pi = (\pi_1, \pi_2, \ldots, \pi_N)$, where $\pi_i = prob(q_1 = i)$.

λ: Indicates the parameters of an HMM, i.e., $\lambda = (P, \pi, B)$.

12.5 SOLUTION TO PROBLEM 1

Problem 1 can be expressed mathematically as follows. Given an HMM $\lambda = (P, \pi, B)$ and a sequence of observations $O = O_1O_2 \ldots O_T$, calculate the probability $p(O_1O_2 \ldots O_T | \lambda)$, i.e., the probability that this sequence was generated by the HMM. (To simplify the mathematical notation, whenever possible, we write the conditional probability $p(O|\lambda)$ and $p(Q|\lambda)$ as $p(O)$ and $p(Q)$ respectively).

12.5.1 A Brute Force Solution

We start by first giving a simple brute force solution. Given a sequence of states $Q = q_1 q_2 \cdots q_T$ the probability of observing a given sequence of observations $O = O_1 O_2 \ldots O_T$, is:

$$p(O|Q) = p(O_1|q_1) p(O_2|q_2) \ldots p(O_T|q_T)$$

$$= b_{q_1}(O_1) b_{q_2}(O_2) \ldots b_{q_T}(O_T).$$

We also have that:

$$p(Q) = \pi_{q_1} P_{q_1, q_2} P_{q_2, q_3} \cdots P_{q_{T-1}, q_T}.$$

Therefore

$$p(O) = \sum_Q p(O|Q) p(Q)$$

$$= \sum_Q \pi_{q_1} b_{q_1}(O_1) P_{q_1, q_2} b_{q_2}(O_2) P_{q_2, q_3} \cdots P_{q_{T-1}, q_T} b_{q_T}(O_T).$$

Therefore, in order to calculate $p(O)$, we have to generate all the possible sequences of states $q_1 q_2 \ldots q_T$ and for each sequence calculate the probability of observing the given sequence of observations $O = O_1 O_2 \ldots O_T$. The final result is obtained by adding up all these probabilities.

For example, let us consider the following HMM: $S = \{1, 2, 3\}$, $v = \{1, 2\}$, $\pi = \{1, 0, 0\}$,

$$P = \begin{pmatrix} 0.5 & 0.5 & 0 \\ 0.2 & 0.2 & 0.6 \\ 0 & 0.4 & 0.6 \end{pmatrix}, \text{ and } B = \begin{pmatrix} 0.8 & 0.2 \\ 0.6 & 0.4 \\ 0.3 & 0.7 \end{pmatrix}.$$

We want to calculate the probability that the sequence of observations $O = 112$, came from this HMM.

Let us consider an arbitrary sequence of states $Q = 123$. Then, we have:

$$p(O|Q) = b_1(1) b_2(2) b_3(2)$$

$$= 0.8 \times 0.6 \times 0.7 = 0.336.$$

$$p(Q) = \pi_1 P_{1, 2} P_{2, 3}$$

$$= 1 \times 0.5 \times 0.6 = 0.3.$$

Hence, the probability of observing the sequence $O = 112$ given $Q = 123$ is: $0.336 \times 0.3 = 0.1008$. We perform this calculation for all possible paths Q and then add up the results.

This brute force method is obviously inefficient and it can only be used for very small numerical cases. Below, we present the forward–backward algorithm, a very efficient procedure for calculating $p(O_1O_2 \ldots O_T)$. This algorithm is also used to solve problems 2 and 3.

12.5.2 The Forward–Backward Algorithm

The algorithm, as its name implies, consists of a forward and a backward algorithm.

a) The Forward Algorithm

Let $\alpha_t(i) = p(O_1O_2 \ldots O_t, q_t = i)$ be the probability that at time t we have observed the sequence $O_1O_2 \ldots O_t$ and the state of the Markov Chain is $q_t = i$. This probability is calculated recursively as follows:

1. *Initialization*

$$\alpha_1(i) = p(O_1, q_t = i) = \pi_i b_i(O_1), \ i = 1, 2, \ldots, N.$$

2. *Recursion*

For $t = 1, 2, \ldots, T-1$

    For $i = 1, 2, \ldots, N$

$$\alpha_{t+1}(i) = p(O_1O_2 \ldots O_t O_{t+1}, q_{t+1} = i)$$

$$= \sum_{j=1}^{N} p(O_1O_2 \ldots O_t, q_t = j) p_{ji} b_i(O_{t+1})$$

$$= \sum_{j=1}^{N} \alpha_t(j) p_{ji} b_i(O_{t+1}).$$

The above recursion can be explained using the diagram in Figure 12.2. We note that the sum $\alpha_t(1)p_{1i} + \cdots + \alpha_t(N)p_{Ni}$ is the probability that we have observed the sequence $O_1O_2 \ldots O_t$ and we have transited to state i at time $t + 1$. Consequently, if we multiply this probability with $b_i(O_{t+1})$, the probability that we observe the value O_{t+1} while in state i, we will obtain the probability that we have observed the sequence of values $O_1O_2 \ldots O_t O_{t+1}$ and we are in state i, which is $\alpha_{t+1}(i)$.

After we run the above recursion, we can calculate the probability $p(O_1O_2 \ldots O_T)$ as follows:

$$p(O_1O_2 \ldots O_T) = \sum_{i=1}^{N} p(O_1O_2 \ldots O_T, q_T = i)$$

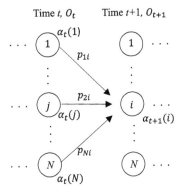

FIGURE 12.2 A diagrammatic view of the forward recursion.

$$= \sum_{i=1}^{N} \alpha_T(i).$$

b) *The Backward Algorithm*

Let $\beta_t(i) = p(O_{t+1}O_{t+2} \ldots O_T | q_t = i)$ be the probability that we will observe the sequence $O_{t+1}O_{t+2} \ldots O_T$ given that at time t the Markov Chain is in state i. This probability is calculated recursively as follows:

1. *Initialization*
   For $i = 1, 2, \ldots, N$

$$\beta_T(i) = p\big(\text{We will observe nothing after } T \,|\, q_T = i\big) = 1.$$

2. *Recursion*
   For $t = T - 1, \ldots, 2, 1$

   For $i = 1, 2, \ldots, N$

$$\beta_t(i) = p\big(O_{t+1}O_{t+2} \ldots O_T | q_t = i\big)$$

$$= \sum_{j=1}^{N} p_{ij} b_j \big(O_{t+1}\big) p(O_{t+2} \ldots O_T | q_{t+1} = j)$$

$$= \sum_{j=1}^{N} p_{ij} b_j \big(O_{t+1}\big) \beta_{t+1}(j).$$

A diagrammatic explanation of this recursion is given in Figure 12.3.

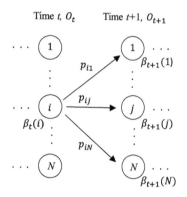

FIGURE 12.3 A diagrammatic view of the backward recursion.

The probability $p(O_1O_2 \ldots O_T)$ is calculated as follows:

$$p(O_1O_2\ldots O_T)=\sum_{i=1}^{N}p(O_1, q_1=i)p(O_2\ldots O_T|q_1=i)$$

$$=\sum_{i=1}^{N}\pi_i b_i(O_1)\beta_1(i).$$

Since $\alpha_1(i) = p(O_1, q_t = i) = \pi_i b_i(O_1)$, $i = 1, 2, \ldots, N$, we have

$$p(O_1O_2\ldots O_T)=\sum_{i=1}^{N}\alpha_1(i)\beta_1(i).$$

In fact, for any t we have the following more general expression:

$$p(O_1O_2\ldots O_T)=\sum_{i=1}^{N}p(O_1\ldots O_t, q_t=i)p(O_{t+1}\ldots O_T|q_t=i)$$

$$=\sum_{i=1}^{N}\alpha_t(i)\beta_t(i).$$

We observe that for $t = T$, and since $\beta_T(i) = 1$, $i = 1, 2, \ldots, N$, the above expression becomes the same as the one obtained from the forward algorithm, i.e.,

$$p(O_1O_2\ldots O_T)=\sum_{i=1}^{N}\alpha_T(i).$$

If we are only interested in the calculation of $p(O_1O_2 \ldots O_T)$, then it suffices to use the forward algorithm. The backward algorithm is used in the solution to problem 3.

As an example, we apply the forward–backward algorithm to the HMM used above to demonstrate the brute force method: $S = \{1, 2, 3\}$, $v = \{1, 2\}$, $\pi = \{1, 0, 0\}$,

$$P = \begin{pmatrix} 0.5 & 0.5 & 0 \\ 0.2 & 0.2 & 0.6 \\ 0 & 0.4 & 0.6 \end{pmatrix}, \text{ and } B = \begin{pmatrix} 0.8 & 0.2 \\ 0.6 & 0.4 \\ 0.3 & 0.7 \end{pmatrix},$$

to calculate the probability that $O = 112$ came from this HMM. We first use the forward algorithm (note that zero terms are omitted).

Initialization:

$$\alpha_1(1) = \pi_1 b_1(1) = 1 \times 0.8 = 0.8.$$

$$\alpha_1(2) = \pi_2 b_2(1) = 0 \times 0.6 = 0.$$

$$\alpha_1(3) = \pi_3 b_3(1) = 0 \times 0.3 = 0.$$

$t = 1$:

$$\alpha_2(1) = \alpha_1(1) p_{11} b_1(1) = 0.8 \times 0.5 \times 0.8 = 0.32.$$

$$\alpha_2(2) = \alpha_1(1) p_{12} b_2(1) = 0.8 \times 0.5 \times 0.6 = 0.24.$$

$$\alpha_2(3) = \alpha_1(1) p_{13} b_3(1) = 0.8 \times 0 \times 0.3 = 0.$$

$t = 2$:

$$\alpha_3(1) = \alpha_2(1) p_{11} b_1(2) + \alpha_2(2) p_{21} b_1(2) = 0.32 \times 0.5 \times 0.2 + 0.24 \times 0.2 \times 0.2 = 0.0416.$$

$$\alpha_3(2) = \alpha_2(1) p_{12} b_2(2) + \alpha_2(2) p_{22} b_2(2) = 0.32 \times 0.5 \times 0.4 + 0.24 \times 0.2 \times 0.4 = 0.0832.$$

$$\alpha_3(3) = \alpha_2(2) p_{23} b_3(2) = 0.24 \times 0.6 \times 0.7 = 0.1008.$$

Finally, we have

$$p(112) = \alpha_3(1) + \alpha_3(2) + \alpha_3(3) = 0.0416 + 0.0832 + 0.1008 = 0.2256.$$

Now, let us apply the backward algorithm. We have

Initialization:

$$\beta_3(1) = \beta_3(2) = \beta_3(3) = 1.$$

$t = 2$:

$$\beta_2(1) = p_{11}b_1(2)\beta_3(1) + p_{12}b_2(2)\beta_3(2)$$

$$= 0.5 \times 0.2 \times 1 + 0.5 \times 0.4 \times 1 = 0.3.$$

$$\beta_2(2) = p_{21}b_1(2)\beta_3(1) + p_{22}b_2(2)\beta_3(2) + p_{23}b_3(2)\beta_3(2)$$

$$= 0.2 \times 0.2 \times 1 + 0.2 \times 0.4 \times 1 + 0.6 \times 0.7 \times 1 = 0.54.$$

$$\beta_2(3) = p_{32}b_2(2)\beta_3(2) + p_{33}b_3(2)\beta_3(2)$$

$$= 0.4 \times 0.4 \times 1 + 0.6 \times 0.7 \times 1 = 0.58.$$

$t = 1$:

$$\beta_1(1) = p_{11}b_1(1)\beta_2(1) + p_{12}b_2(1)\beta_2(2)$$

$$= 0.5 \times 0.8 \times 0.3 + 0.5 \times 0.6 \times 0.54 = 0.282.$$

$$\beta_1(2) = p_{21}b_1(1)\beta_2(1) + p_{22}b_2(1)\beta_2(2) + p_{23}b_3(1)\beta_2(3)$$

$$= 0.2 \times 0.8 \times 0.3 + 0.2 \times 0.6 \times 0.54 + 0.6 \times 0.3 \times 0.58 = 0.2172.$$

$$\beta_1(3) = p_{32}b_2(1)\beta_2(2) + p_{33}b_3(1)\beta_3(2)$$

$$= 0.4 \times 0.6 \times 0.54 + 0.6 \times 0.3 \times 0.58 = 0.234.$$

We can now calculate the probability $p(112)$. We have

$$p(112) = \alpha_1(1)\beta_1(1) = 0.8 \times 0.282 = 0.2256.$$

Also, we can verify that

$$p(112) = \sum_{i=1}^{3} \alpha_2(i)\beta_2(i) = \sum_{i=1}^{3} \alpha_3(i)\beta_3(i).$$

We have:

$$p(112) = \alpha_2(1)\beta_2(1) + \alpha_2(2)\beta_2(2)$$

$$= 0.32 \times 0.3 + 0.24 \times 0.54 = 0.2256,$$

and

$$p(112) = \alpha_3(1)\beta_3(1) + \alpha_3(2)\beta_3(2) + \alpha_3(3)\beta_3(3)$$

$$= 0.0416 \times 1 + 0.0832 \times 1 + 0.1008 \times 1 = 0.2256.$$

12.6 SOLUTION TO PROBLEM 2

In problem 2, we want to find the most probable sequence of states that gives rise to a sequence of observations $O_1O_2 \ldots O_T$, given that we know the parameters of the HMM, i.e., $\lambda = (P, \pi, B)$. Below, we describe two solutions to this problem. The first one is a simple heuristic method, i.e., an intuitive non-theoretical method, whereby at each time t, $t = 1, 2, \ldots, T$, we select the most likely state to occur at that time. For this, we use the expressions derived from the forward–backward method. The second solution uses dynamic programming, a mathematical optimization method, to select the most probable sequence of states, and it is known as the Viterbi algorithm.

12.6.1 The Heuristic Solution

Let $O = O_1O_2 \ldots O_T$ be a sequence of observations. Define,

$$\gamma_t(i) = p(q_t = i|O), \ i = 1, 2, \ldots, T.$$

This is the probability that the system at time t is in state i given that we have observed a sequence O. We have

$$\gamma_t(i) = p(q_t = i|O) = \frac{p(q_t = i, O)}{p(O)}$$

$$= \frac{p(O_1 \ldots O_t, q_t = i)p(O_{t+1} \ldots O_T|q_t = i)}{p(O)}$$

$$= \frac{\alpha_t(i)\beta_t(i)}{\sum_{j=1}^{N} \alpha_t(j)\beta_t(j)}.$$

Therefore, for each t, $t = 1, 2, \ldots, T$, we calculate all the probabilities $\gamma_t(i)$, $i = 1, 2, \ldots, N$, and select the state with the highest probability. In this way, we create a sequence of states that gave rise to a set of observations (see Figure 12.4).

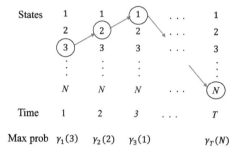

<table>
<tr><td>States</td><td>1</td><td>1</td><td>1</td><td>. . .</td><td>1</td></tr>
<tr><td></td><td>2</td><td>2</td><td>2</td><td>. . . .</td><td>2</td></tr>
<tr><td></td><td>3</td><td>3</td><td>3</td><td>. . .</td><td>3</td></tr>
<tr><td></td><td>⋮</td><td>⋮</td><td>⋮</td><td></td><td>⋮</td></tr>
<tr><td></td><td>N</td><td>N</td><td>N</td><td>. . .</td><td>N</td></tr>
<tr><td>Time</td><td>1</td><td>2</td><td>3</td><td>. . .</td><td>T</td></tr>
<tr><td>Max prob</td><td>$\gamma_1(3)$</td><td>$\gamma_2(2)$</td><td>$\gamma_3(1)$</td><td></td><td>$\gamma_T(N)$</td></tr>
</table>

FIGURE 12.4 Heuristic solution for problem 2.

We note that if some transition probabilities p_{ij} are zero, then it is possible that we may come up with an invalid sequence of states; i.e., the sequence may go from state i at time t to state j at time $t + 1$, whereas in fact the corresponding transition probability p_{ij} is zero. This is because, we chose each state without regard to the previous chosen state and whether p_{ij} is zero or not. However, it is relatively easy to modify this procedure so that to exclude states for which the sequence is invalid.

As an example, we apply this algorithm to the HMM used above: $S = \{1, 2, 3\}$, $v = \{1, 2\}$, $\pi = \{1, 0, 0\}$, $O = 112$,

$$P = \begin{pmatrix} 0.5 & 0.5 & 0 \\ 0.2 & 0.2 & 0.6 \\ 0 & 0.4 & 0.6 \end{pmatrix}, \text{ and } B = \begin{pmatrix} 0.8 & 0.2 \\ 0.6 & 0.4 \\ 0.3 & 0.7 \end{pmatrix},$$

$t = 1$:

$$\gamma_1(1) = \frac{\alpha_1(1)\beta_1(1)}{\sum_{j=1}^{3} \alpha_1(j)\beta_1(j)} = \frac{0.8 \times 0.282}{0.2256} = 1$$

$$\gamma_1(2) = \frac{\alpha_1(2)\beta_1(2)}{\sum_{j=1}^{3} \alpha_1(j)\beta_1(j)} = 0$$

$$\gamma_1(3) = \frac{\alpha_1(3)\beta_1(3)}{\sum_{j=1}^{3} \alpha_1(j)\beta_1(j)} = 0.$$

Therefore, the most likely state that the system was in at time $t = 1$ is state 1.

$t = 2$:

$$\gamma_2(1) = \frac{\alpha_2(1)\beta_2(1)}{\sum_{j=1}^{3} \alpha_2(j)\beta_2(j)} = \frac{0.32 \times 0.3}{0.2256} = 0.426$$

$$\gamma_2(2) = \frac{\alpha_2(2)\beta_2(2)}{\sum_{j=1}^{3}\alpha_2(j)\beta_2(j)} = \frac{0.24\times0.54}{0.2256} = 0.574$$

$$\gamma_2(3) = \frac{\alpha_2(3)\beta_2(3)}{\sum_{j=1}^{3}\alpha_2(j)\beta_2(j)} = 0.$$

Therefore, the most likely state that the system was in at time $t = 2$ is state 2.

$t = 3$:

$$\gamma_2(1) = \frac{\alpha_3(1)\beta_3(1)}{\sum_{j=1}^{3}\alpha_3(j)\beta_3(j)} = \frac{0.0416\times1}{0.2256} = 0.184$$

$$\gamma_2(2) = \frac{\alpha_3(2)\beta_3(2)}{\sum_{j=1}^{3}\alpha_3(j)\beta_3(j)} = \frac{0.0832\times1}{0.2256} = 0.369$$

$$\gamma_2(3) = \frac{\alpha_3(3)\beta_3(3)}{\sum_{j=1}^{3}\alpha_3(j)\beta_3(j)} = \frac{0.1008\times1}{0.2256} = 0.447.$$

Therefore, the most likely state that the system was in at time $t = 3$ is state 3. Consequently, the most likely path is: 123. We note that the denominator in the above calculations of $\gamma_t(i)$ is the same since it is the probability $p(112)$.

12.6.2 The Viterbi Algorithm

The second solution to problem 2 is based on dynamic programming, and it is known as the Viterbi algorithm. The algorithm finds the best sequence $Q = q_1q_2 \dots q_T$ for a given sequence of observations $O = O_1O_2 \dots O_T$ given that we know the parameters $\lambda = (P, B, \pi)$ of an HMM. The algorithm is similar to the forward algorithm for calculating $\alpha_t(i)$, $t = 1, 2, \dots, T$, $i = 1, 2, \dots, N$.

Define

$$\delta_t(i) = \max_{q_1q_2\dots q_{t-1}} p\left(q_1q_2\dots q_{t-1}, q_t = i, O_1O_2\dots O_t\right).$$

$p(q_1q_2 \dots q_{t-1}, q_t = i, O_1O_2 \dots O_t)$ is the probability that a given sequence of states $q_1q_2 \dots q_{t-1}$ ends in state $q_t = i$ at time t and accounts for the observations $O_1O_2 \dots O_t$. $\delta_t(i)$ is the maximum of these probabilities, and it is computed recursively, as follows.

For $t = 1$ we have the following simple expression, for $i = 1, 2, ..., N$:

$$\delta_1(i) = p\big(q_1 = i, O_1\big) = \pi_i b_i\big(O_1\big).$$

For $t = 2$ we have for $i = 1, 2, ..., N$:

$$\delta_2(i) = \max_{q_1} p\big(q_1, q_2 = i, O_1 O_2\big)$$

$$= \max_j \pi_j b_j\big(O_1\big) p_{ji} b_i\big(O_2\big)$$

$$= \max_j \delta_1(j) p_{ji} b_i\big(O_2\big).$$

Likewise, for $t = 3$, we have for $i = 1, 2, ..., N$:

$$\delta_3(i) = \max_{q_1 q_2} p\big(q_1 q_2, q_3 = i, O_1 O_2 O_3\big)$$

$$= \max_{k,j} \pi_k b_k\big(O_1\big) p_{kj} b_j\big(O_2\big) p_{ji} b_i\big(O_3\big).$$

Since $\max_k \pi_k b_k\big(O_1\big) p_{kj} b_j\big(O_2\big) = \delta_2(j)$, we have:

$$\delta_3(i) = \max_j \delta_2(j) p_{ji} b_i\big(O_3\big).$$

Below, we summarize the recursive algorithm.

Let $\psi_t(i)$, $i = 1, 2, ...N$, $t = 1, 2, ..., T$, be the (i, t)th element of a $T \times N$ array. This array is used to store information that permits the selection of the optimum path of states at the end of the execution of the algorithm.

1. *Initialization*

   For $i = 1, 2, ..., N$

$$\delta_1(i) = p\big(q_1 = i, O_1\big) = \pi_i b_i\big(O_1\big)$$

$$\psi_1(i) = 0.$$

2. *Recursion*

   For $t = 2, 3, ..., T$

   For $i = 1, 2, ..., N$

$$\delta_t(i) = \max_j \left[\delta_{t-1}(j) p_{ji} \right] b_i(O_t)$$

$$\psi_t(i) = \operatorname*{argmax}_j \left[\delta_{t-1}(j) p_{ji} \right].$$

3. *Calculation of the optimum path*

The optimum path $Q^* = q_1^* q_2^* \ldots q_T^*$ is selected going backwards using the values $\psi_t(i)$, $i = 1, 2, \ldots N$, $t = 1, 2, \ldots, T$. For $t = T$, we have:

$$q_T^* = \arg\max_j \left[\delta_T(j) \right]$$

and $t = T - 1, T - 2, \ldots, 1$:

$$q_t^* = \psi_{t+1}\left(q_{t+1}^* \right).$$

We demonstrate the Viterbi algorithm using the same HMM as above: $S = \{1, 2, 3\}$, $v = \{1, 2\}$, $\pi = \{1, 0, 0\}$,

$$P = \begin{pmatrix} 0.5 & 0.5 & 0 \\ 0.2 & 0.2 & 0.6 \\ 0 & 0.4 & 0.6 \end{pmatrix}, \text{ and } B = \begin{pmatrix} 0.8 & 0.2 \\ 0.6 & 0.4 \\ 0.3 & 0.7 \end{pmatrix}.$$

We want to calculate the most probable sequence of states that gives rise to the sequence of observations $O = 112$.

$t = 1$

$$\delta_1(1) = \pi_1 b_1(1) = 1 \times 0.8 = 0.8, \ \delta_1(2) = 0, \ \delta_1(3) = 0.$$

$$\psi_1(1) = \psi_1(2) = \psi_1(3) = 0.$$

$t = 2$

$$\delta_2(1) = \max\left\{ \delta_1(1) p_{11}, \ \delta_1(2) p_{21}, \ \delta_1(3) p_{31} \right\} b_1(1)$$

$$= \max\left\{ 0.8 \times 0.5, 0 \times 0.2, 0 \times 0 \right\} \times 0.8 = 0.32.$$

$$\delta_2(2) = \max\left\{ \delta_1(1) p_{12}, \ \delta_1(2) p_{22}, \ \delta_1(3) p_{32} \right\} b_2(1)$$

$$= \max\left\{ 0.8 \times 0.5, 0 \times 0.2, 0 \times 0.4 \right\} \times 0.6 = 0.24.$$

$$\delta_2(3) = \max\left\{\delta_1(1)p_{13}, \delta_1(2)p_{23}, \delta_1(3)p_{33}\right\}b_3(1)$$

$$= \max\left\{0.8\times0,0\times0.6,0\times0.6\right\}\times0.3 = 0.$$

We now proceed to calculate the $\psi_t(i)$ quantities. Note that for this we use the results from the previous iteration. We have

$$\psi_2(1) = \mathrm{argmax}\left\{\delta_1(1)p_{11}, \delta_1(2)p_{21}, \delta_1(3)p_{31}\right\}$$

$$= \mathrm{argmax}\left\{0.8\times0.5,0\times0.2,0\times0\right\}$$

$$= \mathrm{argmax}\left\{0.4,0,0\right\} = 1.$$

$$\psi_2(2) = \mathrm{argmax}\left\{\delta_1(1)p_{12}, \delta_1(2)p_{22}, \delta_1(3)p_{32}\right\}$$

$$= \mathrm{argmax}\left\{0.8\times0.5,0\times0.2,0\times0.4\right\}$$

$$= \mathrm{argmax}\left\{0.4,0,0\right\} = 1.$$

$$\psi_2(3) = \mathrm{argmax}\left\{\delta_1(1)p_{13},\delta_1(2)p_{23},\delta_1(3)p_{33}\right\}$$

$$= \mathrm{argmax}\left\{0.8\times0,0\times0.6,0\times0.6\right\}$$

$$= \mathrm{argmax}\left\{0,0,0\right\} = \left\{1,2,3\right\}.$$

(Note that in the above argmax, all three values are maximum and consequently the solution is a set of values.)

$t = 3$

$$\delta_3(1) = \max\left\{\delta_2(1)p_{11}, \delta_2(2)p_{21}, \delta_2(3)p_{31}\right\}b_1(2)$$

$$= \max\left\{0.32\times0.5,0.24\times0.2,0\times0\right\}\times0.2 = 0.032.$$

$$\delta_3(2) = \max\left\{\delta_1(1)p_{12}, \delta_1(2)p_{22}, \delta_1(3)p_{32}\right\}b_2(2)$$

$$= \max\left\{0.32\times0.5,0.24\times0.2,0\times0.4\right\}\times0.4 = 0.064.$$

$$\delta_3(3) = \max\left\{\delta_1(1)p_{13}, \delta_1(2)p_{23}, \delta_1(3)p_{33}\right\}b_3(2)$$

$$= \max\left\{0.32\times 0, 0.24\times 0.6, 0\times 0.6\right\}\times 0.7 = 0.1008.$$

The $\psi_t(i)$ quantities are as follows:

$$\psi_3(1) = \operatorname{argmax}\left\{\delta_2(1)p_{11}, \delta_2(2)p_{21}, \delta_2(3)p_{31}\right\}$$

$$= \operatorname{argmax}\left\{0.32\times 0.5, 0.24\times 0.2, 0\times 0\right\}$$

$$= \operatorname{argmax}\left\{0.16, 0.048, 0\right\} = 1.$$

$$\psi_3(2) = \operatorname{argmax}\left\{\delta_2(1)p_{12}, \delta_2(2)p_{22}, \delta_2(3)p_{32}\right\}$$

$$= \operatorname{argmax}\left\{0.32\times 0.5, 0.24\times 0.2, 0\times 0.4\right\}$$

$$= \operatorname{argmax}\left\{0.16, 0.048, 0\right\} = 1.$$

$$\psi_3(3) = \operatorname{argmax}\left\{\delta_2(1)p_{13}, \delta_2(2)p_{23}, \delta_2(3)p_{33}\right\}$$

$$= \operatorname{argmax}\left\{0.032\times 0, 0.24\times 0.6, 0\times 0.6\right\}$$

$$= \operatorname{argmax}\left\{0, 0.144, 0\right\} = 2.$$

The $\psi_t(i)$ values are summarized in Table 12.2.

We are now ready to compute the optimum path $q_1^* q_2^* q_3^*$. We have

$$q_3^* = \operatorname{argmax}\left\{0.032, 0.064, 0.1008\right\} = 3$$

$$q_2^* = \psi_3\left(q_3^*\right) = \psi_3(3) = 2.$$

TABLE 12.2 The $\psi_t(i)$ values

$\psi_t(i)$	$t = 1$	$t = 2$	$t = 3$
$i = 1$	0	1	1
$i = 2$	0	1	1
$i = 3$	0	{1,2,3}	2

$$q_1^* = \psi_2\left(q_2^*\right) = \psi_2\left(2\right) = 1.$$

Therefore, the optimum path is 123. We note that this is the same as the path computed using the simple heuristic solution. In general, the two solutions may not be the same, in which case the Viterbi solution is preferable since it is an optimization algorithm.

12.7 SOLUTION TO PROBLEM 3

In problem 3, we want to determine the parameters $\lambda = (P, \pi, B)$ of an HMM given that we have a sequence of observations $O = O_1 O_2 \ldots O_T$. Let Θ be the set of all the parameters to be estimated, i.e., $\Theta = \{p_{ij}, i, j = 1, 2, \ldots, N, b_i(k), i = 1, 2, \ldots, N, k = 1, 2, \ldots, M, \pi\}$. Following the maximum likelihood estimation (MLE) method, Θ can be estimated by maximizing the log of the likelihood $p(O \mid \Theta)$; i.e.,

$$\underset{\Theta}{\text{argmax}} \log\left(p\left(O|\Theta\right)\right).$$

This can be done using a gradient-based method, whereby Θ is updated iteratively, as follows:

$$\Theta^{new} = \Theta^{old} - \gamma \left.\frac{\partial \log\left(p\left(O|\Theta\right)\right)}{\partial \Theta}\right|_{\Theta=\Theta^{old}}$$

where γ is the learning rate (see Section 10.6.1, Chapter 10). The derivatives do not have a closed-form solution and they are calculated iteratively (see [2]). An alternative popular solution is the Baum–Welch algorithm described below. This algorithm is an instance of the expectation-maximization (EM) algorithm (see [3]).

We first define the following quantities. Let $\xi_t(i, j)$ be the probability that the system is in state i at time t and in state j at time $t + 1$. We have

$$\xi_t\left(i, j\right) = p\left(q_t = i, q_{t+1} = j | O\right)$$

$$= \frac{\alpha_t\left(i\right) p_{ij} b_j\left(O_{t+1}\right) \beta_{t+1}\left(j\right)}{p\left(O\right)}.$$

We recall that $p(O) = \sum_{j=1}^{N} \alpha_t\left(j\right) \beta_t\left(j\right)$, and since

$$\beta_t\left(j\right) = \sum_{j=1}^{N} p_{ij} b_j\left(O_{t+1}\right) \beta_{t+1}\left(j\right)$$

we have that

$$p(O) = \sum_{i=1}^{N}\sum_{j=1}^{N}\alpha_t(i)p_{ij}b_j(O_{t+1})\beta_{t+1}(j);$$

i.e., $p(O)$ is the sum of all the numerators in the above expression. For $\xi_t(i,j)$, which guarantees that all the probabilities $\xi_t(i,j)$ sum up to 1. $\xi_t(i,j)$ can also be calculated as follows:

$$\xi_t(i,j) = \frac{\alpha_t(i)p_{ij}b_j(O_{t+1})\beta_{t+1}(j)}{\sum_{i=1}^{N}\sum_{j=1}^{N}\alpha_t(i)p_{ij}b_j(O_{t+1})\beta_{t+1}(j)}.$$

The probability $\gamma_t(i)$ of being in state i at time t given O is

$$\gamma_t(i) = \sum_{j=1}^{N}\xi_t(i,j).$$

We note that the probability $\gamma_t(i)$ can also be interpreted as the expected number of times the system visits state i at time t. Obviously, at time t, the system will visit state i either once or not at all. Therefore, the expected number of times the system visits state i at time t is $1 \times \gamma_t(i) + 0 \times (1 - \gamma_t(i)) = \gamma_t(i)$. In view of this, the expected number of times the system visits state i over the period $[1, T]$ is

$$\sum_{t=1}^{T}\gamma_t(i).$$

The above expression summed from $t = 1$ to $T - 1$ is the expected number of times the system visits state i over the period $[1, T - 1]$, and therefore it is the expected number of times the system transits out of state i over the period $[1, T]$.

Also, the sum

$$\sum_{t=1}^{T-1}\xi_t(i,j)$$

gives the expected number of transitions from state i to state j over the period $[1, T]$. This can be easily shown using the same argument as above. At time t, the number of times the system transits from i to j is either 1 or zero. Therefore, the expected number of transitions from i to j when the system is in state i at time t is: $1 \times \xi_t(i,j) + 0 \times (1 - \xi_t(i,j)) = \xi_t(i,j)$. In view of this, the sum of these expectations from 1 to $T - 1$ is the expected number of transitions from state i to state j over the period $[1, T]$.

The Baum–Welch algorithm is an iterative algorithm, whereby we start with an initial guess of the parameters $\lambda = (P, \pi, B)$ of an HMM, and then we construct a new set of estimates $\bar{\lambda} = (\bar{P}, \bar{\pi}, \bar{B})$ of the HMM parameters. Then, the new estimates are used as the initial guess of the parameters, i.e., we set $\lambda = \bar{\lambda}$, and we repeat the above step. We continue in this manner as long as the probability, also referred to as the likelihood, $p(O|\lambda)$ increases. Given an initial guess of the parameters $\lambda = (P, \pi, B)$, the new estimates are obtained as follows:

$$\bar{\pi}_i = \gamma_1(i), \; i = 1, 2, \ldots, N,$$

$$\bar{P}_{ij} = \frac{\sum_{t=1}^{T-1} \xi_t(i,j)}{\sum_{t=1}^{T-1} \gamma_t(i)}, \; i, j = 1, 2, \ldots, N,$$

$$\bar{b}_i(k) = \frac{\sum_{\substack{t=1 \\ st\, O_t = k}}^{T} \gamma_t(i)}{\sum_{t=1}^{T-1} \gamma_t(i)}, \; k = 1, 2, \ldots, M, \; i = 1, 2, \ldots, N.$$

We note that the following constraints are automatically satisfied.

$$\sum_{i=1}^{N} \bar{\pi}_i = 1, \sum_{j=1}^{N} \bar{P}_{ij} = 1, \; i = 1, 2, \ldots, N, \sum_{k=1}^{M} \bar{b}_i(k) = 1, \; i = 1, 2, \ldots, N.$$

The expression for $\bar{\pi}_i$ is obvious since $\gamma_1(i)$ is the probability of being in state i at time $t = 1$. Now, let us examine the expression for the transition probabilities $\bar{P}_{ij}$, $i, j = 1, 2, \ldots, N$. The numerator gives the expected number of times the system transits from state i to state j over the period $[1, T]$, and the denominator gives the expected number of times the system transits out of state i over the period $[1, T]$. Therefore, their ratio is $\bar{P}_{ij}$. Finally, the expression for $\bar{b}_i(k)$ is the ratio of the expected number of transitions out of the state i while observing the symbol k divided by the expected number of transitions out of state i.

This iterative algorithm converges to a local maximum. This is because the optimization surface is very complex and has multiple maxima. Because of this, an HMM should be trained with different initial values, and the model with the highest probability $p(O|\lambda)$ should be selected.

As an example, we apply one iteration of this algorithm to the same HMM used above: $S = \{1, 2, 3\}$ and $= \{1, 2\}$, $\pi = \{1, 0, 0\}$,

$$P = \begin{pmatrix} 0.5 & 0.5 & 0 \\ 0.2 & 0.2 & 0.6 \\ 0 & 0.4 & 0.6 \end{pmatrix}, \text{ and } B = \begin{pmatrix} 0.8 & 0.2 \\ 0.6 & 0.4 \\ 0.3 & 0.7 \end{pmatrix}.$$

In this case, we assume that we do not know the parameters of the HMM. We only know that the HMM has three states, i.e., $S = \{1, 2, 3\}$, the number of objects is 2, i.e., $v = \{1, 2\}$, and $O = 112$. In order to take advantage of the previously obtained results for this HMM, we use the above exact parameters $\lambda = (P, B, \pi)$ as the initial estimates.

We have

$$\bar{\pi}_1 = \gamma_1(1) = 1, \bar{\pi}_2 = \gamma_1(2) = 0, \text{ and } \pi_3 = \gamma_1(3) = 0.$$

Now, let us estimate $\bar{p}_{12}$. We have

$$\bar{p}_{12} = \frac{\xi_1(1,2) + \xi_2(1, 2)}{\gamma_1(1) + \gamma_2(1)}$$

where

$$\gamma_1(1) + \gamma_2(1) = 1 + 0.426 = 1.426.$$

$$\xi_1(1,2) = \frac{\alpha_1(1) p_{12} b_2(1) \beta_2(2)}{p(O)} = \frac{0.8 \times 0.5 \times 0.6 \times 0.54}{0.2256} = 0.574$$

$$\xi_2(1,2) = \frac{\alpha_2(1) p_{12} b_2(2) \beta_3(2)}{p(O)} = \frac{0.32 \times 0.5 \times 0.4 \times 1}{0.2256} = 0.2837.$$

Hence,

$$\bar{p}_{12} = \frac{0.574 + 0.2837}{1.426} = 0.60.$$

Now, let us estimate $\bar{b}_1(1)$. We have

$$\bar{b}_1(1) = \frac{\gamma_1(1)_{st\ O_1 = 1} + \gamma_2(1)_{st\ O_2 = 1}}{\gamma_1(1) + \gamma_2(1)}.$$

We recall that $\gamma_1(1) = p(q_1 = 1|O)$. Therefore, $\gamma_1(1)_{st\ O_1 = 1}$ is the probability that $q_1 = 1$ given O and given that $O_1 = 1$, i.e., the first observation is 1. Since $O = 112$, we have that $\gamma_1(1)_{st\ O_1 = 1} = \gamma_1(1)$. For the same reason, $\gamma_2(1)_{st\ O_2 = 1} = \gamma_2(1)$. Therefore, $\bar{b}_1(1) = 1$. This means, that $\bar{b}_1(2) = 0$. We have

$$\bar{b}_1(2) = \frac{\gamma_1(1)_{st\ O_1 = 2} + \gamma_2(1)_{st\ O_2 = 2}}{\gamma_1(1) + \gamma_2(1)}.$$

$\gamma_1(1)_{st\ O_1 = 2}$ is the probability that $q_1 = 1$ given O and given that $O_1 = 2$, i.e., the first observation is 2, which is not feasible. Therefore, $\gamma_1(1)_{st\ O_1 = 2} = 0$. Likewise, $\gamma_2(1)_{st\ O_1 = 2} = 0$, and $\bar{b}_1(2) = 0$. The estimation of the remaining $\bar{p}_{ij}$ and $\bar{b}_i(k)$ probabilities is done in the same way.

12.8 SELECTION OF THE NUMBER OF STATES N

We note that in order to determine the parameters $\lambda = (P, \pi, B)$ of an HMM from a sequence of observations $O = O_1 O_2 \ldots O_T$, we need to know the number of states N and the number of objects. The number of objects can be determined by examining the sequence of observations $O = O_1 O_2 \ldots O_T$. The number of states is determined by training the HMM for different values of N, starting from $N = 2$ or from an initial guess in the case where we know where the data comes from. For each value of N we train the HMM and record the achieved maximum likelihood L. Then, we plot L against N, and select the number of states from which L is maximum. For this, we need to have a sufficient number of likelihood values so that we can discern a pattern. Note that the number of parameters that have to be estimated is increased as the number of states increases, and therefore larger samples may be needed. As guidance, one can use the rule of thumb of at least 10 observations for each parameter to be estimated.

An alternative way to selecting the number of states is to use the AIC and BIC criteria, which take into account both the likelihood and the number of parameters to be estimated. They are given by the expressions:

$$AIC = -2\log L + 2p$$

$$BIC = -2\log L + p\log T$$

where L is the likelihood, T is the number of observations, and p is the total number of independent parameters, given by the expression:

$$p = M^2 + NM - 1,$$

where N is the number of states and M is the number of objects. We train the HMM for different values of N and plot the AIC and BIC as a function of N. We select the value of N for which the AIC or BIC is the lowest.

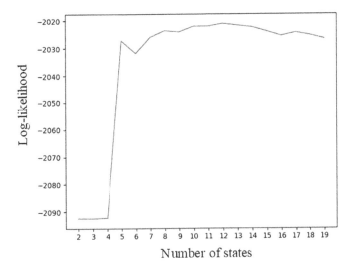

FIGURE 12.5 Log-likelihood vs number of states.

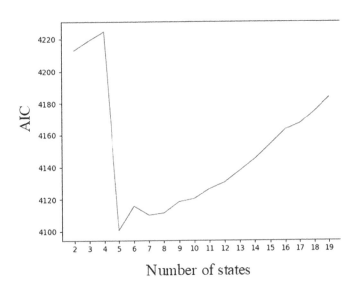

FIGURE 12.6 *AIC* vs number of states.

In practice, one should plot all three curves, i.e., *L*, *AIC*, and *BIC* as a function of the number of states *N*, and then make an educated decision. An example is shown in Figures 12.5–12.7. Figure 12.5 gives the log-likelihood as a function of the number of states. Figures 12.6 and 12.7 give the *AIC* and *BIC*, respectively, as a function of the number of states. Examining these three figures, it seems that 5 states is a good choice.

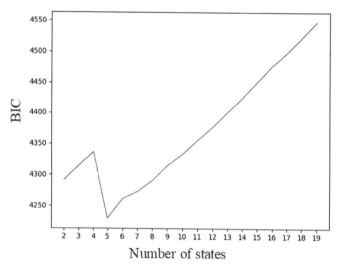

FIGURE 12.7 *BIC* vs number of states.

12.9 FORECASTING O_{T+t}

If an HMM is used as a forecasting tool, then we want to predict the next value O_{T+1} after we have trained the HMM on a sequence of observations $O = O_1O_2 \ldots O_T$. The probability of observing the value k at time $t = T + 1$, given $O = O_1O_2 \ldots O_T$, is

$$p(O_{T+1} = k|O) = \frac{\sum_{i=1}^{N} a_T(i) \sum_{j=1}^{N} p_{ij} p(O_{T+1} = k|q_{T+1} = j)}{p(O)}$$

$$= \frac{\sum_{i=1}^{N} a_T(i) \sum_{j=1}^{N} p_{ij} b_j(k)}{\sum_{j=1}^{N} a_T(i)}.$$

The numerator of the above expression is derived as follows. We recall that $a_T(i)$ is the probability that the HMM is in state i at time $t = T$ given that we have observed $O = O_1O_2 \ldots O_T$. Consequently, the probability that the HMM is in state i at time $t = T$, and the value k is observed at time $t = T + 1$ is $a_T(i)\sum_{j=1}^{n} p_{ij} b_j(k)$. Summing up this term over all i's gives the numerator of the above expression, which is the probability that we observe the value k at time $t = T + 1$, given that we have observed $O = O_1O_2 \ldots O_T$.

Using the above expression we can obtain the entire forecasting distribution $p(O_{T+1} = k|O)$, $k = 1, 2, \ldots, M$, for time $t = T + 1$. From this distribution, we can calculate measures, such as the mean, the median, and prediction interval at different levels of confidence such as 80%. The latter, is an interval $[a, b]$ that contains, 80% of the distribution; i.e., a and b are the 10% and 90% quantiles, respectively. Similar prediction intervals can be constructed for other confidence levels, such as the 90%.

For the above expression $p(O_{T+1} = k|O)$ can be easily extended so that to forecast any future value $O_{T+\tau}$, $\tau \geq 1$. We have

$$p(O_{T+\tau} = k|O) = \frac{\sum_{i=1}^{N} a_T(i) \sum_{j=1}^{N} p_{ij}^{(\tau)} b_j(k)}{\sum_{j=1}^{N} a_T(i)},$$

where $p_{ij}^{(\tau)}$ is the transition probability of going from state i to state j in τ steps.

12.10 CONTINUOUS OBSERVATION PROBABILITY DISTRIBUTIONS

So far, we assumed that the observations come from a discrete set of symbols $\{v_1, v_2, ..., v_M\}$, and therefore the probability distributions $b_i(k)$, $k = 1, 2, ..., M$, of observing symbol v_k when the system is in state $i = 1, 2, ..., N$, is also discrete. However, there are many cases where the observations are continuous and in this case the associated pdf is also continuous. It is possible to discretize an observation density, but this approach introduces inaccuracies due to the approximations introduced when discretizing a continuous variable.

HMMs with continuous observation densities have been analyzed assuming that the densities are normally distributed with parameters dependent on the state i; i.e., $b_i(x)$, the pdf of observing a value x when in state i, follows a normal distribution with mean μ_i and variance σ_i^2, $i = 1, 2, ..., N$.

More generally, $b_i(x)$ can be represented by a mixture of m normal distributions, each with different mean and variance. These m distributions are weighed with m weights that depend on the state i; i.e.,

$$b_i(x) = \sum_{j=1}^{m} c_{ij} f_{X_{ij}}(x), \, i = 1, 2, ..., N,$$

where c_{ij}, $j = 1, 2, ..., m$, are the weights, and $f_{X_{ij}}(x)$ is normally distributed, i.e., $f_{X_{ij}}(x) \sim N\left(\mu_{ij}, \sigma_{ij}^2\right), j = 1, 2.., m$, where the mean μ_{ij} and variance σ_{ij}^2 depend on the state i. In order for $b_i(x)$ to be a pdf, $c_{ij} \geq 0$, $j = 1, 2, ..., m$, $i = 1, 2, ..., N$, and the sum of weights have to add up to 1, i.e.,

$$\sum_{j=1}^{m} c_{ij} = 1, \, i = 1, 2, ..., N.$$

This guarantees that $b_i(x)$ is normalized, i.e.,

$$\int_{-\infty}^{+\infty} b_i(x) dx = 1, \, i = 1, 2, ..., N.$$

In the most general case, $f_{Xij}(x)$ can be a log-concave or elliptically symmetric density. Also, the observations can be vectors rather than scalars, as assumed above.

The forward–backward algorithm, the Viterbi algorithm, and the Baum–Welch estimation procedure described in the previous sections have been extended to the case of continuous observation densities (see [1,2]).

12.11 AUTOREGRESSIVE HMMS

An underlying assumption in the HMMs examined above, is that the successive observations are not correlated. This is because at time t we select a value k with probability $b_i(k)$ which depends only on the state i that the HMM is currently in, and not on the previous observation or sequence of observations. The only dependency introduced is that the current state at time t depends on the previous state at time $t-1$. It does not depend, however, on the sequence of states it went though before t. This dependency only on the previous state, is an underlying assumption of Markov Chains.

There are real-life cases where the successive observations are correlated, such as the air temperature measured every minute. One way of taking into account the correlation between successive observations is to use an autoregressive HMM, notated as AR(p)-HMM. This is an interesting class of HMMs with many applications, whereby the value observed at a given time t is given by an AR(p) process whose parameters depend on the state of the system.

For instance, let us consider an HMM with three states $i = 1, 2, 3$. For each state i we define an AR(p) process with the same order p as follows. For simplicity, we assume that $p = 1$, i.e., the autoregressive model is given by the expression: $X_t = \delta + a_1 X_{t-1} + \varepsilon_t$. Then, we define the following three AR(1) models:

$$X_t = \delta^1 + a_1^1 X_{t-1} + \varepsilon_t^1$$

$$X_t = \delta^2 + a_1^2 X_{t-1} + \varepsilon_t^2$$

$$X_t = \delta^3 + a_1^3 X_{t-1} + \varepsilon_t^3$$

where the superscript indicates the state of the HMM, and $\varepsilon_t^i \sim N(0, \sigma_i^2)$, $i = 1, 2, 3$. The AR(3)-HMM process works as follows. Let us assume that at time $t = 0$ the HMM is in state 1. Then the first observation X_1 is generated using the AR(1): $X_1 = \delta^1 + \varepsilon_1^1$, where $X_0 = 0$. At time $t = 2$ we assume that the HMM shifts to state 3, and in this case, the second value X_2 is generated using the AR(1): $X_2 = \delta^3 + a_1^3 X_1 + \varepsilon_2^3$. At time $t = 3$ we assume that the HMM shifts to state 2 and $X_3 = \delta^2 + a_1^2 X_2 + \varepsilon_3^2$, and so on.

The observations can also be vectors, as opposed to scalars assumed above, and in this case the HMM is notated as a VAR(p)-HMM. For further details on AR(p)-HMMs (see [2]).

EXERCISES

1. The probability of rain tomorrow is 0.8 if it is raining today, and the probability it is clear tomorrow is 0.9 if it is clear today.

   a. Construct the one-step transition matrix

   b. Find the steady-state probabilities

2. The males in an animal population are classified as aggressive or passive. It is observed that 70% of the male offspring of aggressive males are themselves aggressive, while 50% of the offspring of passive males are aggressive. In the long run, assuming that these percentages remain the same, what proportion of the males in this animal population will be aggressive?

3. Let X_t be the number of jobs awaiting execution at a CPU at time t. If a job is executing at time t, it will complete its execution at time $t + 1$ with a probability β. Between the time t and $t + 1$ only one job can arrive with probability α. Set-up the Markov Chain and compute the steady-state probabilities.

4. Consider the HMM example used in the chapter: $S = \{1, 2, 3\}$, $v = \{1, 2\}$, $\pi = \{1, 0, 0\}$,

$$P = \begin{pmatrix} 0.5 & 0.5 & 0 \\ 0.2 & 0.2 & 0.6 \\ 0 & 0.4 & 0.6 \end{pmatrix}, \text{ and } B = \begin{pmatrix} 0.8 & 0.2 \\ 0.6 & 0.4 \\ 0.3 & 0.7 \end{pmatrix}.$$

   Calculate the probability $p(O)$ that the sequence of observations $O = 221$ came from this HMM using the brute force method.

5. Consider the same HMM as in the above exercise 4. Apply the forward–backward algorithm to determine the probability $p(O)$ that the sequence of observations $O = 221$ came from this HMM. ($p(O)$ should be the same as $p(O)$ calculated in exercise 4!)

6. Consider the HMM described in exercise 4. Apply the Viterbi algorithm to determine the most likely sequence of states that gave rise to the sequence of observations $O = 221$. Compare this with the solution obtained using the simple heuristic method for problem 2.

HMM PROJECT

The objective is to solve the three problems associated with HMMs with discrete observation probability distributions for a given data set. For this analysis you can use any software, such as, Python, R, MatLab, and SAS, or a combination of functions from different software.

Data Set Generation

The data set will be generated by simulating the following HMM: $S = \{1, 2, 3, 4\}$, $v = \{1, 2, 3\}$, $\pi = \{1, 0, 0, 0\}$,

$$P = \begin{pmatrix} 0.4 & 0.4 & 0.1 & 0.1 \\ 0.2 & 0.3 & 0.3 & 0.2 \\ 0.1 & 0.1 & 0.4 & 0.4 \\ 0.4 & 0.3 & 0.2 & 0.1 \end{pmatrix}, \text{ and } B = \begin{pmatrix} 0.8 & 0.1 & 0.1 \\ 0.3 & 0.4 & 0.3 \\ 0.1 & 0.2 & 0.7 \\ 0.2 & 0.5 & 0.3 \end{pmatrix}.$$

A sequence of observations can be generated as follows.

$t = 1$:

- The HMM is in state q_1 = 1.

- Draw a pseudo-random number r and use the probabilities $b_1(k)$, k = 1, 2, 3, to determine observation O_1; i.e.,

  o if $r \leq 0.8$, then O_1 = 1

  o If $0.8 < r \leq 0.9$, then O_1 = 2.

  o Otherwise, O_1 = 3.

$t = 2$:

- The HMM shifts to state j, j = 1, 2, 3, 4. For this, draw a pseudo-random number r and use the branching probabilities corresponding to state 1 to determine the next state q_2; i.e.,

  o If $r \leq 0.4$, then q_2 = 1.

  o If $0.4 < r \leq 0.8$, then q_2 = 2.

  o If $0.8 < r \leq 0.9$, then q_2 = 3.

  o Otherwise, q_2 = 4.

- Having determined the state, use matrix B to determine O_2 as described above for t = 1.

Continue in this manner until you have generated 1000 observations, i.e., t = 1000. Remember to also record the generated sequence of states.

Task 1 Estimate $p(O|\lambda)$

Given that you know the parameters of the HMM, calculate the probability $p(O\,|\,\lambda)$ that the sequence of observations O = 123312331233 came from the HMM.

Task 2 Estimate the Most Probable Sequence Q

Given that you know the parameters of the HMM, calculate the most probable sequence of states Q that gave rise to the sequence of observations O = 123312331233.

Task 3 Train the HMM

Now forget that you know the parameters of the HMM, and use an HMM solver to estimate its parameters $\lambda = (P, \pi, B)$ using the 1000 generated observations.

Task 3.1 Number of States is Known

Assuming that the HMM has four states, i.e., $S = \{1, 2, 3, 4\}$, and that the number of objects is 3, i.e., $v = \{1, 2, 3\}$. Give the estimated parameters in one or more tables along with their corresponding p-values. Compare your results against the actual parameters $\lambda = (P, \pi, B)$ of the HMM used to generate the data. Discuss your results.

Task 3.2 Number of States is not Known

Assume now that you do not know the number of states of the HMM, but you know that the number of objects is 3, i.e., $v = \{1, 2, 3\}$. Train different HMMs each with a different number of states, starting with an HMM with two states, and for each HMM calculate the likelihood, *AIC*, and *BIC*. Plot these three quantities as a function of the number of states. Keep increasing the number of states until you begin to discern a pattern in each of the three plots. Select the best HMM. Discuss your results.

Note that the number of parameters increases as the number of states increases. A rule of thumb is that for each parameter, you need at least 10 observations. As you increase the number of states, you may require more than 1000 observations. In this case, simply generate additional observations as described above.

REFERENCES

1. L.R. Rabiner. "A Tutorial on Hidden Markov Models and Selected Applications in Speech Recognition", *Proceedings of the IEEE* 77: 257–286 (1989).
2. J.P. Coelho, T.M. Pinho, and J. Boaventura-Cunha. *Hidden Markov Models: Theory and Implementation Using Matlab*, CRC Press, 2019.
3. C. Bishop. *Pattern Recognition and Machine Learning*, Springer, 2006.

Appendix A: Some Basic Concepts of Queueing Theory

I N THIS APPENDIX, WE PRESENT SOME SIMPLE BUT IMPORTANT CONCEPTS from queueing theory that one needs to be aware of when studying the performance of systems where customers have to queue for service. In an IoT setting, customers are messages send to a server by sensors, messages send by a server to actuators, IP packets transmitted between two devices, jobs running on a CPU, etc. A service station is typically a transmitter that transmits packets out of a device or a CPU processing messages and other user processes. An example of a queueing system is the re-certification of IoT devices described in Section 3.2.1.

The simplest queueing system is a single queue served by a single server, commonly known as the *single server queue*, as shown in Figure A.1. A customer arrives from outside, joins the queue, waits for its turn to get served, and upon service completion departs from the server. In order to define a single server queue, we need to specify (a) the distribution of the inter-arrival time, (b) the distribution of the service time, and (c) the service discipline. The inter-arrival time can be assumed to be constant, or exponential, or a mixture of exponentials. Examples of a mixture of exponentials are the Erlang, the hyper-exponential distribution, and other more complex combinations of exponential distributions, such as the phase-type distribution. The inter-arrival time distribution could also be any arbitrary distribution defined only by its mean and variance. Likewise, the service time distribution can be any of the above-mentioned distributions. The service discipline defines the order in which customers in the queue are served. Common disciplines are the first-in-first-out (FIFO), round robin, and static priorities. FIFO is the typical discipline used in many queueing situations we encounter in life, such as, a check-out counter in a super market, and waiting for a teller in a bank. In round robin, customers are served in a FIFO manner, but they are only given a slice of service, that is, a short amount of service. If the service of a customer is not completed at the end of a slice of service, the customer is placed at the end of the queue where it waits its turn for another slice. The customer cycles through the queue in this manner until it is eventually fully processed. This discipline is used for executing processes on a CPU. Finally, statistic priorities are used to give different priorities to different classes of customers. For instance, let us consider a server that processes messages from sensors and issues commands to actuators. We assume that there are two types of

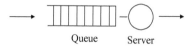

Queue   Server

FIGURE A.1 The single server queue.

messages, real-time (RT) and non-real-time (nonRT). RT messages have to be processed as fast as possible since they represent tasks that need to be executed in real time. nonRT messages represent non-real-time tasks and therefore they are not time constrained. The server maintains two queues, an RT queue and a nonRT queue. Messages in the RT queue have pre-emptive priority over those in the nonRT queue; i.e., they are executed first, and they can also interrupt the execution of a nonRT message so that they are processed. In this Appendix, we shall consider only the single server queue.

Now, let us see how a queue is formed. Let us assume the simplest case where the inter-arrival time is constant equal to 10 minutes and the service time is also constant equal to 6 minutes. Let us say that at time $t = 0$ the queue is empty and the server is idle. At time $t = 1$, a customer arrives and since the server is idle it goes straight into service. Its service will be completed at time 7 and the next arrival will occur at time 11. Obviously, the second customer will not have to wait since it will find the server idle. In general, as the system evolves, an arriving customer will never have to wait since the previous one has departed before its arrival. Now, let us reverse the situation and let us assume that the inter-arrival time is 6 minutes and the service time is 10 minutes. As before, we assume that at time $t = 0$ the queue is empty and the server is idle, and at time $t = 1$ the first customer arrives. Now, we see that the next customer will arrive at time 7 while the service of the first customer will be completed at time 11. As the system evolves, more and more customers will accumulate in the queue, and as a result the queue will grow infinitely big.

In general, the inter-arrival times and service times are not constant and they vary around a mean. Let us assume that the inter-arrival times are exponentially distributed with mean 10 and the service times are also exponentially distributed with mean 6. In this case, we can still see the formation of a queue, because of the variability of the inter-arrival and service times. For instance, a queue will be formed if a number of successive service times happen to be long while at the same time the successive inter-arrivals happen to be short. This formation of a queue does not continue for a long time, and eventually the queue becomes empty and the server idle.

An example of queue formation is shown in Figure A.2. On the x-axis, the up-arrows indicate the arrival of customers, and the down-arrows the departure of customers. On the y-axis, the number of customers in the queue, including the one in service, is indicated. The time during which the server is busy working is known as the *busy period*, and the time during which the server is idle is known as the *idle period*. These two periods alternate continuously, and depending upon the amount of work the server has to do, they have different lengths. This alternation between the two periods is an indication that the queue is *stable*.

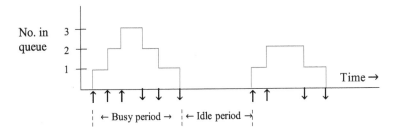

FIGURE A.2 A stable queue.

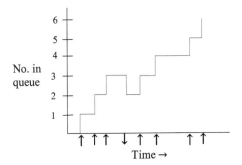

FIGURE A.3 An example of an unstable queue.

When a queue is *unstable*, the number of customers in the queue grows continuously, as shown in Figure A.3; that is, the server never becomes idle. Unstable queues are detrimental to the performance of a queueing-based system. For instance, if an IoT server receives too many messages from a population of sensors, then its queues become unstable, and as a result messages are dropped, or they take an extremely long period of time to get processed.

Of interest, therefore, is how can we design a stable queueing system? Going back to the single server queue example, how can we tell if it is stable or not? Fortunately, this is very easy to verify by examining the mean inter-arrival time and the mean service time. Let $1/\lambda$ and $1/\mu$ be the mean arrival time and the mean service time, respectively. Then, λ and μ are the arrival rate and the service rate, respectively. The rate is defined as the average number of occurrences per unit time. For instance, if $1/\lambda = 10$ minutes, then $\lambda = 1/10 = 0.1$ arrivals per minute, or 60×0.1 arrivals per 60 minutes, i.e., $\lambda = 6$ per hour. This can be also calculated by observing that on the average we have an arrival every 10 minutes, and therefore we have 6 arrivals per hour; likewise, we have for the service time. For instance, if the average service time is $1/\mu = 5$ minutes, then the service rate $\mu = 1/5 = 0.2$ per minute, or $\mu = 12$ per hour. We note that 12 per hour is the maximum capacity of the server; i.e., it cannot serve more than 12 customers on the average per hour. Consequently, in order for the single server queue to be stable, the arrival rate has to be less that the maximum service capacity, i.e.,

$$\lambda < \mu.$$

This is the condition for stability of the single server queue. As λ increases, the busy periods begin to get longer and the idle periods shorter. For $\lambda = \mu$ the queue is neither stable nor unstable; that is, it has excessively long busy periods followed by extremely short idle periods. For $\lambda > \mu$, the queue is unstable. That is, it grows infinitely big, as shown in Figure A.3.

The same happens if we fix the arrival rate and change the service rate. That is, as μ decreases and gets closer to λ, the busy periods begin to get longer and the idle periods shorter, and eventually the queue becomes unstable when $\mu < \lambda$.

The stability condition can be written as follows:

$$\frac{\lambda}{\mu} < 1$$

or,

$$\rho < 1,$$

where $\rho = \lambda/\mu$. ρ is referred to as the *traffic intensity* and it is the utilization of the server. Utilization is defined as the average percent of time that the server is busy working, and it varies between zero and 1. Let us assume that $\lambda = 6$ per hour and $\mu = 12$ per hour. Then, on the average the server works half the time, since it can process on the average 12 customers per hour, but only 6 customers per hour arrive on the average. If we measure the server's utilization on say, a 5-minute basis, we will see that it fluctuates above and below 0.5. However, its long-term average will be 0.5.

An equivalent alternative way of looking at the utilization is to re-write ρ as follows: $\rho = \lambda \left(1/\mu\right)$ i.e., it is equal to the (arrival rate) × (mean service time). For instance, if the arrival rate is 6 per hour and the mean service time is 5 minutes, or 1/12 hours, then the server works for $6 \times \left(1/12\right) = 0.5$; i.e., it is busy 50%.

The quantity $1 - \rho$ is the fraction of time that the server is idle. Since the server can only be idle when there are no customers in the queue, this quantity is also the fraction of time that the system is empty, i.e., the queue is empty and the server is idle.

The arrival and service rates of a single server queue should be such that the utilization of the server is less than 1.

Another related quantity of interest is the *throughput* of the server. This is defined as the average number of completed jobs per unit time; i.e., the average number of customers that depart from the queue per unit time. Let the mean service time in a single server queue be 5 minutes. If the server has always work to do, i.e., there is always a customer waiting in the queue to be serviced, then the number of customers it can service per hour is 12. This is the *maximum* throughput. However, the throughput is not always 12 per hour. For instance, if the arrival rate is 6 customers per hour, then only 6 customers depart on the average. In this case, the throughput is 6 per hour. Likewise, if the arrival rate is 8 customers per hour, then the throughput is 8 per hour. If the arrival rate is increased to, say, 14 customers per

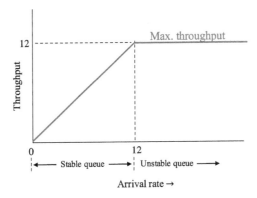

FIGURE A.4 Throughput vs arrival rate.

hour, then the queue becomes unstable (or saturated). In this case the throughput is 12 per hour, since on the average it cannot process more customers. Therefore, the throughput is equal to the arrival rate as long as the queue is stable, and equal to the maximum capacity of the server when the queue is unstable. Figure A.4 shows the server's throughput as a function of the arrival rate.

Example: Let the mean inter-arrival time be 5 minutes, i.e., the arrival rate is 1/5 = 0.2 per minute, and let the mean service time be 2 minutes. Then, we can obtain the following performance measures:

- Server utilization: $0.2 \times 2 = 0.4$ or 40%.

- Percent of time the server is idle: $1 - 0.40 = 0.60$, i.e. 60% of the time.

- Percent of time that the system is empty, i.e., there is no one waiting or being served: 0.60 (equal to the percent of time the server is idle).

Of interest is also the average number of customers in the system, L, that is those in the queue plus the one in service. In general, L as a function of λ has the shape shown in Figure A.5. We observe that L tends to infinity as λ tends to μ.

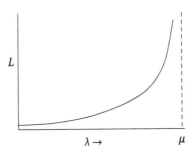

FIGURE A.5 L vs λ.

We can obtain a theoretical expression for L using queueing theory. For instance, in the case where the inter-arrival time and the service time are exponentially distributed, we have

$$L = \frac{\rho}{1-\rho}.$$

Finally, the mean waiting time of a customer in the system, i.e., queueing and being served, W, is also of interest. There is a simple relationship, known as Little's relation, between W, L, and λ; i.e.:

$$\lambda W = L.$$

For instance, let us assume that the rate of arrival is 6 per hour, and that the average number of customers waiting in the system is 3. Then a customer spends an average of 1/2 an hour in the system, i.e., waiting and been served.

Appendix B: Maximum Likelihood Estimation (MLE)

IN THIS APPENDIX, WE GIVE A BRIEF OVERVIEW OF THE maximum likelihood estimation (MLE) method, examine its relation with Bayesian inference and the least squares method, and derive the log-likelihood function of the Gaussian MA(1) and AR(1) models.

B.1 THE MLE METHOD

The MLE method is a popular technique for estimating the parameters of a probability distribution using a set of observed data. The optimum parameters are obtained by maximizing the likelihood, i.e., the probability, that the set of the observed data follows this probability distribution. More specifically, let $f(x)$ be a probability distribution that describes a certain system, and let Θ be the set of unknown parameters of $f(x)$, where $\Theta = \{ \theta_1, \theta_2, ..., \theta_x \}$. Let us also assume that we have n observations $x_1, x_2, ..., x_n$ from this system. Then, the parameters Θ can be estimated by maximizing the likelihood that these n observations follow $f(x)$. The likelihood is notated as $L(\Theta; x)$, where Θ indicates the set of unknown parameters and x the set of observations. Assuming that the observations are independent from each other, we have

$$L\left(\Theta; x\right) = f\left(x_1; \Theta\right) f\left(x_2; \Theta\right) ... f\left(x_n; \Theta\right),$$

where $f(x_i; \Theta)$ indicates the probability of observing x_i given the set of parameters Θ. Therefore, we are looking for the set of values of Θ that maximizes $L(\Theta; x)$, i.e.,

$$\operatorname*{argmax}_{\Theta} L\left(\Theta; x\right) = \operatorname*{argmax}_{\Theta} \prod_{i=1}^{n} f\left(x_i; \Theta\right).$$

For computational convinience, we maximize the natural log of the likelihood, called the *log-likelihood*. Let $l(\Theta; x) = \log\left(L(\Theta; x)\right)$. If $l(\Theta; x)$ is differentiable in Θ, then we can obtain the equations:

$$\frac{\partial l\left(\Theta; x\right)}{\partial \theta_1} = 0, \quad \frac{\partial l\left(\Theta; x\right)}{\partial \theta_2} = 0, ..., \quad \frac{\partial l\left(\Theta; x\right)}{\partial \theta_k} = 0.$$

We can estimate the k parameters θ_1, θ_2, ..., θ_κ if the above equations can be solved explicitly. See for instance Section 5.1 (Chapter 5), where we obtained the parameters of the simple linear regression. In general, numerical optimization is used since these equations are not easy to solve.

There are many different techniques for numerical optimization, such as the gradient-descent method, the Nelder–Mead method, the Newton method, the genetic algorithm, and the simulated annealing technique. It is possible that the numerical solution may lead to a local maximum and this is why different initial values should be used. Note that the natural log function is negative for values less than one and positive for values greater than one. So, it is possible that the optimum value of $l(\Theta; x)$ is negative.

Below, we give two examples of the MLE technique. In the first example, we estimate the parameter p of a binomial probability distribution, and in the second example we estimate the mean and variance of a normal distribution.

We consider a Bernoulli experiment, where the outcome of each independent trial is 1 or 0 with probability p and $1 - p$, respectively. Let X_i be a random variable that indicates the outcome of the ith trial, i.e., X_i takes the values 0 and 1. Then, the probability that it is equal to x_i is

$$f\left(x_i; p\right) = p^{x_i} \left(1-p\right)^{1-x_i}.$$

The above expression is a compact way of saying that the probability that $x_i = 1$ is p, and the probability that $x_i = 0$ is $1 - p$. Given the outcomes from n trials, x_1, x_2, ..., x_n, the likelihood $L(p; x)$ is

$$L\left(p; x\right) = f\left(x_1; p\right) f\left(x_2; p\right)...f\left(x_n; p\right),$$

where x indicates the n outcomes. The above expression can be written as follows:

$$L\left(p; x\right) = p^{x_1} \left(1-p\right)^{1-x_1} p^{x_2} \left(1-p\right)^{1-x_2} \cdots p^{x_n} \left(1-p\right)^{1-x_n}$$

or

$$L\left(p; x\right) = p^{\sum_{i=1}^{n} x_i} \left(1-p\right)^{n-\sum_{i=1}^{n} x_i}.$$

Taking logs, we have

$$\log\left(L\left(p; x\right)\right) = \left(\sum_{i=1}^{n} x_i\right) \log\left(p\right) + \left(n - \sum_{i=1}^{n} x_i\right) \log\left(1-p\right).$$

In order to find p that maximizes the above expression we set the derivative of $\log(L(p; x))$ with respect to p to zero. We have

$$\frac{d\log\left(L\left(p;x\right)\right)}{dp} = \left(\sum_{i=1}^{n}x_i\right)\frac{1}{p} + \left(n - \sum_{i=1}^{n}x_i\right)\frac{1}{1-p} = 0.$$

We recall that $d\log(x) = 1/x$, and $d\log(f(x)) = f'(x)/f(x)$. Solving for p we obtain

$$p = \frac{1}{n}\sum_{i=1}^{n}x_i;$$

i.e., if we have 5 trials with the outcome 1,0,1,1,0, then $p = 3/5$. This result is hardly surprising since this is the way one would compute p, but it simply validates the MLE technique.

In the second example, we show how the MLE method can be used to estimate the mean and variance of a normal distribution. Let $x_1, x_2, \ldots, x_n$ be n independent observations from a normal distribution $N(\mu, \sigma^2)$, where μ and σ^2 are not known. Then, the probability that x_i comes from the distribution $N(\mu, \sigma^2)$ is

$$f\left(x_i; \mu, \sigma^2\right) = \frac{1}{\sqrt{2\pi\sigma^2}}\exp\left\{-\frac{1}{2}\frac{\left(x_i - \mu\right)^2}{\sigma^2}\right\}.$$

Since the observations are independent from each other, the likelihood that they all come from the normal distribution $N(\mu, \sigma^2)$ is:

$$L\left(\mu, \sigma^2; x\right) = f\left(x_1, x_2, \ldots, x_n; \mu, \sigma^2\right) = \prod_{i=1}^{n}f\left(x_i; \mu, \sigma^2\right)$$

$$= \left(\frac{1}{2\pi\sigma^2}\right)^{n/2}\exp\left\{-\frac{\sum_{i=1}^{n}\left(x_i - \mu\right)^2}{2\sigma^2}\right\}.$$

We want to find μ and σ^2 that maximize the log of $L(\mu, \sigma^2; x)$, given by the expression:

$$\log\left(L\left(\mu, \sigma^2; x\right)\right) = -\frac{n}{2}\log\left(2\pi\sigma^2\right) - \frac{1}{2\sigma^2}\sum_{i=1}^{n}\left(x_i - \mu\right)^2.$$

For this we set the derivatives of $\log(L(\mu, \sigma^2; x))$ with respect to μ and to σ to zero and then solve for μ and σ. For the first derivative, we have

$$\frac{d\log\left(L\left(\mu, \sigma^2; x\right)\right)}{d\mu} = -\frac{1}{2\sigma^2}\sum_{i=1}^{n}\left(-2\left(x_i - \mu\right)\right) = \frac{1}{\sigma^2}\sum_{i=1}^{n}\left(x_i - \mu\right) = 0$$

or

$$\sum_{i=1}^{n}(x_i - \mu) = \left(\sum_{i=1}^{n}x_i\right) - n\mu = 0$$

or

$$\mu = \frac{1}{n}\sum_{i=1}^{n}x_i.$$

Setting the derivative of $\log(L(\mu, \sigma^2; x)$ with respect to σ to zero, gives

$$\frac{d\log\left(L\left(\mu, \sigma^2; x\right)\right)}{d\sigma} = -\frac{n}{\sigma} + \frac{1}{\sigma^3}\sum_{i=1}^{n}(x_i - \mu)^2 = 0$$

or

$$\frac{1}{\sigma^2}\sum_{i=1}^{n}(x_i - \mu)^2 = n$$

or

$$\sigma^2 = \frac{1}{n}\sum_{i=1}^{n}(x_i - \mu)^2.$$

We observe that the MLE technique gives the same results as we would have obtained using their expected values.

B.2 RELATION OF MLE TO BAYESIAN INFERENCE

The MLE method coincides with the most probable Bayesian estimator given a uniform prior distribution of the parameters Θ. We have

$$p(\Theta|x) = \frac{p(x|\Theta)p(\Theta)}{p(x)},$$

where $p(\Theta)$ is the prior distribution of the parameters Θ, and for which we assume that it is uniformly distributed, i.e., $p(\theta_i) = 1/k$, $i = 1, 2, \ldots, k$. The probability $p(x)$ is the probability of observing the values $x = \{x_1, x_2, \ldots, x_n\}$ averaged over all Θ parameters. $p(\Theta)$ is independent of Θ since we assumed it is uniformly distributed, and so is $p(x)$. Therefore, in order to obtain the parameters Θ by maximizing $p(\Theta|x)$ it suffices to maximize $p(x|\Theta)$, which is the same as in MLE.

B.3 MLE AND THE LEAST SQUARES METHOD

Let us consider the simple linear regression model $Y = a_0 + a_1 X_1 + \varepsilon$, where $\varepsilon \sim N(0, \sigma^2)$, described in Section 5.1 (Chapter 5). We will apply the MLE method to obtain the coefficients a_0 and a_1. We recall that

$$y_i \sim N\left(a_0 + a_1 x_i, \sigma^2\right),$$

and consequently

$$L\left(a_0, a_1, \sigma^2; y_1, y_2, \ldots, y_n\right) = f\left(y_1, y_2, \ldots, y_n; a_0, a_1, \sigma^2\right)$$

$$= \prod_{i=1}^{n} \frac{1}{\sqrt{2\pi\sigma^2}} \exp\left\{ -\frac{\left(y_i - a_0 - a_1 x_i\right)^2}{2\sigma^2} \right\}$$

$$= \left(\frac{1}{2\pi\sigma^2}\right)^{n/2} \exp\left\{ -\frac{\sum_{i=1}^{n}\left(y_i - a_0 - a_1 x_i\right)^2}{2\sigma^2} \right\}.$$

Therefore, maximizing the above likelihood for a given σ is equivalent to

$$\min \sum_{i=1}^{n}\left(y_i - a_0 - a_1 x_i\right)^2$$

which is the same maximization problem as in the least squares method (see Section 5.1, Chapter 5). This means that the least squares method is equivalent to the MLE method for estimating a_0 and a_1. This provides an additional justification for the use of the sum of squared errors for fitting a linear regression model, since there are different error functions that could be used.

B.4 MLE OF THE GAUSSIAN MA(1)

So far, we have assumed that the observations $x_1, x_2, \ldots, x_n$ are i.i.d. (independent and identically distributed). This allows us to express the likelihood function as the product of the probability density functions of the individual observations; i.e.,

$$L\left(\Theta; x\right) = f\left(x_1; \Theta\right) f\left(x_2; \Theta\right) \ldots f\left(x_n; \Theta\right),$$

where $f(x_i; \Theta)$ is the probability of observing x_i given the set of parameters Θ. However, this is not the case for a time series where the successive observations are correlated. In this section, we show how the parameters of a Gaussian MA(1) model can be estimated using MLE. In the following section, we apply MLE to a Gaussian AR(1) model.

The MA(1) model was analyzed in Section 6.3 (Chapter 6). It is given by the expression

$$X_t = \mu + \varepsilon_t + \theta_1 \varepsilon_{t-1},$$

where μ is the mean of the time series, θ_1 is the weight, and ε_t, and ε_{t-1} are errors normally distributed with mean 0 and variance σ^2. The mean of X_t is

$$E(X_t) = E(\mu + \varepsilon_t + \theta_1 \varepsilon_{t-1}) = \mu,$$

since $E(\varepsilon_t) = E(\varepsilon_{t-1}) = 0$. Now, we note that both ε_t and ε_{t-1} are normally distributed. Therefore, their sum is also normally distributed with a mean equal to the sum of their means and a variance equal to the sum of their variances. In view of this, $\varepsilon_t + \theta_1 \varepsilon_{t-1} \sim N(0, \sigma^2 + \theta_1^2 \sigma^2)$. Therefore, X_t is normally distributed with mean μ and variance equal to $\sigma^2(1+\theta_1^2)$.

Now, let us consider n successive random variables $X_1, X_2, \dots, X_n$, where X_i is the ith estimate defined by the above MA(1) model. We assume that their joint distribution follows a multivariate normal distribution; i.e., the probability of observing $x = (x_1 \, x_2, \dots x_n)$ is

$$f_{X_1, X_2, \dots, X_n}(x; \mu, \Sigma) = \frac{1}{\sqrt{(2\pi)^n \det(\Sigma)}} e^{-\frac{1}{2}(x-\mu)^T \Sigma^{-1}(x-\mu)},$$

where $x = (x_1, x_2, \dots, x_n)^T$, $\mu = (\mu_1, \mu_2, \dots, \mu_n)^T$, and Σ is the covariance matrix of $X_1, X_2, \dots, X_n$, given by the expression:

$$\Sigma = \sigma^2 \begin{pmatrix} 1+\theta_1^2 & \theta & & & & \\ \theta & 1+\theta_1^2 & \theta & & & \\ & \theta & 1+\theta_1^2 & \theta & & \\ & & & & \ddots & \\ & & & & \theta & 1+\theta_1^2 \end{pmatrix}.$$

The log-likelihood of $f_{X_1, X_2, \dots, X_n}(x; \Theta)$ is

$$\log L(x_1, x_2, \dots, x_n; \Theta) = \log\left(\frac{1}{\sqrt{(2\pi)^n \det(\Sigma)}} e^{-\frac{1}{2}(x-\mu)^T \Sigma^{-1}(x-\mu)} \right)$$

$$= -\frac{n}{2}\log(2\pi) - \frac{1}{2}\sum_{i=1}^{n}\log(d_{ii}) - \frac{1}{2}\sum_{i=1}^{n}\frac{\tilde{x}_i}{d_{ii}},$$

where

$$d_{ii} = \sigma^2 \frac{1 + \theta^2 + \theta^4 + \cdots + \theta^{2i}}{1 + \theta^2 + \theta^4 \cdots + \theta^{2(i-1)}}$$

and

$$\tilde{x}_i = x_i - \frac{\theta\left(1 + \theta^2 + \theta^4 + \cdots + \theta^{2i}\right)}{1 + \theta^2 + \theta^4 \cdots + \theta^{2(i-1)}}\tilde{x}_{i-1}.$$

The parameters μ, θ_1 and σ^2 are estimated by maximizing the above log-likelihood numerically.

B.5 MLE OF THE GAUSSIAN AR(1)

The AR(1) mode, studied in Section 6.4.1 (Chapter 6), is given by the expression:

$$X_t = \delta + a_1 X_{t-1} + \varepsilon_t,$$

where $\varepsilon_t \sim N(0, \sigma^2)$. The following expression hold: $E(X_t) = \delta/(1 - a_1)$, $\mathrm{Var}(X_t) = \sigma^2/(1 - a_1^2)$, and $\mathrm{Cov}(X_t, X_{t-k}) = a_1^k \sigma^2 / (1 - a_1^2)$.

Now, let us consider n successive random variables X_1, X_2, ..., X_n, where X_i is the ith estimate defined by the above AR(1) model. We assume that the joint distribution follows a multivariate normal distribution; that is, the probability of observing $x_1, x_2, \ldots x_n$ is

$$f_{X_1, X_2, \ldots, X_n}\left(x; \mu, \Sigma\right) = \frac{1}{\sqrt{(2\pi)^n \det(\Sigma)}} e^{-\frac{1}{2}(x-\mu)^T \Sigma^{-1}(x-\mu)}$$

where $x = (x_1, x_2, \ldots, x_n)^T$, $\mu = (\mu_1, \mu_2, \ldots, \mu_n)^T$, and Σ is the covariance matrix of X_1, X_2, ..., X_n, given by the following expression:

$$\Sigma = \frac{\sigma^2}{1 - a_1^2} \begin{pmatrix} 1 & a_1 & \cdots & a_1^{n-1} \\ a_1 & 1 & \cdots & a_1^{n-1} \\ \vdots & \vdots & & \vdots \\ a_1^{n-1} & a_1^{n-2} & \cdots & 1 \end{pmatrix}.$$

The log-likelihood of $f_{X_1, X_2, \ldots, X_n}(x; \Theta)$ is

$$\log L\left(x_1, x_2, \ldots, x_n; \Theta\right) = \log\left(\frac{1}{\sqrt{(2\pi)^n \det(\Sigma)}} e^{-\frac{1}{2}(x-\mu)^T \Sigma^{-1}(x-\mu)}\right)$$

$$= -\frac{n}{2}\log\left(2\pi\right) - \frac{1}{2}\log\left(\left|\Sigma^{-1}\right|\right) - \frac{1}{2}\left(x - \mu\right)^{T}\Sigma^{-1}\left(x - \mu\right).$$

The parameters $\Theta = \{\delta, a_1, \sigma^2\}$ are estimated by maximizing the above log-likelihood numerically.

Index

Page numbers in *italic* indicate a figure and page numbers in **bold** indicate a table on the corresponding page.

Milton Keynes UK
Ingram Content Group UK Ltd.
UKHW050454071024
449327UK00015B/369

9 780367 686314